Mechanophotonics for Organic Photonic Integrated Circuits

Online at: https://doi.org/10.1088/978-0-7503-5336-6

IOP Series in Advances in Optics, Photonics and Optoelectronics

SERIES EDITOR

Professor Rajpal S Sirohi Consultant Scientist

About the Editor

Rajpal S Sirohi is currently working as a faculty member in the Department of Physics, Alabama A&M University, Huntsville, AL, USA. Prior to this, he was a consultant scientist at the Indian Institute of Science, Bangalore, and before that he was Chair Professor in the Department of Physics, Tezpur University, Assam. During 2000–11 he was an academic administrator, being vice-chancellor to a couple of universities and the director of the Indian Institute of Technology, Delhi. He is the recipient of many international and national awards and the author of more than 400 papers. Dr Sirohi is involved with research concerning optical metrology, optical instrumentation, holography, and speckle phenomena.

About the series

Optics, photonics, and optoelectronics are enabling technologies in many branches of science, engineering, medicine, and agriculture. These technologies have reshaped our outlook and our ways of interacting with each other, and have brought people closer together. They help us to understand many phenomena better and provide deeper insight into the functioning of nature. Further, these technologies themselves are evolving at a rapid rate. Their applications encompass very large spatial scales, from nanometers to the astronomical scale, and a very large temporal range, from picoseconds to billions of years. This series on advances in optics, photonics, and optoelectronics aims to cover topics that are of interest to both academia and industry. Some of the topics to be covered by the books in this series include biophotonics and medical imaging, devices, electromagnetics, fiber optics, information storage, instrumentation, light sources, CCD and CMOS imagers, metamaterials, optical metrology, optical networks, photovoltaics, free-form optics and its evaluation, singular optics, cryptography, and sensors.

About IOP ebooks

The authors are encouraged to take advantage of the features made possible by electronic publication to enhance the reader experience through the use of color, animation, and video and by incorporating supplementary files in their work.

Do you have an idea for a book you'd like to explore?

For further information and details of submitting book proposals, see iopscience.org/books or contact Ashley Gasque at Ashley.gasque@iop.org.

A full list of titles published in this series can be found here: https://iopscience.iop.org/bookListInfo/series-on-advances-in-optics-photonics-and-optoelectronics.

Mechanophotonics for Organic Photonic Integrated Circuits

Rajadurai Chandrasekar
School of Chemistry, University of Hyderabad, Hyderabad, India

IOP Publishing, Bristol, UK

ISBN 978-0-7503-5336-6 (ebook)
ISBN 978-0-7503-5334-2 (print)
ISBN 978-0-7503-5337-3 (myPrint)
ISBN 978-0-7503-5335-9 (mobi)

DOI 10.1088/978-0-7503-5336-6

Version: 20240701

IOP ebooks

British Library Cataloguing-in-Publication Data: A catalogue record for this book is available from the British Library.

Published by IOP Publishing, wholly owned by The Institute of Physics, London

IOP Publishing, No.2 The Distillery, Glassfields, Avon Street, Bristol, BS2 0GR, UK

US Office: IOP Publishing, Inc., 190 North Independence Mall West, Suite 601, Philadelphia, PA 19106, USA

Dedicated to my Family

Contents

Preface

Mechanophotonics for Organic Photonic Integrated Circuits is a unique book that delves into the emerging interdisciplinary field of mechanophotonics, where mechanical micromanipulation converges with nanophotonics of molecular optical microcrystals. The book offers comprehensive insights into designing, microfabricating, and integrating molecular flexible crystals into photonic components, ultimately leading to the development of prototype monolithic or hybrid photonic circuits using atomic force microscopy manipulation techniques.

The initial two chapters serve as a solid foundation, covering the fundamental aspects of light, the physics and chemistry of organic materials, nonlinear optics, mechanical properties of molecular crystals, the characteristics of various photonic components, and photonic integrated circuits (PICs). These chapters cater to a diverse audience, including chemists, physicists, and materials scientists and engineers interested in the field of mechanophotonics.

Chapter 3 is dedicated to detailing various micromechanical manipulation operations employed in mechanophotonics, accompanied by selected examples to provide a practical understanding of the techniques involved. This chapter serves as a valuable resource for researchers and practitioners seeking a deeper understanding of the mechanical manipulation processes integral to the field.

In chapter 4, the book showcases recent advancements in organic crystal-based PIC components and presents prototype PICs constructed using flexible organic crystals. This section offers a comprehensive overview of the progress made in this innovative field and is a must-read for those interested in cutting-edge developments within mechanophotonics.

The final chapter introduces a novel area known as crystal photonics foundry, where crystals are precisely milled into various photonic components using focused ion beams. The integration of these fabricated crystal components into circuits adds another layer of innovation to the field, highlighting the versatility and potential applications of mechanophotonics.

In summary, this book is a valuable resource that not only educates readers on the foundational concepts but also provides in-depth insights into the practical aspects and recent advancements in this evolving discipline. Whether you are a researcher, scientist, or engineer, this book offers a comprehensive exploration of mechanophotonics and its transformative potential in the realm of photonic integrated circuits.

Acknowledgments

I am deeply grateful for the unwavering support provided by my family—Marina, Gitika, and Sujit—throughout the entire process of producing this book. Their encouragement and understanding have been instrumental in bringing this project to fruition.

I would like to extend my earnest appreciation to my doctoral students and postdocs, whose invaluable assistance played a crucial role in the preparation of this book.

Special thanks to my colleague, Prof. N Viswanathan, from the School of Physics at the University of Hyderabad, for proofreading the introductory chapters.

Author biography

Professor Rajadurai Chandrasekar

Born in the historic city of Madurai, Professor Chandrasekar has earned distinction as a prominent materials scientist. Currently, he holds the position of Professor at the School of Chemistry and associated with Centre for Nanotechnology at the University of Hyderabad, India. He earned his MSc degree from the American College, Madurai, in 1997, and later obtained an MTech degree from the Indian Institute of Technology, Delhi, in 1998. Subsequently, he accomplished his PhD at the Max Planck Institute for Polymer Research, Mainz, in 2004, followed by a four-year post-doctoral stint at the Institute of Nanotechnology, Karlsruhe Institute of Technology. His multifaceted background reflects a comprehensive approach to research and education in the interdisciplinary fields of optics and materials science.

Chandrasekar has published over 130+ peer-reviewed papers and 5 patents in the area of materials chemistry and physics, particularly on the nanophotonic properties of molecular solids. With over 16 years of experience in academia, he has made significant original contributions to the field of organic photonic integrated circuits. Recently, he initiated two new research fields namely, (i) *Mechanophotonics*—a domain that develops and exploits mechanical micromanipulation of single crystal photonic modules towards circuits, and (ii) *Crystal photonic foundry*—a novel approach involving focused ion beam milling of organic single crystals to produce various organic single crystal photonic modules.

He has made original contributions to the field of passive organic optical waveguides, organic photochromic crystal modulators, linear and nonlinear organic optical cavities, organic directional couplers, chiral organic cavities, organic interferometry, and developed various mechanical micromanipulation operations using AFM cantilever tip for the construction of several prototype organic single crystal photonic circuits.

Chandrasekar is recognized as a Fellow of distinguished academic societies, including the Indian Academy of Sciences (2020), the National Academy of Sciences, India (2023), the Optical Society of India (2018), and the Royal Society of Chemistry (2019).

He is on the editorial advisory board of various materials science journals viz. *Small*, *Advanced Optical Materials*, *Materials Advances*, *Journal of Materials Chemistry C*, and *ACS Applied Engineering Materials*. He served as a guest editor for a special section titled '*Organic Photonic Integrated Circuits and Components*' in the *Advanced Optical Materials* journal.

List of symbols

A	Amplitude
A	Proportional limit
A'	Coefficient
a_B	Bohr radius
a_{ex}	Exciton Bohr radius
a	cavity size
$\mathbf{B}$	Magnetic field
c	Speed of light in vacuum
d	Deformation value
d_c	Distance between the charges
d_i	diameter of the inner arc
$\mathbf{D}$	Electric field density
D	Propagation distance
D	Cavity diameter
E	Energy
$\mathbf{E}$	Electric field
$\mathbf{E}(t)$	Time dependent electric field
$\mathbf{E}_g$	Band gap
E_a	Electric field of atoms
E_B	Exciton binding energy
E_g	Initial transition state
$\mathrm{E_l}$	Light's electric field strength
E_f	Final transition state
$\mathbf{F}$	Field
F	Force
$\boldsymbol{F}$	Finesse
h	Planck's constant
H	Horizontal polarization
$\hat{\mathbf{H}}$	Hamiltonian
$\mathbf{H_c}$	Contact hardness
H_I	Indentation hardness
I_a	Light intensity corresponding to an atomic field
I	Intensity
I_E	Elasticity index
I_{tip}	FL intensities of out-coupled at the tip of the crystal
I_{body}	FL intensities of light incidence region
I	Intensity of the guided signal spots
I_o	PL intensity at the excited spot
I_{cir}	Circulating optical intensity
J	Current density
$\mathbf{k}$	Wave vector
k_o	Free space wave number
k	Wavenumber
k_B	Boltzmann constant
L	Waveguide length OR crystal of length
L	Cavity length
l	Polar mode number

m	Mode number
m^*	Effective mass
m_0	Rest mass
M	Azimuthal mode number
M1 and M2	Light-reflecting mirrors
n	Refractive index
n_{eff}	Effective refractive index
n_{S}	Substrate refractive index
$n_{\perp}$	Refractive index along and parallel direction
$n_{\parallel}$	Refractive index along and perpendicular direction
N	Atoms per unit volume
N_0	molecular density
N_{A}	Avogadro's number
$\mathbf{P}$	Polarization
$\mathbf{P}(t)$	Time dependant polarization
P_{in}	Input power
P_{out}	Output power
P_{cir}	Circulating optical power
P_{loss}	Power loss
P	Momentum
p	unit vector
p	Angular mode number
Q	Quality factor
Q_{mat}	Material absorption loss
Q_{ss}	Scattering loss
Q_{rad}	Radiation loss
Q_{ex}	External coupling loss
R_{ID}	Resistance to irreversible deformation
R	Light reflectivity
R	Radius
r_{d}	radial distance
r_{o}	Equilibrium interatomic separation
r_{c}	Radius of curvature
r	Radial mode number
s	constant scaling factor
T	Thickness of the crystal
$\mathbf{T}$	Temperature
T	Power transmission ratio
T_{rt}	Round trip time
t_{f}	Thickness of the polymer film
t_{c}	Thickness of the crystal
U	Potential energy
V	Vertical polarization
V_{eff}	Effective mode volume
w	Width of the waveguide
W	Width
$\mathbf{Y}$	Young's modulus
Y_{I}	Indentation modulus
$\mathbf{Y_c}$	Young's modulus of the crystal
$\mathbf{Y_f}$	Young's modulus of the polymer

z	light propagation distance in the material
$\boldsymbol{\alpha}$	Attenuation coefficient
α, β	Amplitudes of quantum states
α, β, and γ	One, two, and three-photon absorption coefficients
α'	Optical loss coefficient
β_{pm}	Propagation constant
Γ	Full-width at half-maximum
ε_o	Dielectric permittivity of the vacuum
ε	Relative dielectric permittivity
ϵ	Extension/compression ratio
ζ	Strain
ζ_B	Bending strain of the bilayer model
η	Polymer thickness
θ_i	Light's incident angle
θ_c	Critical angle
θ	Angle, Angular parameter
κ	Waveguides wavenumber
λ	Wavelength
λ_c	Cut-off wavelength
μ_o	Magnetic permeability of the vacuum
μ	Relative magnetic permeability
$\boldsymbol{\mu}$	Dipole moment
$\boldsymbol{\Delta\mu}_{g,\,f}$	The change in the final static dipole moment relative to the ground state
ν	Frequency
π	3.141 59……
$\pi\cdots\pi$	Pi–Pi stacking of molecules
ρ	Charge density
$\boldsymbol{\rho}$	Density
ρ_{laser}	Critical superfluid density
σ	Stress
$\sigma_2(\lambda)$	TPA cross-section
τ_P	Photon lifetime
v_P	Phase velocity
v_g	Group velocity
φ	Cone angle of the indenter
χ	Electric susceptibility
$\chi^{(2)}$	Second-order electric susceptibility
$\chi^{(3)}$	Third-order electric susceptibility
Ψ	Wave function
ω	Frequency of the light
∇^2	Second-order Laplacian operator
$\hbar$	Reduced Planck's constant
Φ_{FL}	Quantum yield

List of abbreviations

ADF	Add-drop-filter
AFM	Atomic force microscopy
ANN	Artificial neural network
o-BCB	2,2′-((1*E*,1′*E*)-1,4-phenylenebis(ethene-2,1-diyl))-dibenzonitrile
BFIMP	2,4-Dibromo-6-(((2-bromo-5-fluorophenyl)imino)methyl)phenol
BMPIN	(*E*)-2-bromo-6-(((4-methoxyphenyl)imino)methyl)-4-nitrophenol
BPEA	9,10-Bis(phenylethynyl)anthracene
BPIN	(*E*)-1-(4-bromo)iminomethyl-2-hydroxyl-naphthalene
BPP	4,4′-Bis(2,6-di(1H-pyrazol-1-yl)pyridin-4-yl)biphenyl
BPyIN	(*E*)-1-(((5-bromopyridin-2-yl)imino)methyl)naphthalen-2-ol
BTD	4,7-Bis(4-octylthiophen-2-yl)benzo[c][1,2,5]thiadiazole
$BTD2CF_3$	(*Z*)-2-(3,5-bis(trifluoromethyl)phenyl)-3-(7-methoxybenzo[c][1,2,5]thiadiazol-4-yl)acrylonitrile
CB	Conduction band
CD	Circular dichroism
CF_3OMe	(*Z*)-3-(3′,5′-bis(trifluoromethyl)-[1,1′-biphenyl]-4-yl)-2-(4-methoxyphenyl) acrylonitrile
CMOS	Complementary metal–oxide semiconductor
CPIN	(*E*)-1-(4-(chloro)phenyl) iminomethyl-2-hydroxyl-naphthalene
CROW	Coupled-resonator-optical waveguides
CT	Charge transfer
$Cz\text{-}2CF_3$	(*Z*)-3-(4-(9H-carbazol-9-yl)phenyl)-2-(3,5-bis(trifluoromethyl)phenyl) acrylonitrile
DBA	9,10-Dibromoanthracene
DC	Directional coupler
DPIN	(*E*)-1-(4-(dimethylamino)phenyl)iminomethyl-2-hydroxyl-naphthalene
DPBT	2-(*N,N*-diethylanilin-4-yl)-4,6-bis(3,5-dimethylpyrazol-1-yl)-1,3,5-triazine
DPPDO	1,5-Diphenyl-1,4-pentadien-3-one
DR	Disk resonator
DSB	Distyrylbenzene
DTPz	Dithieno[3,2-a:2′,3′-c]phenazine
EP	Exciton–polariton
ET	Energy transfer
FDTD	Finite-difference-time-domain
FESEM	Field emission scanning electron microscopy
FET	Field-effect transistor
FIB	Focused ion beam
FL	Fluorescence
FLIM	FL-lifetime imaging
FLM	Fiber loop mirror
F–P	Fabry–Pérot
FSR	Free spectral range
F-TIR	Frustrated total internal reflection
FWHM	Full-width at half-maximum
HDC	Hybrid directional coupler
HOADF	Hybrid organic-add-drop filter
IM	Interferometer

IPIN	(*E*)-1-(4-(iodo)phenyl)iminomethyl-2-hydroxyl-naphthalene
ITO-PET	Indium tin oxide-coated polyethylene terephthalate
LED	Light emitting diode
LMAIS	Liquid metal alloy ion sources
MIR	Mid-infrared
MOSFET	Metal–oxide-semiconductor field-effect transistor
NA	Numerical aperture
NDIPH	*N*,*N′*-bis(*p*-tolyl)-1,4,5,8-naphthalene diimide
NLO	Nonlinear optical
NLO-CD	Nonlinear optical circular dichroism
OPIC	Organic photonic integrated circuit
PIC	Photonic integrated circuit
PL	Photoluminescence
2-PL	Two-photon luminescence
3-PL	Three-photon luminescence
RR	Ring resonator
SAA	2,2′-((1*E*,1′*E*)-hydrazine-1,2-diylidenebis(methaneylylidene))diphenol
SAED	Selected area electron diffraction
SHG	Second harmonic generation
SOI	Silicon-on-insulator
SW	Spiral waveguide
TC	Thiocyanine
TDSB	1,4-Dimethoxy-2,5-di[4-(methylthio)styryl]benzene
TE	Transverse electric
TEM	Transmission electron microscopy
THG	Third harmonic generation
THz	Terahertz
TIR	Total internal reflection
TM	Transverse magnetic
TPA	Two-photon absorption
TPL	Two-photon luminescence
TPI	2,4,5-Triphenylimidazole
TTO	9,9,9′,9′-Tetramethyl-9H,9′H,9′H-[2,2′:7,2′-terfluoren]-9′-one
UOH1	University of Hyderabad 1
VB	Valence band
VLC	Visible light communication
WDM	Wavelength division multiplexer
WGM	Whispering-gallery-mode

IOP Publishing

Mechanophotonics for Organic Photonic Integrated Circuits

Rajadurai Chandrasekar

Chapter 1

Introduction

1.1 The nature of light waves

Light is an electromagnetic wave that can undergo reflection, diffraction, refraction, interference, polarization, loss, etc (figure 1.1(a)). Generally, a wave can be understood in terms of its frequency (ω), amplitude (A) and wavelength (λ). Here, λ is related to the wave vector (**k**). In an isotropic medium, **k** is pointing towards the direction of wave propagation or phase motion. The magnitude of **k** in three dimensions is related to the angular wavenumber, $k = \frac{2\pi}{\lambda}$, and the latter parameter denotes a change in spectral phase ($kx - \omega t$) per unit length. The positive and negative values of **k** describe forward and backward propagating waves, respectively. The velocity at which a sinusoidal wave's constant phase travels is known as *phase velocity*, $v_{\rm P}$.

$$y(x, t) = \text{Asin}\,(kx - \omega t) \tag{1.1}$$

$$kx - \omega t = \text{Constant} \Rightarrow \frac{d(kx - \omega t)}{dt} = 0 \Rightarrow k\frac{dx}{dt} - \omega = 0 \Rightarrow \frac{dx}{dt} = \frac{\omega}{k} = v_{\rm P} \tag{1.2}$$

The velocity of a group of waves (with different frequencies and amplitude) traveling as a pocket or pulse is called the *group velocity*, $v_{\rm g}$. A wave group or wave packet emerges due to the interference between the waves. The resultant amplitude variation defines the wave group shape (figure 1.1(b)). If the $v_{\rm P}$ of the individual waves varies; they do not propagate together (called *dispersion*). Consequently, $v_{\rm g}$ will be different from the $v_{\rm P}$ of waves that make the group. The speed and direction of energy flow is given as

$$v_{\rm g} = \frac{d\omega}{d\mathbf{k}} \tag{1.3}$$

doi:10.1088/978-0-7503-5336-6ch1

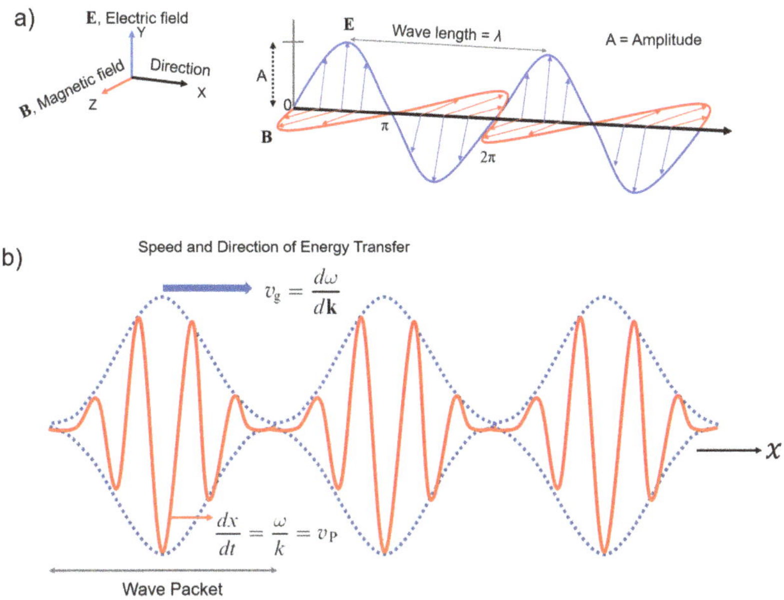

Figure 1.1. (a) A propagating electromagnetic wave. (b) Phase and group velocity of a plane wave.

1.1.1 The Maxwell and Helmholtz equations

The understanding of the origin of the electromagnetic wave starts from the famous Maxwell's four equations which are given below.

$$\nabla \cdot \mathbf{E} = \frac{\rho}{\varepsilon_o} \quad \text{(Gauss' law for electric fields)} \tag{1.4}$$

$$\nabla \cdot \mathbf{B} = 0 \quad \text{(Gauss' law for magnetic fields)} \tag{1.5}$$

$$\nabla \times \mathbf{E} = \left(-\frac{\partial \mathbf{B}}{\partial t}\right) \quad \text{(Faraday's law)} \tag{1.6}$$

$$\nabla \times \mathbf{B} = \mu_o J + \mu_o \varepsilon_o \left(\frac{\partial \mathbf{E}}{\partial t}\right) \quad \text{(Ampere − Maxwell law)} \tag{1.7}$$

Here, ε_o is the dielectric permittivity ($\varepsilon_o \approx 8.854 \times 10^{-12}$ F m$^{-1} \approx (\frac{1}{36\pi}) \times 10^{-9}$ C V^{-1} m^{-1}) and μ_o is magnetic permeability ($\mu_o \approx 4\pi \times 10^{-7}$ V s A^{-1} m^{-1}) of the vacuum. **E** and **B** are the electric and magnetic fields, respectively.

In equations (1.4) and (1.7), when charge density (ρ) and current density (J) are zero, then Maxwell's equation in *vacuum* can be written as,

$$\nabla \cdot \mathbf{E} = 0 \tag{1.8}$$

$$\nabla \cdot \mathbf{B} = 0 \tag{1.9}$$

$$\nabla \times \mathbf{E} = \left(-\frac{\partial \mathbf{B}}{\partial t}\right) \tag{1.10}$$

$$\nabla \times \mathbf{B} = \mu_o \varepsilon_o \left(\frac{\partial \mathbf{E}}{\partial t}\right) \tag{1.11}$$

These equations demonstrate the existence of electromagnetic waves in a vacuum and are normally used for optical fields as it is usually not generated by free currents or free charges.

In the coupled differential equations (1.6) and (1.7), we have two unknown parameters, **E** and **B**. For example, application of $\nabla\times$ (curl) on both sides of equation (1.6) and rearrangement provides

$$\nabla \times \nabla \times \mathbf{E} = \nabla \times \left(-\frac{\partial \mathbf{B}}{\partial t}\right) \tag{1.12}$$

$$\nabla \times \nabla \times \mathbf{E} = -\frac{\partial}{\partial t}\ \nabla \times \mathbf{B} \tag{1.13}$$

Insertion of the curl of **B** (from equation (1.11)) in the right side of equation (1.13)

$$\nabla \times \nabla \times \mathbf{E} = -\frac{\partial}{\partial t}\left(\mu_o \varepsilon_o \frac{\partial \mathbf{E}}{\partial t}\right) \tag{1.14}$$

$$\nabla \times \nabla \times \mathbf{E} = -\mu_o \varepsilon_o \frac{\partial^2 \mathbf{E}}{\partial t^2} \tag{1.15}$$

The left side of the equation (1.15) can be solved by applying $\nabla \times \nabla \times \mathbf{F} = \nabla(\nabla \cdot \mathbf{F}) - \nabla^2\,\mathbf{F}$.

$$\nabla(\nabla \cdot \mathbf{E}) - \nabla^2\,\mathbf{E} = -\mu_o \varepsilon_o \frac{\partial^2 \mathbf{E}}{\partial t^2} \tag{1.16}$$

Application of equation (1.8), i.e., $\nabla\mathbf{E} = 0$ in *vacuum* in equation (1.16) and removal of the sign provides the *wave equation* for **E** in vacuum,

$$\nabla^2 \mathbf{E} = \mu_o \varepsilon_o \frac{\partial^2 \mathbf{E}}{\partial t^2} \tag{1.17}$$

Similarly, we can write the *wave equation* for **B** as,

$$\nabla^2 \mathbf{B} = \mu_o \varepsilon_o \frac{\partial^2 \mathbf{B}}{\partial t^2} \tag{1.18}$$

These wave equations are linear, second-order, and homogeneous partial differential equations describing the propagation of a field from one location to another (a propagating wave). Further, these equations also provide the *velocity* of the propagating wave as well, as shown below.

Equations (1.17) and (1.18) are analogous to the general form of the wave equation for field, **F**

$$\nabla^2 \mathbf{F} = \frac{1}{v_P^2} \frac{\partial^2 \mathbf{F}}{\partial t^2} \tag{1.19}$$

Now,

$$\frac{1}{v_P^2} = \mu_o \varepsilon_o \tag{1.20}$$

Rearranging equation (1.20) and substituting the values of μ_o and ε_o provides equation (1.21), the phase velocity in free space, and the solution to wave equations (1.17) and (1.18) propagates with the speed of light (c).

$$v_P = \frac{1}{\sqrt{\mu_o \varepsilon_o}} = c = 2.99792458 \times 10^8 \text{ m s}^{-1} \tag{1.21}$$

Therefore, a pair of wave equations describing monochromatic (single frequency) light can be written as,

$$\nabla^2 \mathbf{E} = \frac{1}{c^2} \left(\frac{\partial^2 \mathbf{E}}{\partial t^2} \right) \tag{1.22}$$

$$\nabla^2 \mathbf{B} = \frac{1}{c^2} \left(\frac{\partial^2 \mathbf{B}}{\partial t^2} \right) \tag{1.23}$$

where ∇^2 is the second-order Laplacian operator, which is described as $\nabla^2 = \left(\frac{\partial^2}{\partial x^2} + \frac{\partial^2}{\partial y^2} + \frac{\partial^2}{\partial z^2} \right)$ in Cartesian coordinates.

The two variables in an electromagnetic wave, i.e., both **E** and **B**, are vector quantities and are transverse to the direction of propagation, and mutually perpendicular and coupled (figure 1.1(a)). In free space, their magnitudes are related as

$$\mathbf{E} = c\mathbf{B} \tag{1.24}$$

In the following equation (1.25), the left side provides the electric field vector components, x, y, z in 3D *space* and the right side provides the corresponding *time*-dependent components.

$$\begin{pmatrix} \frac{\partial^2 \mathbf{E}_x}{\partial x^2} + \frac{\partial^2 \mathbf{E}_x}{\partial y^2} + \frac{\partial^2 \mathbf{E}_x}{\partial z^2} \\ \frac{\partial^2 \mathbf{E}_y}{\partial x^2} + \frac{\partial^2 \mathbf{E}_y}{\partial y^2} + \frac{\partial^2 \mathbf{E}_y}{\partial z^2} \\ \frac{\partial^2 \mathbf{E}_z}{\partial x^2} + \frac{\partial^2 \mathbf{E}_z}{\partial y^2} + \frac{\partial^2 \mathbf{E}_z}{\partial z^2} \end{pmatrix} = \frac{1}{c^2} \begin{pmatrix} \frac{\partial^2 \mathbf{E}_x}{\partial t^2} \\ \frac{\partial^2 \mathbf{E}_y}{\partial t^2} \\ \frac{\partial^2 \mathbf{E}_z}{\partial t^2} \end{pmatrix} \tag{1.25}$$

(Field's space components) (Field's time components)

The three non-coupled partial differential equations of second-order can be written as given below.

$$\frac{\partial^2 \mathbf{E}_x}{\partial x^2} + \frac{\partial^2 \mathbf{E}_x}{\partial y^2} + \frac{\partial^2 \mathbf{E}_x}{\partial z^2} = \frac{1}{c^2}\frac{\partial^2 \mathbf{E}_x}{\partial t^2} \tag{1.26}$$

$$\frac{\partial^2 \mathbf{E}_y}{\partial x^2} + \frac{\partial^2 \mathbf{E}_y}{\partial y^2} + \frac{\partial^2 \mathbf{E}_y}{\partial z^2} = \frac{1}{c^2}\frac{\partial^2 \mathbf{E}_y}{\partial t^2} \tag{1.27}$$

$$\frac{\partial^2 \mathbf{E}_z}{\partial x^2} + \frac{\partial^2 \mathbf{E}_z}{\partial y^2} + \frac{\partial^2 \mathbf{E}_z}{\partial z^2} = \frac{1}{c^2}\frac{\partial^2 \mathbf{E}_z}{\partial t^2} \tag{1.28}$$

The solution to the above equations gives the electric field of the wave at any location and at any time.

$$\mathbf{E} = \begin{pmatrix} \mathbf{E}_x(x, y, z, t) \\ \mathbf{E}_y(x, y, z, t) \\ \mathbf{E}_z(x, y, z, t) \end{pmatrix} \tag{1.29}$$

Similarly, for the magnetic field components as

$$\mathbf{B} = \begin{pmatrix} \mathbf{B}_x(x, y, z, t) \\ \mathbf{B}_y(x, y, z, t) \\ \mathbf{B}_z(x, y, z, t) \end{pmatrix} \tag{1.30}$$

From all three components of **E** and **B** fields obtained from equations (1.29) and (1.30), we can calculate the entire *temporal* and *spatial* behavior of an electromagnetic wave.

Maxwell's equation in a chargeless and current less *medium* when $\rho = 0$, and $J = 0$ equations (1.17) and (1.18) can be rewritten as:

$$\nabla^2 \mathbf{E} = \mu\mu_o \varepsilon\varepsilon_o \frac{\partial^2 \mathbf{E}}{\partial t^2} \tag{1.31}$$

$$\nabla^2 \mathbf{B} = \mu\mu_o \varepsilon\varepsilon_o \frac{\partial^2 \mathbf{B}}{\partial t^2} \tag{1.32}$$

where μ and ε are the relative magnetic permeability and relative dielectric permittivity, respectively.

With a plane wave solution, k can be written as:

$$k = \omega\sqrt{\mu\mu_o \varepsilon\varepsilon_o} \tag{1.33}$$

Thus,

$$v_P = \frac{\omega}{k} = \frac{1}{\sqrt{\mu\mu_o \varepsilon\varepsilon_o}} \tag{1.34}$$

In media,

$$v_P = \frac{c}{n} \tag{1.35}$$

where n is the refractive index of the medium and generally,

$$n = \sqrt{\varepsilon\mu} \tag{1.36}$$

Using equations (1.34) and (1.35), one can write equations (1.31) and (1.32) as follows:

$$\nabla^2\mathbf{E} - \frac{n^2}{c^2}\left(\frac{\partial^2\mathbf{E}}{\partial t^2}\right) = 0 \tag{1.37}$$

$$\nabla^2\mathbf{B} - \frac{n^2}{c^2}\left(\frac{\partial^2\mathbf{B}}{\partial t^2}\right) = 0 \tag{1.38}$$

Introducing the scalar E($\mathbf{r}$) function results in $\mathbf{E}(\mathbf{r},t) = \mathrm{p}\ \mathrm{E}(\mathbf{r},t)$, where p is the unit vector. For a plane monochromatic wave with a frequency, ω, the time-dependent part of equation (1.37) is:

$$\mathrm{E}(\mathbf{r},\ t) = \mathrm{E}(\mathbf{r})\, e^{i\omega t} \tag{1.39}$$

Hence, equation (1.37) can be written in the scalar form of the time independent function as:

$$\nabla^2\mathrm{E}(\mathbf{r}) + \frac{n^2(\mathbf{r})}{c^2}\omega^2\mathrm{E}(\mathbf{r}) = 0 \tag{1.40}$$

or

$$\nabla^2\mathrm{E}(\mathbf{r}) + k^2\mathrm{E}(\mathbf{r}) = 0,\ k = \frac{n(\mathbf{r})}{c}\omega \tag{1.41}$$

Equation (1.41) is known as the scalar *Helmholtz equation*, which provides the spatial distribution of the electric field of a light propagating with a frequency in a medium with refractive index $n(\mathbf{r})$. Similar consideration also applies to $\mathbf{B}(\mathbf{r},t)$.

If n is not a function of frequency, the materials are referred to as dispersion-less, and $v_P = v_g$. However, dispersion exists for almost all materials, i.e., $n(\omega)$, and therefore $v_P \neq v_g$.

1.1.2 The resemblance between the Helmholtz and Schrödinger equations

In quantum or wave mechanics, the state of a single electron is described as a wave function, denoted as Ψ, which depends on time and three coordinates (x,y,z). The Ψ of an electron can be determined by solving the steady-state *Schrödinger* equation

$$\hat{\mathbf{H}}\Psi = E\Psi \tag{1.42}$$

where $\hat{\mathbf{H}}$ is the Hamiltonian operator and E is particle (constant) energy.

For a 1D space with a constant potential energy U_o and $-\frac{\hbar^2}{2m}\frac{d^2\Psi(x)}{dx^2}$ is kinetic energy, equation (1.42) can be written as:

$$-\frac{\hbar^2}{2m}\frac{d^2\Psi(x)}{dx^2} + U_o\Psi(x) = E\Psi(x) \tag{1.43}$$

where $\hbar$ is the reduced Planck's constant, $\hbar = \frac{h}{2\pi}$.

Dividing equation (1.43) by $\frac{\hbar^2}{2m}$

$$\frac{d^2\Psi(x)}{dx^2} + \frac{2m}{\hbar^2}(E - U_o)\Psi(x) = 0 \tag{1.44}$$

and by introducing wavenumber notation,

$$\frac{d^2\Psi(x)}{dx^2} + k^2\Psi(x) = 0 \tag{1.45}$$

$$k^2 = \frac{2m}{\hbar^2}(E - U_o) \tag{1.46}$$

$$\frac{\hbar^2k^2}{2m} = E - U_o = \frac{P^2}{2m} \tag{1.47}$$

where P is momentum and $E - U_o$ is the particle kinetic energy. Now, the mathematical resemblance between electromagnetic waves (equation (1.40)) and quantum particles can be seen in the following pair of equations,

$$\nabla^2\mathrm{E}(\mathbf{r}) + \frac{n^2(\mathbf{r})}{c^2}\omega^2\mathrm{E}(\mathbf{r}) = 0 \tag{1.40}$$

$$\frac{\hbar^2}{2m}\nabla^2\,\Psi(\mathbf{r}) + [E - U(\mathbf{r})]\Psi(\mathbf{r}) = 0 \tag{1.48}$$

For a 1D problem, both equations (1.40) and (1.48) take the following form,

$$\frac{d^2\mathrm{E}(x)}{dx^2} + \frac{n^2(x)}{c^2}\omega^2\mathrm{E}(x) = 0 \tag{1.49}$$

$$\frac{d^2\Psi(x)}{dx^2} + \frac{2m}{\hbar^2}[E - U(x)]\Psi(x) = 0 \tag{1.50}$$

Here, the substitution, $\frac{n^2(x)}{c^2}\omega^2 \equiv \frac{2m}{\hbar^2}[E - U(x)] = k^2$ fuses equations (1.49) and (1.50).

For space-independent refractive index $n(x) = n_o \equiv$ constant, and constant potential $U(x) = U_o \equiv$ constant. The wavenumber, k governing electromagnetic wave and quantum particle is given as:

$$k_{\mathrm{EM}} = \frac{n_o\omega}{c} \tag{1.51}$$

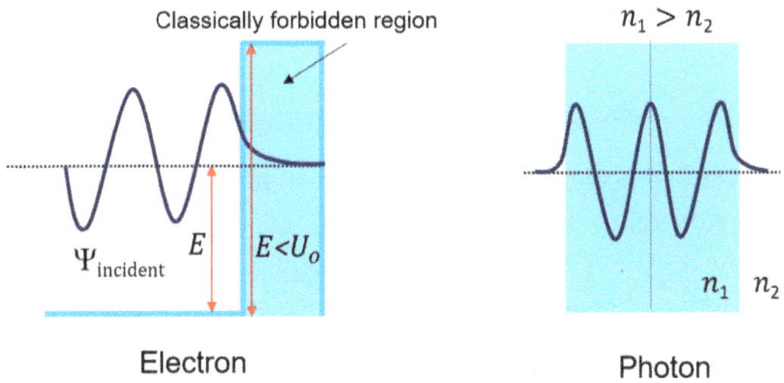

Figure 1.2. Escape of electron and photon into energetically forbidden classical regions.

and

$$k_{\text{QM}} = \frac{1}{\hbar}\sqrt{2m(E - U_{\text{o}})} \tag{1.52}$$

The expression of k points out that space variation in *refractive index* may result in an analogous effect in space variation in *potential.* Equations (1.51) and (1.52) describe the velocity (v) of waves.

$$v_{\text{EM}} = \frac{c}{n} \tag{1.53}$$

and

$$v_{\text{QM}} = \sqrt{\frac{2(E - U_{\text{o}})}{m}} \tag{1.54}$$

The remarkable parallelism between optical and quantum phenomena is evident through several shared characteristics, including:

- The wave-like behavior exhibited by both light and electrons, highlighting the wave-particle duality inherent in quantum mechanics.
- The presence of evanescent waves, which can extend into classically prohibited regions, defying classical expectations (see section 2.1.1).
- The occurrence of wave interference phenomena, over barriers and within potential wells, underscoring the interference patterns that can manifest in quantum systems (figure 1.2).

1.2 The photon nature of light

The electromagnetic wave description of light is not adequate to describe the absorption and emission of a material. The photon nature of light involved in these processes clarifies the quantum mechanical transitions of material between its energy levels. The particle nature of a photon provides momentum (P) which is determined by its wavelength or **k**.

The de Broglie wavelength is

$$\lambda = \frac{h}{P} \tag{1.55}$$

The total energy of a body in terms of its rest mass (m_0) and momentum in a relativistic dispersion relation is expressed as:

$$E^2 = P^2c^2 + m_0^2c^4 \tag{1.56}$$

As for photon $m_0 = 0$, equation (1.56) becomes,

$$E = Pc \tag{1.57}$$

We know that $E = h\nu$; therefore,

$$\lambda = \frac{h}{P} = \frac{h}{\frac{E}{c}} = \frac{h}{\frac{h\nu}{c}} = \frac{c}{\nu} \tag{1.58}$$

In free space:
- the speed of the photon is $c = \lambda\nu$, which is expected for a harmonic wave and it is consistent with the wave theory;
- the energy of the photon is, $E = h\nu = \hbar\omega = Pc$;
- the momentum of the photon is $P = \frac{h\nu}{c} = \frac{h}{\lambda} = \hbar k$.

The energy of a photon of λ in free space can be calculated as:

$$h\nu = \frac{hc}{\lambda} = \frac{1239.8}{\lambda \text{ (in nm)}} \text{ eV} \tag{1.59}$$

For example, the photon energy for a λ of 400 nm is 3.09 eV, and for 1550 nm it is 0.8 eV, which is a typical bandgap of many semiconductors.

1.3 The polarization states of light

The polarization state of light characterizes the electromagnetic field's capacity to oscillate in multiple directions during its propagation. In the majority of cases, light is considered unpolarized due to the random orientation of its electric field vectors. The concept of polarization in electromagnetic waves pertains to the direction of motion of the electric vector (**E**) tip, as depicted in figure 1.3. Conventionally, the electric vector is invariably orthogonal to the direction of wave propagation, but its specific orientation can be anywhere within the plane that is perpendicular to this direction. This fundamental property of light polarization has significant implications for various optical and photonic phenomena.

In *linear* or *plane polarization*, the electric field tip of light oscillates solely within a single plane, and this plane is perpendicular to the direction of propagation (z-axis). Linear polarization can be further categorized into two primary orientations: horizontal (H) and vertical (V). Linear polarization in light is achieved using polarizers, which selectively transmit light waves based on their electric vector

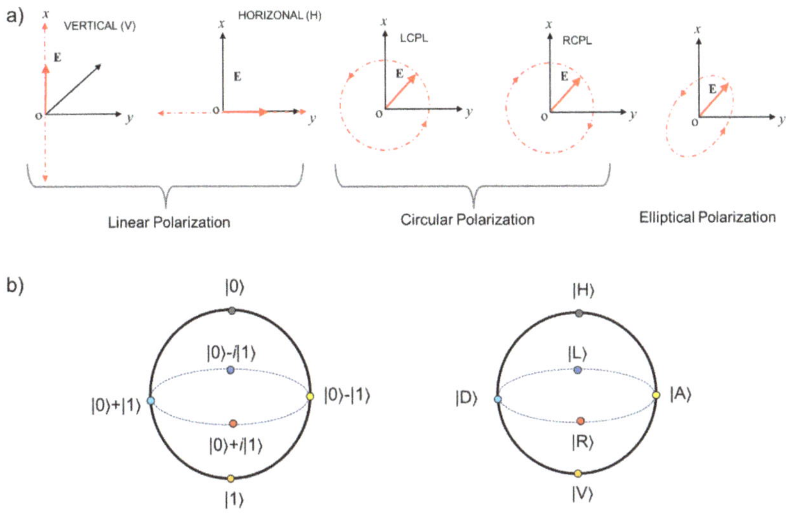

Figure 1.3. (a) Polarization states of light's electric field, **E**. The dotted lines show the movement of the tip of the electric field vector with time. (b) The states of a qubit on the Bloch sphere (Poincaré sphere in optics) for quantum information processing. The circles denote the **E** tip pointing directions.

orientation. The amplitude of the transmitted light (I_t) through a polarizer is related to incident light (I_i) by the relation: $I_t = I_i \cos^2\theta$. Here, I_i is proportional to E^2, and θ is the angle between the polarizer's transmission axis and the light's polarization direction. For example, when $\theta = 0°$ (or 180°), I_t is at its maximum, and for $\theta = 90°$ (or 270°) the value of $I_t = 0$.

Circularly polarized light (CPL) exhibits a unique property in which the electric field tip traces a circular path over time. Depending upon the direction of rotation (handedness), this polarization is called right circularly polarized light (RCPL) and left circularly polarized light (LCPL). This particular polarization state is characterized by two electric field components of equal amplitude, which are orthogonal to each other and maintain a 90° phase difference. This leads to a rapid rotation of the electric field vector with a constant magnitude. The CPL can be produced by passing a linearly polarized light through a quarter wave ($\lambda/4$) plate at an angle of 45° relative to the optical axis.

In *elliptical polarization*, the magnitude and direction of the electric field vector undergo constant variation. This state is defined by two perpendicular electric field components with differing amplitudes and a 90° phase difference between them.

Unpolarized light can be produced by the combination of vertical and horizontal linear polarization components or LCPL and RCPL. This results in partially polarized light due to the statistical superposition of an unpolarized component and a fully polarized one.

The interaction of light at the interface between two materials is influenced by its polarization state. Additionally, the absorption and scattering of light within materials are polarization-dependent phenomena. In the case of anisotropic materials, the refractive index is also polarization-dependent, leading to variations in the speed of light propagation and consequent phase shifts, such as the transformation from linear to circular polarization. Notably, when plane-polarized light traverses through optically active media, it experiences a rotation in its polarization state.

Polarization states of light play a pivotal role in quantum photonic technologies, particularly in quantum processing and quantum communication [1, 2]. These polarization degrees of light offer a valuable means to encode quantum bits (or qubits), which are the quantum counterparts of classical binary bits. A qubit, denoted as $|\psi\rangle$, can exist in one of two fundamental states, $|0\rangle$ or $|1\rangle$, or in a linear superposition of these basis states, which can be expressed as

$$|\psi\rangle = \alpha|0\rangle + \beta|1\rangle \tag{1.60}$$

The symbols α and β represent the amplitudes of quantum states (complex numbers), and they determine the probability of detecting the photon in a particular state during an experiment. These amplitudes are subject to the normalization rule, which stipulates that the sum of the squared magnitudes of α and β must equal 1:

$$|\alpha|^2 + |\beta|^2 = 1 \tag{1.61}$$

In contrast to the deterministic measurement of classical bits, quantum mechanics provides outcomes with probabilistic characteristics. When an experiment is conducted to measure the quantum state $|\psi\rangle$, the result 0 is obtained with a probability of $|\alpha|^2$, and the result 1 is obtained with a probability of $|\beta|^2$. This implies that the state of $|\psi\rangle$ can collapse to either $|0\rangle$ or $|1\rangle$ based on the outcome of the measurement, in accordance with the probabilistic nature of quantum phenomena [3].

A single qubit in a superposition, denoted as $|\psi\rangle = \alpha|0\rangle + \beta|1\rangle$, can be equivalently represented in matrix form as a two-dimensional vector, with the amplitudes α and β functioning as the components of this vector.

$$|\psi\rangle = \begin{pmatrix}\alpha\\ \beta\end{pmatrix} \tag{1.62}$$

Now the states can be represented as

$$|0\rangle = \begin{bmatrix}1\\ 0\end{bmatrix} \text{ and } |1\rangle = \begin{bmatrix}0\\ 1\end{bmatrix}$$

For example, a qubit (0 or 1) can be encoded using the horizontal ($|\mathrm{H}\rangle = |0\rangle$) and vertical ($|\mathrm{V}\rangle = |1\rangle$) polarization of a single photon, i.e., $|\psi\rangle = \alpha|\mathrm{H}\rangle + \beta|\mathrm{V}\rangle$.

An arbitrary set of states for a single qubit can be visualized on the Poincaré (or Bloch) sphere, representing horizontal ($|0\rangle$) and vertical ($|1\rangle$), diagonal ($|0\rangle+|1\rangle$), anti-diagonal ($|0\rangle-|1\rangle$), left circular ($|0\rangle-i|1\rangle$), and right circular ($|0\rangle+i|1\rangle$) (figure 1.3(b)). The north and south poles of the sphere correspond to measurable states ($|0\rangle$) and ($|1\rangle$). In other regions of the sphere, the qubit exists in superposition states. Along the equator of the sphere, the qubit may probabilistically collapse to either ($|0\rangle$) or ($|1\rangle$) with a 50:50 probability upon measurement. Experimentally, the polarization manipulation (encoding and decoding) can be achieved using waveplates and polarization beam splitters.

Furthermore, the creation of quantum entangled states is attainable through nonlinear optical (NLO) crystals. The principles of superposition and entanglement [4, 5] form the foundation for advanced quantum computing capabilities.

1.4 The chemistry and physics of materials

Photon absorption via interband transition is one of the exciting forms of light–matter interactions in semiconductors. This transition gives rise to the generation of an electron in the conduction band (CB) and a hole in the valence band (VB) (figure 1.4(a)). The electrostatic Coulomb attractive forces between them create a bound hole–electron pair (akin to a hydrogen-like bound state) commonly referred to as an '*exciton*'. The energy necessary for the creation of an *exciton* is defined by the band gap energy ($\mathbf{E}_g$). The excitons have lighter effective mass (m^*) than that of bare electrons ($\approx 0.2\ m_e$).

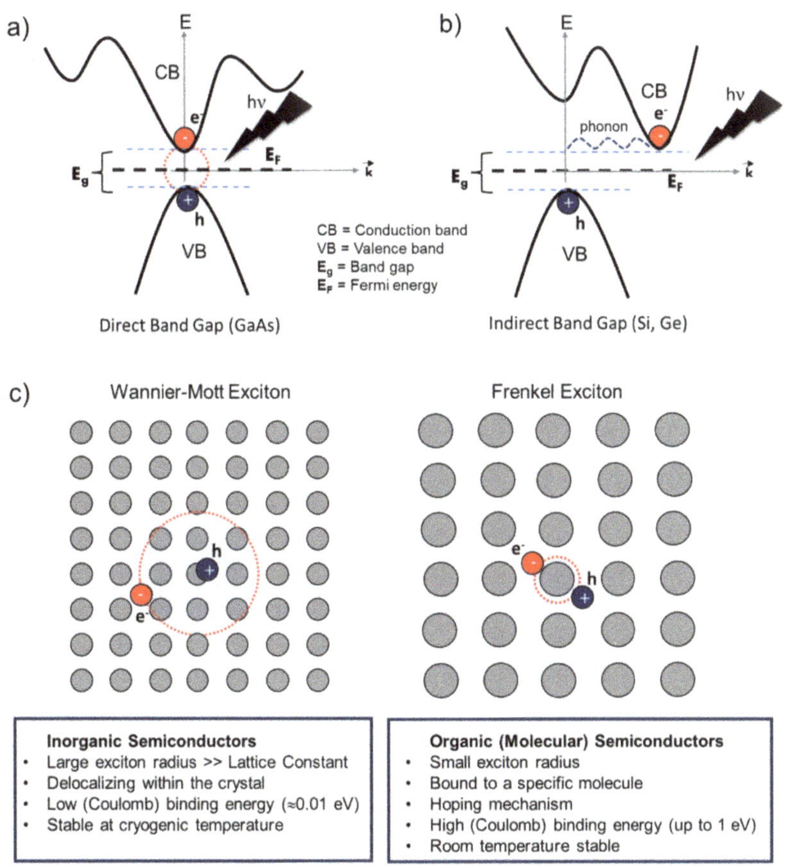

Figure 1.4. (a and b) The direct and indirect band structures, respectively. (c) Qualitative illustration of the Wannier–Mott and Frenkel excitons formed in inorganic and organic semiconductors, respectively.

Analogous to the hydrogen atom with a Bohr radius (a_B)[1] of 0.053 nm, an exciton created within a material is characterized by the '*exciton Bohr radius*' denoted as a_{ex}

$$a_{ex} = \frac{4\pi\hbar^2\varepsilon\varepsilon_o}{e^2 m^*} \tag{1.63}$$

where ε and ε_o are the material's relative electric permittivity and free space electric permittivity, respectively, $\hbar$ is the reduced Planck's constant, and e is the electron charge (1.602×10^{-19} C).

$$a_{ex} \approx a_B \varepsilon \frac{m_e}{m^*} \tag{1.64}$$

The effective mass of the electron (m_e^*) and hole (m_h^*) can be calculated by fitting the parabolic functions around the CB or the VB using the relation,

$$m^* = \pm\hbar^2[d^2E\mathbf{k}/d\mathbf{k}^2]^{-1} \tag{1.65}$$

where E is the energy of an electron at wave vector $\mathbf{k}$ in that band.

The a_{ex} value varies for different materials, serving as a useful metric to evaluate the degree of confinement in semiconducting nanomaterials. For example, the values of ε, m_{eff}, and a_B for PbS are 17.3, 0.05, and 18 nm, respectively, and for CdS 8.9, 0.16, and 3 nm, respectively.

The presence of the electron–hole pair interactions results in a stabilizing energy, thereby lowering the energy of the pair compared to that of unbound electrons and holes. These excitons can either recombine into a photon via radiative recombination or thermally dissipate their energy non-radiatively. The excitons formed in inorganic and organic semiconductor materials exhibit variations in their exciton binding energies (E_B). The E_B is related to a_B by

$$E_B = \frac{e^2}{2\varepsilon a_B} \tag{1.66}$$

For a non-magnetic material, $\mu \approx 1$. Therefore, equation (1.36) can be written as:

$$n = \sqrt{\varepsilon} \tag{1.67}$$

The relative permittivity or *dielectric constant* [note: it is a function of frequency, $\varepsilon(\omega)$] is a dimensionless number that is, in general, complex-valued.

Since ε is the complex function with the real and imaginary parts, $\varepsilon = \varepsilon' + i\varepsilon''$ and so the value of n is also a complex function[2], $\varepsilon = n^2 = (n' + i\,n'')^2$.

---

[1] The Bohr radius is defined as the radius of the maximum probability for finding the ground state electron in the hydrogen atom; $a_B = \frac{4\pi\varepsilon_0\hbar^2}{m_e e^2} \approx 0.5292$ Å, or 0.053 nm, or 53 pm, or 5.3×10^{-11} m.
Here $\hbar$ is Planck's constant divided by 2π, m_e is the mass of electron, and e corresponds to electron charge.

[2] The direct measurement of the complex refractive index poses a challenge. Extracting this information from parameters like reflectance, transmittance, or angles of refraction for polarized light using conventional theories is not straightforward. However, alternative methods, including the Kramers–Kronig transform method [6], ellipsometry [7], or the reflection-transmission method [8], prove effective in deriving the complex refractive index.

For homogenous (isotropic) dielectric materials, the electric field densities **D** is given by,

$$\mathbf{D} = \varepsilon\, \mathbf{E} \tag{1.68}$$

Further,

$$\varepsilon = \varepsilon_{o}\,(1 + \chi) \tag{1.69}$$

where χ is the *electric susceptibility* of the material.

Additionally, combining equations (1.68) and (1.69) provides equation (1.70),

$$\mathbf{D} = \varepsilon\,\mathbf{E} = \varepsilon_{o}(1 + \chi)\,\mathbf{E} = (\varepsilon_{o}\mathbf{E} + \varepsilon_{o}\chi\mathbf{E}) = \varepsilon_{o}\,\mathbf{E} + \mathbf{P} \tag{1.70}$$

The n is related to χ by the following relation,

$$n = \sqrt{\varepsilon} = \sqrt{1 + 4\pi\chi} \tag{1.71}$$

Here, χ represents the degree of polarization (**P**) of the electron cloud of an atom in response to applied light's electric field and it is also called the first-order susceptibility tensor. For organic and inorganic semiconductors, ε is in the range of 3 and 6–12, respectively.

For *anisotropic* materials, based on ε in the x-, y-, and z-directions, the relation in matrix form can be written as:

$$\begin{bmatrix} \mathbf{D}_x \\ \mathbf{D}_y \\ \mathbf{D}_z \end{bmatrix} = \begin{bmatrix} \varepsilon_{x,x} & \varepsilon_{x,y} & \varepsilon_{x,z} \\ \varepsilon_{y,x} & \varepsilon_{y,y} & \varepsilon_{y,z} \\ \varepsilon_{z,x} & \varepsilon_{z,y} & \varepsilon_{z,z} \end{bmatrix} \begin{bmatrix} \mathbf{E}_x \\ \mathbf{E}_y \\ \mathbf{E}_z \end{bmatrix} \tag{1.72}$$

For NLO materials, ε depends on the magnitude applied **E** of the light.

In the band structure of inorganic semiconductors, we observe discrete energy bands characterized by a VB and an empty CB. The schematic images of the direct and indirect band gaps are given in figures 1.4(a) and (b). In contrast, organic semiconductors exhibit a band structure consisting of molecular orbitals, with the energy gap being the separation between the highest occupied molecular orbital (HOMO) and the lowest unoccupied molecular orbital (LUMO). The dispersions for both HOMO/LUMO are relatively flat.

For inorganic semiconductors, the small binding energy of excitons (often referred to as Wannier–Mott excitons) limits strong light–matter coupling, particularly at room temperature (RT). For example, in GaAs, the binding energy is approximately 10 meV, which is about three times smaller than that of hydrogen atoms [9]. In contrast, materials such as GaN and ZnO exhibit slightly higher exciton binding energies, approximately 20 and 60 meV, respectively [10, 11].

The larger value of ε in inorganic semiconductors increases the dielectric screening within the lattice, leading to a reduction in the Coulomb interaction between the electrons and holes and consequently a decrease in the E_B. The formation of Wannier–Mott excitons [12] is also facilitated by the small electron m^*. Consequently, these excitons have a larger radius of the order of 10 nm, (larger than the lattice spacing). The binding energies of these excitons are about $E_B \approx 10$–20 meV, much lower than thermal energy at RT, i.e. $k_B\mathbf{T} = 25.6$ meV (where k_B is the Boltzmann constant = $8.617\ 3303 \times 10^{-5}$ eV K^{-1} and **T** is the temperature in Kelvin units); therefore, they are unstable at RT. These excitons can be stabilized at cryogenic temperature.

Organic semiconductors exhibit charge-transfer (CT) characteristics, with their excitons being of the Frenkel type, as illustrated in figure 1.4(c) [13, 14]. Molecules with electron donor and acceptor groups, so-called push–pull molecules, show effective CT characteristics. In these excitons, the electron and hole are situated in the same molecule making the a_B small. This proximity provides strong binding energy of the order of 0.1–1.0 eV, which is much higher than inorganic semiconductors [15]. As a result, Frenkel excitons remain stable at RT exhibiting a lifetime of a few picoseconds and m^* in the order of $\approx 0.0001\ m_e$, when they are confined.

As a consequence, these excitons exhibit strong coupling with light when trapped within a high finesse optical resonator (to be discussed in section 2.2.1.1) [16], giving rise to a new class of hybrid states known as *exciton-polaritons* (EPs) [17, 18]. EPs represent quasi-particles, embodying a dual nature of part-light and part-matter. However, there is a trade-off associated with their high E_B, which reduces the effective refractive index (n_{eff}) of organic crystals to approximately 1.6–1.8 (although still higher than that of SiO_2), as described by equations (1.66) and (1.67).

In CT excitons, the charge carriers are typically located on adjacent or nearest-neighbor molecules. These unique characteristics of excited states (excitons) distinguish organic semiconductors fundamentally from their inorganic counterparts [15]. It is anticipated that organic materials will contribute to significant advancements in nanophotonic applications, encompassing areas such as nonlinear optics, light emission, waveguiding, cavities, lasing, and photonic circuitry.

Another very important development of organic molecules in photonics is their use as micro/nano scale lasers. For lasers, population inversion of the gain medium is considered an essential prerequisite. Nevertheless, for many potential applications, the properties of the emitted light such as coherence, directionality, and monochromaticity are imperative, not the principle of operation of the laser.

1.4.1 Population inversion-less lasers

In 1996, Yamamoto *et al* theoretically predicted the first *population inversion-less* EP lasers [19]. One of the distinguishing features of EPs is their exceedingly low m^*, of the order of 10^4 times smaller than the bare electron. These EP particles are bosonic in nature [20, 21] and their very light (m^*) mass allows the realization of phase transition even at modest particle densities forming condensate with long-range order (spatial coherence). In a strong coupling regime, i.e., if the rate at which photons escape from the optical cavity is smaller than the rate of production of photons from the excitons and vice-versa, the EPs can form a condensate (like atomic Bose–Einstein condensates (BECs) [20, 21], and superfluids) or coherent state at low density and high temperature.

Formation of EP has been demonstrated in GaAs [22, 23], CdTe [24, 25], GaN [8], ZnO [26], and melt-grown anthracene single crystals [27]. The EP condensate is useful for producing population inversion-less lasers. Standard light lasers based on stimulated emission require the attainment of population inversion. On the other hand, in semiconductor (quantum well) lasers this population inversion condition is fulfilled by the *exciton Mott transition* criterion:

$$\rho_{\text{laser}} \geqslant 1/(a_{\text{ex}})^2 \tag{1.73}$$

where ρ_{laser} is the critical superfluid density of bosons. For example, the substitution of the values of a_{ex} for GaN and ZnO into equation (1.73) meant for population inversion condition, we find $\rho_{\text{laser}}(\text{GaN}) \approx 1.3 \times 10^{13}\ \text{cm}^{-2}$ and $\rho_{\text{laser}}(\text{ZnO}) \approx 3.1 \times 10^{13}\ \text{cm}^{-2}$, respectively [28]. This condition allows the achievement of coherent light waves without meeting the requirement of population inversion.

Unlike electrons, the EPs are chargeless and therefore, not that much influenced by the environment. The strong coupling is achieved between excitons in bulk or quantum wells, and photons trapped inside a high finesse optical resonator [16]. Advanced research over recent years has shown that condensation of EPs in micro-optical cavities [21] or resonators represents a fundamentally different and technologically more efficient way of generating coherent light. Experimentally, EP condensates were produced around 10 K in GaAs and CdTe materials. At cryogenic temperature, the EP lifetime of GaAs is an order of magnitude higher than the organic materials. Remarkably, GaN, ZnO, and organic semiconductors show polariton condensates at higher temperatures, including RT [8, 15, 26].

Recently, Cuves *et al* proved the quantum nature of EPs by exciting them with an entangled photon [29]. This result demonstrates that EP emission can be used for quantum cryptography and quantum information processing applications.

1.5 Nonlinear optical (NLO) properties

Naturally, the interaction of light and matter is very weak. Therefore, the interaction of ambient light or continuous wave laser with matter falls in the linear optical regime. The order and magnitude of light–matter interaction can be assessed by comparing the light's electric field strength (E_{l}) to the electric field (E_{a}) of atoms of the material [30–32]. The field within the matter reflects the interactions between charged particles within the materials. Sunlight has a typical E_{l} of $\approx$600 V m^{-1}, while the typical atomic field in the matter is $E_{\text{a}} \approx 5.24 \times 10^{11}$ V m^{-1}, which can be numerically estimated from the following equation [31]:

$$E_{\text{a}} = \frac{e}{4\pi\varepsilon_{\text{o}}\, a_{\text{o}}^2} \tag{1.74}$$

As one can see clearly, $E_{\text{a}} \ggg \text{E}_{\text{l}}$. This is expected because the atomic field at the scale of individual atoms is typically much stronger than the macroscopic electric fields associated with sunlight. Therefore, any interacting light only to some extent polarizes the electrons in the material from their normal position. In other words, the light–matter interaction is very feeble.

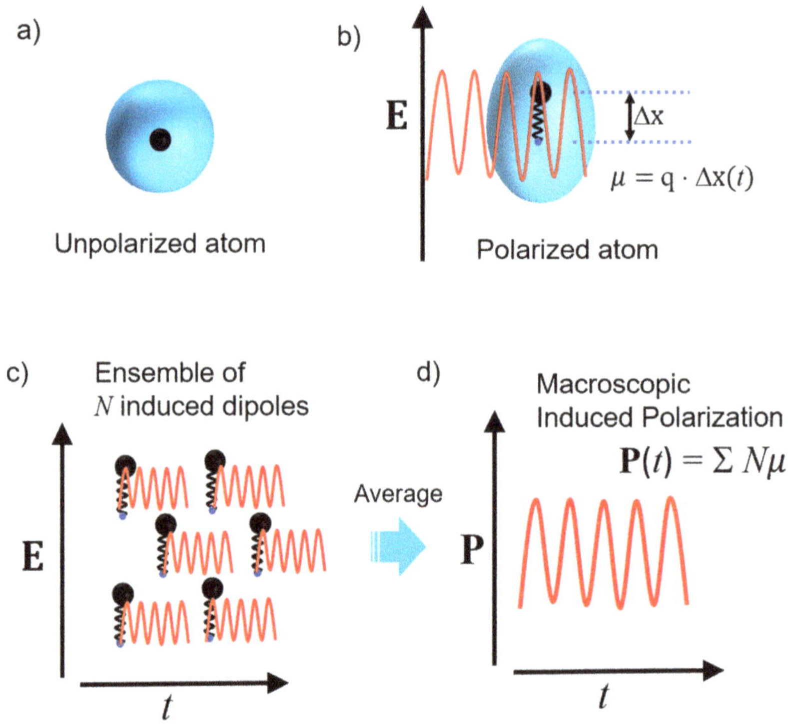

Figure 1.5. Schematic illustration of (a) an unpolarized atom, (b) electron cloud distortion in an atom by the light's electric field (**E**), and formation of an electric dipole moment ($\boldsymbol{\mu}$). (c) Ensemble of N induced dipoles and (d) the resultant macroscopic induced polarization. The oscillating dipole radiates in the direction of incident light perpendicular to the oscillating vector.

NLO light–matter interactions are observed at very high optical intensities once $E_a \approx E_l$. The required light intensity (I_a) corresponding to E_a can be calculated from the following equation [31]:

$$I_a = \tfrac{1}{2}\, c\varepsilon_o (E_a)^2 = 3.5 \times 10^{20}\ W\ \text{m}^{-2} = 3.5 \times 10^{16}\ W\ \text{cm}^{-2} \tag{1.75}$$

Typically, the light intensity of this order is provided by the pulse lasers.

In an atom, light's electric field **E** creates a separation of charges in the direction collinear to it due to the movement of the nucleus in the **E** direction and the electron cloud in the opposite direction (figures 1.5(a) and (b)). This polarization produces an oscillating dipole $\boldsymbol{\mu}$ for each atom in the material pointing in the direction of light's **E**. For bulk material, the resultant optical response is given as induced macroscopic polarization, $\mathbf{P}(t) = \Sigma\, N\, \boldsymbol{\mu}$, where N is atoms per unit volume.

The concept of optical nonlinearity of materials can be understood better by considering the dependence of $\mathbf{P}(t)$ on the strength of $\mathbf{E}(t)$, where t is time. For optically linear materials, **P** is a linear function of **E**, i.e., **P** oscillates at the same frequency as **E**. This can be described by the relationship (in SI units):

$$\mathbf{P}(t) = \varepsilon_o \chi^{(1)} \mathbf{E}(t) \tag{1.76}$$

where $\chi^{(1)}$ is called the material's linear electric susceptibility. For simplicity, $\mathbf{P}(t)$ and $\mathbf{E}(t)$ are taken as scalar quantities.

At very high applied electric fields, **P** is a nonlinear function of **E**. This leads to the appearance of the second, third, and higher orders of the NLO properties. For NLO materials (responding instantaneously), their *macroscopic* (bulk level)[3] optical response can be given as a power series in the field strength.

$$\mathbf{P}(t) = \varepsilon_o[\chi^{(1)}\,\mathbf{E}(t) + \chi^{(2)}\,\mathbf{E}^2(t) + \chi^{(3)}\,\mathbf{E}^3(t) + \cdots] \tag{1.77}$$

$$\mathbf{P}(t) = \mathbf{P}^{(1)}(t) + \mathbf{P}^{(2)}(t) + \mathbf{P}^{(3)}(t) + \cdots \tag{1.78}$$

where $\chi^{(2)}$ and $\chi^{(3)}$ are the second- and third-order NLO susceptibilities, respectively. Mostly, NLO susceptibilities depend upon the frequencies of the applied fields. It is also assumed that the medium reacts instantaneously (without any loss and dispersion); therefore, NLO susceptibilities are taken as constants.

The second- and third-order NLO polarizations, $\mathbf{P}^{(2)}(t)$ and $\mathbf{P}^{(3)}(t)$, respectively, can be written as:

$$\mathbf{P}^{(2)}(t) = \varepsilon_o \chi^{(2)}\,\mathbf{E}^2(t) \tag{1.79}$$

$$\mathbf{P}^{(3)}(t) = \varepsilon_o \chi^{(3)} \mathbf{E}^3(t) \tag{1.80}$$

Therefore, their optical processes are different. Under non-resonant excitation conditions $\chi^{(2)}$ is in the order of $\chi^{(1)}/E_a$. For condensed matter, $\chi^{(1)}$ is of the order of unity, because $\chi^{(1)}$ is the product of the number density of atoms [which is of the order of $(a_o)^{-3}$] and polarizability of atoms [which is of the order of $(a_o)^3$].

Therefore, the magnitude of $\chi^{(2)}$ is $1/E_a \cong 1.94 \times 10^{-12}$ m V^{-1}. Similarly, $\chi^{(3)}$ is in the order of $\chi^{(1)}/(E_a)^2 \cong 3.78 \times 10^{-24}$ m^2 V^{-2}.

$$\chi^{(1)} > \chi^{(2)} > \chi^{(3)}$$

Similarly, their emitted intensities, I, vary as

$$I^{(1)} > I^{(2)} > I^{(3)}$$

The time-varying polarization can produce new components of the electromagnetic field. Therefore, the wave equation described in equation (1.37) in NLO media takes the form as [31]

$$\nabla^2\mathbf{E} - \frac{n^2}{c^2}\left(\frac{\partial^2\mathbf{E}}{\partial t^2}\right) = \frac{1}{\varepsilon_o c^2}\left(\frac{\partial^2\mathbf{P}^{\mathrm{NLO}}}{\partial t^2}\right) \tag{1.81}$$

[3] The *macroscopic* material polarization phenomenon can be traced back to the *microscopic* molecular level; the polarization intensity is represented by the dipole moment (μ) induced by the electric field.
$\mu(t) = \alpha\mathbf{E}(t) + \beta\mathbf{E}^2(t) + \gamma\mathbf{E}^3(t) + \cdots$, where α is polarizability, β and γ are the first and second hyperpolarizability, respectively.

The $\frac{\partial^2 \mathbf{P}^{NLO}}{\partial t^2}$ component in equation (1.81) is, in a way, like the *Larmor formula* used in electrodynamics, which essentially explains the production of electromagnetic radiation during time-dependent acceleration of nonrelativistic charges (electron or proton or an ion, whose velocity is smaller than the speed of light).

1.5.1 Second harmonic generation (SHG)/frequency doubling/up-conversion

Two important NLO phenomena commonly observed in organic materials are second harmonic generation (SHG) and two-photon luminescence (TPL) [30–32]. In the SHG process, the NLO material destroys two photons of frequency ω, and simultaneously generates a photon of frequency 2ω, in a single quantum mechanical process (figure 1.6(a)). For example, two 800 nm (infrared) photons combine to simultaneously produce 400 nm (violet) photons. SHG or *frequency doubling* or *up-conversion* is an energy and momentum-conserved NLO process.

To understand this process further, let us imagine a laser beam of strength **E** (with $I_a \approx 10^{16}$ W cm^{-2}) represented as $\mathbf{E}(t) = \mathrm{E_o} \cos \omega t$ incident on a non-centrosymmetric crystal for which $\chi^{(2)} \neq 0$. The ensuing NLO polarization that is created in the same crystal is given as:

$$\mathbf{P}^{(2)}(t) = \frac{1}{2}\varepsilon_o \chi^{(2)} \mathrm{E}_o^2 + \frac{1}{2}\varepsilon_o \chi^{(2)} \mathrm{E}_o^2 \cos 2\omega t \tag{1.82}$$

There are two terms in equation (1.82) for $\mathbf{P}^{(2)}(t)$, the first is a non-frequency contribution known as *optical rectification*, which leads to the creation of a static electric field in nonlinear crystals, useful for the production of terahertz (THz; 1 THz = 10^{12} Hz) radiation. The second term represents generation of 2ω (SHG).

As mentioned before, only non-centrosymmetric crystals (lack of inversion symmetry) exhibit SHG, while third-order NLO interactions can occur in both centrosymmetric and non-centrosymmetric media.

For centrosymmetric (in other words, having inversion symmetry) crystals, $\chi^{(2)} = 0$, as a result, no SHG is produced as per equation (1.82). Take an example of a cubic symmetry lattice, if we replace x-, y-, z-directions by $-x$, $-y$, $-z$,

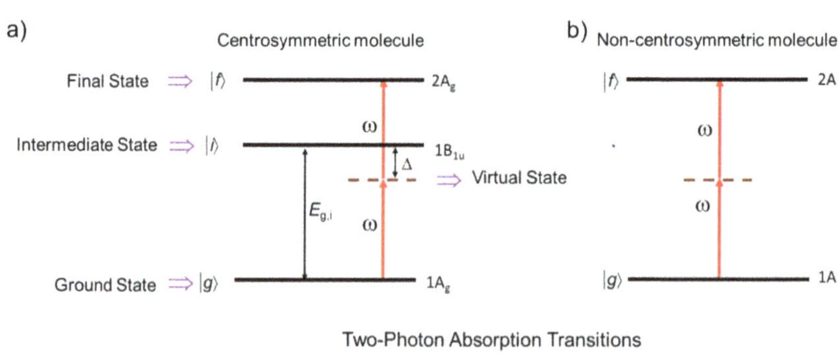

Figure 1.6. Graphical representations of energy level diagram of two-photon transitions in (a) centrosymmetric and (b) non-centrosymmetric molecules. The states are labeled for D_2h and C_2 symmetry, respectively.

respectively, the lattice is not distinguishable. If $\mathbf{E}_x$ is applied in the $+x$-direction, with a corresponding $\mathbf{P}_x$ in the $+x$-direction, due to inversion, it adds a minus sign to $\mathbf{E}_x \rightarrow -\mathbf{E}_x$ and $\mathbf{P}_x \rightarrow -\mathbf{P}_x$. Once the term $\mathbf{E}_x^2$ is involved, the sign of $\mathbf{P}_x$ will not reverse when $\mathbf{E}_x$ reverse its sign. This shows that $\mathbf{P}_x$ must be an odd function of $\mathbf{E}_x$ comprising $\mathbf{E}_x$, $\mathbf{E}_x^3$, ... terms, not $\mathbf{E}_x^2$. Therefore, for a centrosymmetric lattice, $\chi^{(2)} = 0$.

In addition to high nonlinear susceptibility, *phase matching* is also important to increase the NLO efficiency. Phase matching occurs when the fundamental $\mathbf{k}(\omega)$ and SH beam, $\mathbf{k}(2\omega)$ are equal (i.e. $\Delta\mathbf{k} = 0$), that is, momentum is conserved.

Therefore, the efficiency of SHG from the non-centrosymmetric material is related to equation (1.83)

$$\Delta\mathbf{k} = \mathbf{k}(2\omega) - 2\,\mathbf{k}(\omega) \approx 0 \tag{1.83}$$

In crystalline (birefringent) materials, $\mathbf{k}$ is a vector quantity, which means that the phase matching occurs only when $n(\omega) = n(2\omega)$. It shows that the speed of fundamental and second harmonic waves should be equal [31]. More discussion on SHG in birefringence materials can be found elsewhere (see also section 2.1.6).

Generally, during the irradiation of fundamental IR radiation, molecular crystals produce both SHG and TPL. Therefore, the use of a longer wavelength of fundamental IR radiation which falls away from the TPA region produces only SHG. Both SHG and TPL are two-photon mediated processes, and their intensities are quadratically dependent on the laser pump power. The slope value of the log–log plot of NLO signal intensity versus the pump power directly provides information on whether the transition is a 2-photon or 3-photon absorption process (see section 2.1.6 and figures 2.9(f) and (i)).

1.5.2 Two-photon absorption (TPA)

TPA is an NLO process that takes place due to the simultaneous absorption of two photons by a molecule and the subsequent transition from the ground to the excited state (figure 1.6). TPL is exhibited by both centrosymmetric and non-centrosymmetric materials, though the optical selection rules for both types of materials are different. Mostly, TPL shows up at the single-photon PL band region where the absorption at the doubled frequency of the pump radiation is sufficient. However, TPA is several orders of magnitude weaker than one-photon absorption. TPA is a $\chi^{(3)}$ process and Göppert-Mayer [33] experimentally demonstrated this process using a laser.

The two transition steps that take place from the initial (E_g) and final (E_f) states in different time sequences in a TPA process is given as:

$$E_f - E_g = 2\,h\nu \tag{1.84}$$

When a laser light of intensity (I) passes through an optical material of thickness z, it undergoes attenuation, which can be written as:

$$\frac{dI(z)}{dz} = -\alpha I(z) - \beta I^2(z) - \gamma I^3(z) - \cdots, \tag{1.85}$$

Here, α, β, and γ, are 1-, 2-, and 3-photon *absorption coefficients* of the transmitting material medium.

TPA at a specific frequency is proportional to $\text{Im}(\chi^{(3)})$, as per the following equation [34]:

$$\beta(\omega) = \frac{3\omega}{2n^2c^2\varepsilon_0}\text{Im}(\chi^{(3)}) \tag{1.86}$$

The TPA wavelength is always different from one-photon absorption. Therefore, when there is no one-photon absorption or linear absorption ($\alpha = 0$) at the laser wavelength λ, we have,

$$\frac{dI(z)}{dz} = -\beta\, I^2(z) \tag{1.87}$$

Equation (1.87) indicates that the TPA probability of molecules is proportional to the square of the local light intensity. Therefore,

$$I(z, \lambda) = \frac{I_0(\lambda)}{1 + \beta(\lambda)I_0(\lambda)z} \tag{1.88}$$

where $I_0(\lambda)$ is the incident light intensity with a top-hat pulse shape and z is the light propagation length in the material.

$\beta(\lambda)$ can be further expressed as

$$\beta(\lambda) = \sigma'_2(\lambda)N_0 = \sigma'_2(\lambda)N_A\, d_0 \times 10^{-3}\ (\text{in units of cm GW}^{-1}) \tag{1.89}$$

Here $\sigma'_2(\lambda)$ is the molecular TPA cross-section in units of $\text{cm}^4\ \text{GW}^{-1}$ that characterizes average TPA per molecule, N_0 is the molecular density in units of $1/\text{cm}^3$, N_A is Avogadro's number, and d_0 is the molar concentration of the absorbing molecules in units of M.

The alternative expression for TPA cross-section is

$$\sigma_2(\lambda) = \sigma'_2(\lambda)h\nu\ (\text{in cm}^4/(\text{photon/s}) \text{ or cm}^4\text{ s units}) \tag{1.90}$$

Here, simply, 1 Göppert-Mayer (GM) = 10^{-50} cm^4 s, as the usually measured TPA cross-section is from 10^{-51} to 10^{-46} cm^4 s [34].

The NLO transmission for a TPA medium with optical path length l_0 can be written as

$$T(I_0, \lambda) = \frac{I(l_0, \lambda)}{I_0(\lambda)} = \frac{1}{1 + \beta(\lambda)I_0(\lambda)l_0} \tag{1.91}$$

For a plane-polarized light, the value $\sigma_2(\lambda)$ for a transition from ground state to final state at the Lorentzian line-shaped TPA band maximum (in SI units) [31, 35, 36] is given by:

$$\sigma_2(\lambda)_{\max} = \frac{2\pi h\nu^2L^4}{\varepsilon_0^2n^2c^2}\left(\frac{1}{\Gamma}\right)S_{f,g} \tag{1.92}$$

where $S_{f,g} = \left[\sum_i \frac{\langle\mu_{g,i}\,\mu_{i,f}\rangle}{\langle E_{g,i} - h\nu\rangle}\right]^2$.

Here, $E_{g,\,i}$ is the ground state and an intermediate state (i) energy gap, and Γ is the full-width at half-maximum of the TPA or 2PA band in energy units. For the overall orientation of the molecule, the average is 1/5, if we have a co-parallel $\mu_{g,\,i}$ and $\mu_{i,\,f}$, as it is valid for most of the TPA chromophores. Therefore, without the vector notations, an equation for this situation can be written as:

$$S_{f,\,g} = \frac{1}{5}\left[\underset{\text{(Dipolar part)}}{\left(\frac{\Delta\mu_{g,\,f}\mu_{g,\,f}}{h\nu}\right)^2} + \underset{\text{(Two-photon part)}}{\sum\nolimits_{i\neq f,\,g}\left(\frac{\mu^2_{g,\,i}\mu^2_{i,\,f}}{(E_{g,\,i} - h\nu)^2}\right)}\right] \tag{1.93}$$

where $\Delta\mu_{g,\,f}$ is the change in the final static dipole moment relative to the ground state. In equation (1.93), the first term is related to the dipolar part and the second is the two-photon part [37].

For centrosymmetric molecules, all dipole moments are static and hence the dipolar part is zero. Therefore, $\sigma_2(\lambda)$ is majorly influenced by the interaction of the ground state with two excited states. In such a situation, equation (1.93) can be simplified to an approximate form:

$$\sigma_2(\lambda)_{\max} \approx C\frac{\mu^2_{g,\,i}\mu^2_{i,\,f}}{((E_{g,\,i}/h\nu) - 1)^2\Gamma} \tag{1.94}$$

where C is the constant. In this model, there are three essential states proposed with alternating symmetry. The *gerade* (symmetric) wave functions for the states are $|g\rangle$ (ground state) and $|f\rangle$ (final state), and for *ungerade* it is $|i\rangle$ (intermediate state). For a linear molecule with D_2h symmetry, the states are given in figure 1.6(a) [38, 39]. The electric dipole allowed one-photon transition for $g\leftrightarrow i$ and $i\leftrightarrow f$. In contrast, these two transitions are out of resonance with the optical frequency (ν) for TPA. However, a non-stationary (virtual) state exists (for about 5 fs) due to the mixing of $|g\rangle$ and $|i\rangle$ states, only when a molecule is induced by the electromagnetic field of the first photon [40].

This transient presence of $|i\rangle$ state with *ungerade* parity allows the second photon to induce a molecular dipole transition to the $|f\rangle$ state. Therefore, unlike one-photon transitions, here the selection rule is reversed and $1A_g\leftrightarrow 2A_g$ two-photon transition is allowed for all centrosymmetric chromophores.

For non-centrosymmetric dipolar chromophores, the $g\leftrightarrow f$ is electric dipole allowed and the dipolar part of equation (1.93) is non-zero (figure 1.6(b)) and the contribution of the two-photon term is smaller. Here, $|f\rangle$ takes the role of $|i\rangle$ and both 1PA and TPA show this transition.

Unlike $\mu_{i,\,f}$, the values of $\mu^2_{g,\,i}\mu^2_{g,\,f}$ are proportional to one-photon oscillator strength, which can be determined from the optically linear absorption spectrum. Therefore, the design of TPA chromophores needs guidance from the calculated transition moments. The following points can be considered to maximize the TPA cross-section of a chromophore [41–43].

- The incorporation of electron donor and acceptor groups in a conjugated molecule enhances $\mu_{g,i}$, $\mu_{i,f}$, $\mu_{g,f}$, and/or $\Delta\mu_{g,f}$, and subsequently increases the displacement of charge during the transition from donor HOMO to acceptor LUMO.
- Long π-conjugated planar molecules increase the value of $\mu_{g,i}$, $\mu_{i,f}$, $\mu_{g,f}$, and/or $\Delta\mu_{g,f}$.
- The presence of strong 1PA close to TPA pump wavelength in centrosymmetric molecules enhances $\sigma_2(\lambda)$ when Δ value is small ($\Delta = E_{g,i} - h\nu$). When $\Delta = 0$, TPA is usually not observed due to the overlap with 1PA. Therefore, a sharp 1PA or smaller Γ is required.
- The presence of narrow 1PA and TPA bands. As the area of the TPA band is governed by $S_{f,g}$, a narrow TPA band (it means smaller Γ) increases the $\sigma_2(\lambda)_{\max}$ as per equation (1.94).

TPA is shown by organic molecules, organic liquids and liquid crystals, conjugated polymers, fullerenes, coordination and organometallic compounds, and porphyrins, and metalloporphyrins.

1.5.3 Molecular building blocks for NLO materials

The design of NLO molecules involves tailoring the molecular structure to exhibit specific NLO properties. Key considerations in the design process include the enhancement of dipole moments, control of electron distribution, and achieving suitable symmetries. In-depth investigations into the design of NLO molecules have resulted in identifying important factors that augment molecular hyperpolarizability given in the footnote 3 on pp 1–18. Here are some principles and strategies involved in the design of NLO molecules:

- *Push–pull architecture*: In push–pull architecture, the molecule consists of an electron donor and an acceptor group separated by a π-conjugated bridge. This design induces a CT transition, leading to enhanced NLO effects, for example, TPL.
- *Enhanced dipole moments*: The dipole moment of the molecule can be increased by introducing electron-donating and electron-withdrawing groups. Asymmetry in the electron distribution enhances the molecular response to an external electric field. The significant electronegativity difference between the atoms in these compounds imparts electron density transfer from the donor (electron-rich) to acceptor (electron-deficient) groups leading to partial positive and negative charge development on the donor and acceptor groups of the molecule. As a result of these intramolecular electron transfers, a strong dipole moment (μ in Debye unit) develops even in the ground state. The magnitude of μ is directly proportional to the magnitude of charge separation ($\mu = q \times d_c$, where q is the charge and d_c is the distance between the charges).
- *Conjugation length*: The NLO response can be improved by increasing the conjugation length between the donor–acceptor groups and/or by increasing the strength of these groups. Extension of the conjugation path length or

several connections of the molecular system to 2D or 3D configurations increase the electronic delocalization and 2PA responses. Longer conjugated systems often result in stronger NLO effects.

- *Planarity*: Planarity of molecules allows efficient electron delocalization.
- *Symmetry considerations*: The design of molecules that can crystallize in specific symmetries facilitates special NLO processes. For example, molecules with non-centrosymmetric crystal structures exhibit SHG.
- *Polymeric and hybrid approaches*: The use of polymeric or hybrid materials, combining organic and inorganic components, can provide enhanced structural stability and tailored NLO properties.
- *Octupolar structures*: Design of molecules with octupolar symmetry to induce third-order nonlinear optical effects. These structures often involve three donor-π-acceptor moieties.
- *Tailoring functional groups*: The introduction of specific functional groups enhances the molecule's NLO properties. Examples include –CN, $–NO_2$, –F, –COOH, sulfonyl, triflyl (CF_3SO_2), arylcarbonyl (ArCO–), aldehyde (CHO), and phosphonate groups known for their electron-withdrawing characteristics, and $–NH_2$, $–NMe_2$, –OR, *N*-substituted carbazoles, pyrroles, etc for electron-donating properties.
- *Conformational flexibility*: The design of molecules that allows for conformational flexibility can influence the ease of molecular alignment, crucial for some NLO processes.
- *Co-crystallization:* Organic electron donor and acceptor compounds can be co-crystallized into a single-phase material using hydrogen bonds, halogen bonds, π–π stacking, and charge-transfer interactions are known to form CT complexes exhibiting NLO activity.
- *Computational chemistry*: Quantum chemical calculations play a crucial role in unraveling electronic structures, forecasting static and dynamic hyperpolarizabilities, estimating excited-state dipole moments and polarizabilities, and determining higher-order NLO properties.

The design of NLO molecules often involves a combination of theoretical predictions, computational modeling, and experimental synthesis to achieve the desired NLO effects for specific applications in areas such as photonics, telecommunications, and laser technology. The representative NLO molecules with various donor and acceptor groups are given in figure 1.7.

Molecules DMNA, INBP, and ANS are NLO active due to their push–pull construction [44–46]. Nano/microcrystals of a highly zwitterionic diaminodicyanoquinodimethane (BCADQ) molecule, synthesized in the colloidal state through a reprecipitation-digestion method, exhibited notable optical limiting characteristics and nonlinear absorption/scattering effects. The *optical limiting*[4] response was found

[4] Optical limiting: Optical limiting materials restrict the transmission of light as laser intensity rises, turning opaque when surpassing a specific threshold.

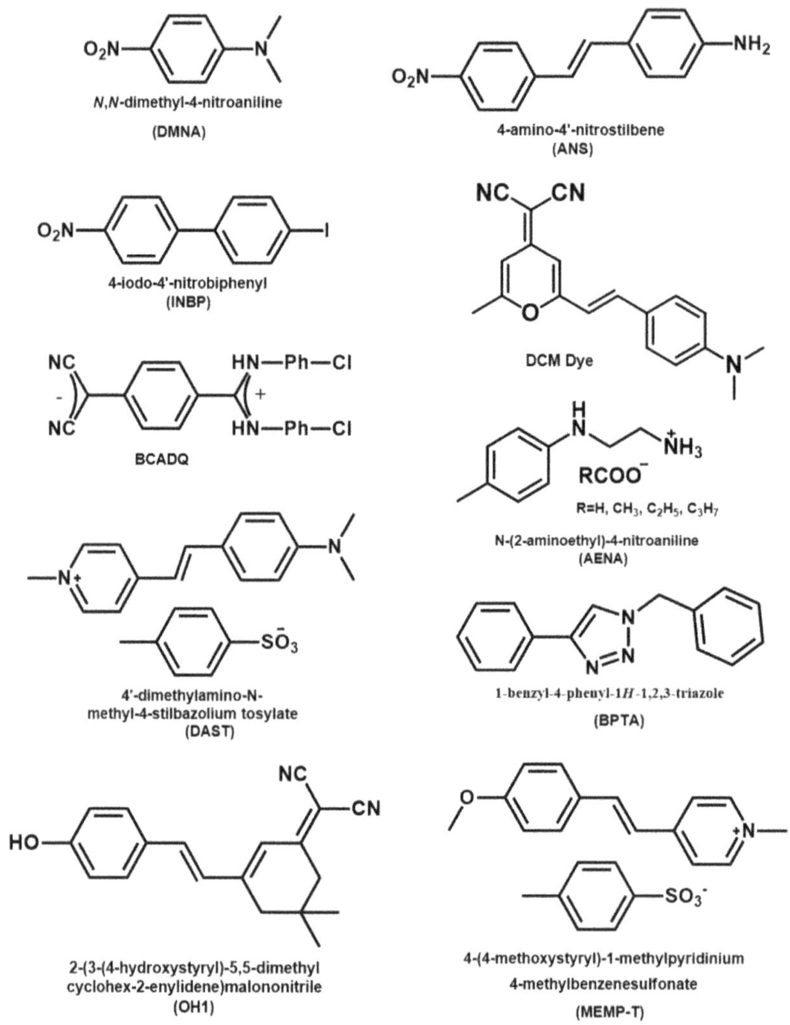

Figure 1.7. Representative examples of NLO molecules.

to be contingent upon the size distribution of the molecular crystals within the colloid [47].

DAST, an organic salt, features a positively-charged stilbazolium cation and a negatively-charged tosylate anion, illustrated in figure 1.7. The crystal packing is achieved through potent Coulomb interactions between these oppositely-charged molecules. The stilbazolium cation, known for its high efficiency as an NLO chromophore, adopts an acentric structure. Concurrently, the tosylate counterion is strategically employed to induce non-centrosymmetric crystallization. DAST exhibits an extremely high NLO susceptibility, $\chi^{(2)} = 580 \pm 30$ pm V^{-1} at 1.54 μm [48].

The OH1 molecule exhibits a substantial dipole moment, with a value of $\mu_g = 3.44 \times 10^{-29}$ Cm. Featuring an extended π-conjugated electron system, it further demonstrates a considerable hyperpolarizability of $\beta_z = 765 \times 10^{-40}$ m^4 V^{-1}, as determined in a chloroform solution. The CT characteristics of OH1 enable the generation of a macroscopic $\chi^{(2)}$, measuring 240±20 pm V^{-1} for SHG at a fundamental wavelength of 1.9 μm [49].

The red-emitting DCM dye possesses strong electron-donating and accepting groups with a very high dipole moment (6.1 D in the ground state). Murzina *et al* reported the generation of TPL from self-assembled DCM dye microspheres that exhibit an *optical cavity* effect (which will be discussed in chapter 2; section 2.2) [50].

Radhakrishnan *et al* introduced a series of crystals featuring achiral remote functionalized NLO chromophore, *N*-(2-aminoethyl)-4-nitroaniline, combined with a homologous series of carboxylic acids. This study revealed the significant influence of supramolecular interactions on the solid-state SHG. Quantum chemical computations on the supramolecular clusters validated the dominant influence of cooperative interactions on molecular hyperpolarizability [51].

Leveraging azide–alkyne click chemistry, Bu *et al* synthesized the compound 1-benzyl-4-phenyl-1,2,3-triazole (BPTA) featuring motifs that allow n–π transitions. This approach resulted in the formation of two polymorphs of BPTA crystals characterized by non-centrosymmetric packing. These polymorphs not only demonstrate robust second-order NLO responses but also exhibit high thermal stability and broad optical transparency [52].

Nelson *et al* reported the generation of tunable multi-cycle THz pulses with center frequencies ranging from 0.3 to 0.8 THz using the organic nonlinear crystal HMQ-TMS [2-(4-hydroxy-3-methoxystyryl)-1-methylquinolinium 2-4,-6-trimethylbenzenesulfonate]. This was achieved through collinearly phase-matched optical rectification of temporally shaped 800 nm pulses [53].

A detailed review of multiphoton absorbing materials, molecular designs, characterizations, and applications can be found in the review of Prasad *et al* [34]

1.6 Mechanical properties of organic crystals

The mechanical properties of crystalline materials dictate the ability of materials to be shaped in a desirable geometry. Among mechanical properties, *hardness, brittleness, elasticity, stiffness,* and *plasticity* are important for the utilization of organic crystals for photonic circuit applications. The capacity of the crystals to resist permanent shape change, or penetration, scratching, or cutting due to external stress (σ) is known as hardness. The brittleness of a material indicates how easily a crystal gets fractured when an external load or force is applied to it. Brittle crystals observe less energy from the stress and get fractures without substantial strain (ζ). It is important to remember that the brittleness of certain materials is temperature dependent. Elasticity is a property when a crystal changes its shape due to applied load and will regain its original shape after the load removal. Young's modulus (**Y**) or modulus of elasticity is a parameter used to measure the elasticity of the crystal. Highly elastic materials show higher **Y** measured in N μm^{-2}. Stiffness (usually

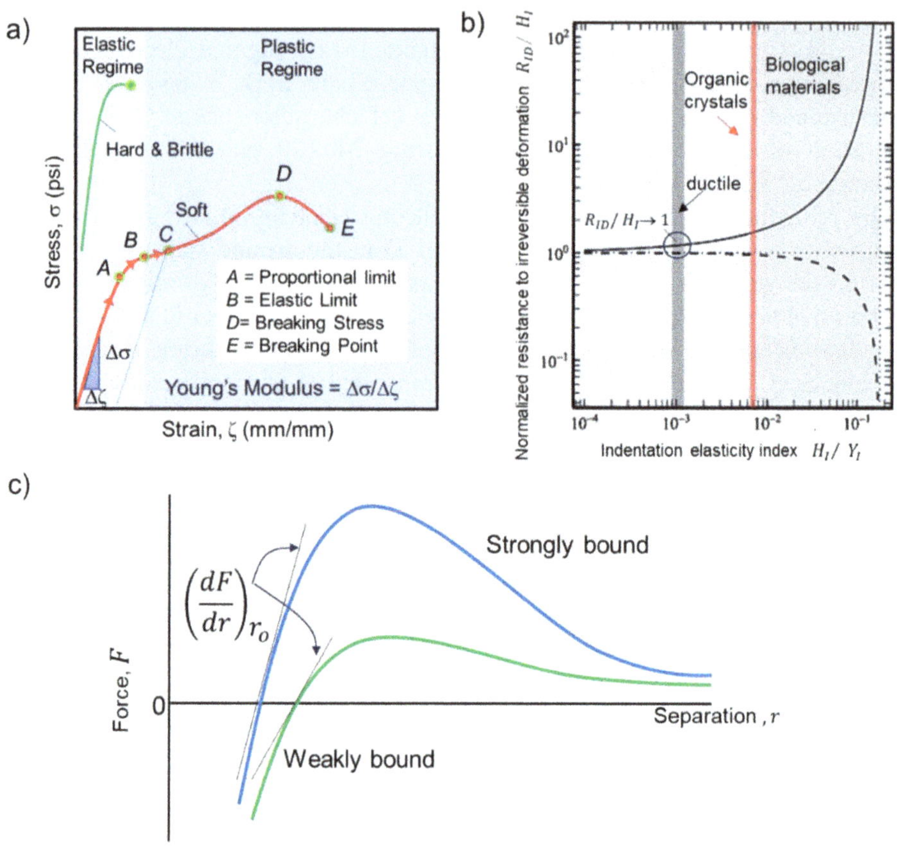

Figure 1.8. (a) The plot depicts the relationship between stress–strain (σ–ζ) of solid materials. [165] John Wiley & Sons. (2020). (b) A plot of normalized resistance to irreversible deformation versus elasticity index to understand the hardness ($\mathbf{H_c}$) of various materials. Reprinted from [54], Copyright (2017), with permission from Elsevier. (c) Force versus interatomic separation for weakly and strongly bonded atoms. The slope of the curve at r_0 is proportional to the magnitude of Young's modulus (**Y**).

measured in N μm^{-1}) denotes the ability of materials to resist deformation when load is applied. Plastic materials are opposite to elastic materials, they show permanent deformation when load is applied/removed.

The mechanical behavior of crystals also can be understood from their stress–strain, (σ)–(ζ), relationship (figure 1.8(a)). Most of the crystals, as per Hooke's law, show a linear σ–ζ relationship till a specific σ value, called proportional limit (*A*). The linear section of the slope of $\Delta\sigma/\Delta\zeta$ provides the value of **Y** to understand the stiffness or the resistance to elastic deformation of the crystal. For materials such as cast iron, concrete, and several polymers, the elastic portion of the σ–ζ is nonlinear. After *A*, the elastic limit or yield point (*B*) is that which provides the elastic crystal, i.e., the crystals that revert to their original dimension upon removal of σ. Hard and brittle materials break in this elastic regime. At point *C*, permanent deformation or plasticity occurs when the load is removed. From point *C* to *D* (called breaking stress), large permanent deformation is shown by the crystal even for a small σ. From *D* to breaking point *E*, even for a low σ, the crystal shows large deformation like viscous fluid and ultimately breaks at this point.

Usually, nanoindentation tests are performed to assess mechanical attributes such as **Y** and contact hardness (**H**$_c$) of molecular crystals. Currently, these two parameters are the measurable quantities of resistance to elastic or plastic deformation of molecular crystals. The indentation hardness is measured by denting a crystal with an external hard and sharp calibrated Berkovich diamond indenter tip (a three-sided pyramidal indenter). During the experiment, load or force is exerted on the sample and the displacement, which is the distance the indenter travels into a material, is measured. The load and displacement data together with the area (the interaction of the indenter with the sample) provide mechanical properties, viz. elastic modulus and hardness.

In organic crystals, hardness is related to the ability of the intermolecular forces to slip when there is less friction between molecules. However, the high **Y** value makes the crystals brittle. For organic crystals, **Y** versus **H**$_c$ show a nearly linear relationship; however, their ratio differs across materials due to their varying amounts of relative irreversible deformation. In the Sakai model [54], the *elastic component* is characterized by Y_I and the *irreversible component* is characterized by the *resistance to irreversible deformation* (R_{ID}).

$$R_{ID} = \frac{H_I}{(1 - \sqrt{\frac{H_I}{Y_I}}\sqrt{\frac{2}{\tan(\varphi)}})^2} \tag{1.95}$$

where φ is the equivalent cone angle of the indenter.

The value of R_{ID} can be estimated from the *elasticity index* (I_E) and the latter term reflects the type of deformation.

$$I_E = H_I / Y_I \tag{1.96}$$

where H_I is the indentation hardness, and Y_I is the indentation modulus. However, the R_{ID} is not only dependent on I_E but also φ.

In figure 1.8(b), the solid line (equation (1.95) with $\varphi = 19.7°$ for a Berkovich indenter) approaches 1 in the limit of small I_E, while it diverges as it approaches maximum probable value.

In other words, when $I_E \rightarrow 0$, H_I converges to resistant to irreversible deformation, i.e., $R_{ID}/H_I \rightarrow 1$ (left side of figure 1.8(b)). As a result, the indentation tests sensibly estimate small values of I_E. For example, ductile metals I_E is about 0.001 (see the dark gray area in figure 1.8(b)). Larger values of I_E indicate dominant reversible deformation (see the right side of figure 1.8(b)). The light gray area specifies the range of values observed for biological materials ($I_E = 0.5$). The I_E value of organic crystals is 0.04, which is near to biological materials and between ductile ($I_E = 0.001$) and brittle materials (I_E up to 0.3).

Based on the magnitude of the **Y** value, crystals are classified into *intermediate stiff* (6–16 GPa), *complaint* (4–6 GPa), and *super-complaint* (<4 GPa). The **H**$_c$ of *medium-hard, soft,* and *super-soft* crystals are in the range of 0.26–0.72, 0.15–0.26, and <0.15 GPa, respectively [55]. More than 90% of the brittle molecular crystals exhibit **Y/H$_c$** values below 35, whereas crystals with higher **Y/H$_c$** values demonstrate that they are plastic. The magnitude of **Y** hinges on parameters such as intermolecular hydrogen bond strength and their orientation, molecular packing, and crystal voids. However, the ease of irreversible slip of molecular layers commands **H$_c$** [56]. If the intermolecular interactions are similar in all directions, those crystals are characteristically brittle.

Naumov and Spackman *et al* have generated a global mechanical properties profile by analyzing **Y** and **H$_c$** data of hundreds of organic crystals reported over the past three decades in comparison to different materials including metals and alloys (figure 1.9) [57]. The analysis shows that the elasticity of organic crystals possessing

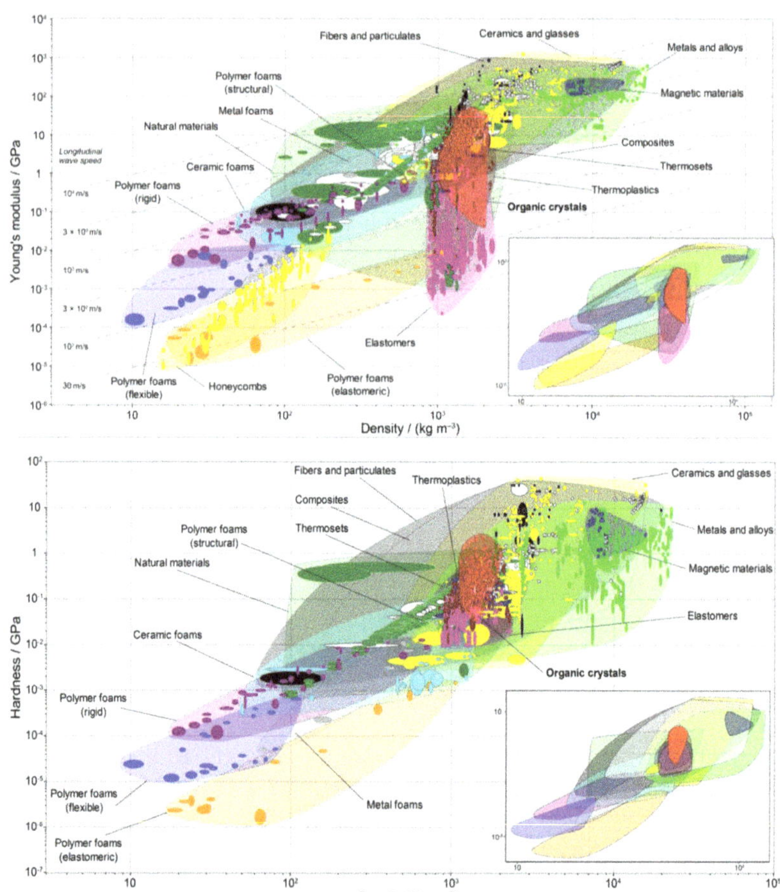

Figure 1.9. Young's modulus–density (**Y**–ρ) and hardness–density (**H$_c$**–ρ) materials space for organic crystals and other classes of materials. The opaque bubbles represent the range of values for performance indices of particular materials, while the larger translucent envelopes enclose points for a given class of materials. A more simplified distribution of the classes is represented in the subplot in the bottom right corner. The family of organic crystals is outlined with a red envelope. The plots in panels are on a double log scale. The ρ values were calculated from the respective crystal structures. [57] John Wiley & Sons. (2022).

Y>20 GPa is similar to soft metals and alloys and is superseded by rigid metals, ceramics/ceramic composites, glasses, and magnetic materials. To use organic crystals to fabricate photonic circuit components their **Y** values should be moderate so that reversible deformation can be obtained even under high stresses. They found that the typical **Y** value of organic crystals is in the range of 1–25 GPa. Amongst these organic crystals, 8% of them exhibit **Y**<1 GPa and 8% show **Y**>25 GPa.

The $\mathbf{H_c}$ of the organic crystals is in the range of 0.04–1.67 GPa. As organic compounds have carbon and other light elements as substantial contributors to their masses, they naturally occupy a slender density range (1–2 kg m^{-3}) in the global **Y**–**ρ** and $\mathbf{H_c}$–**ρ** spaces. This critical analysis shows that indeed organic crystals can be considered as a distinct materials class as they bridge rigid/dense materials (e.g. metals, ceramics, and composites) and mechanically acquiescent/lightweight materials (e.g. polymers, elastomers, and foams), although they intersect with a few biogenic materials.

1.6.1 Flexible molecular crystals and their types

The general notion about crystals' mechanical properties is that they are hard and brittle (figure 1.10). Defying this view, some crystals show unusual mechanical compliance in response to applied mechanical force, the so-called *flexible crystals*. These crystals are generally classified into two types, namely *plastic* and *elastic*, depending upon their response to mechanical force. Upon release of the force, plastic crystals display permanent deformation (bending and shearing) owing to

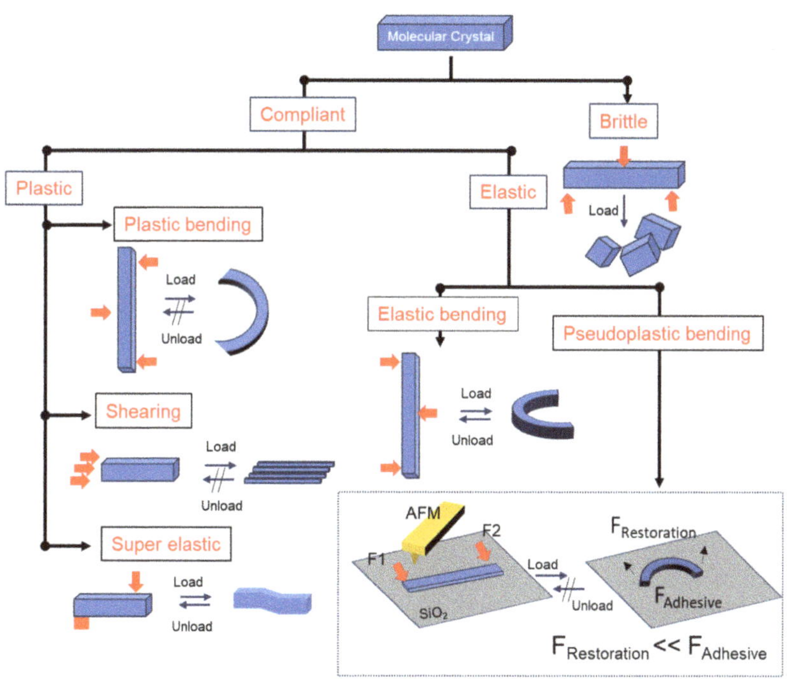

Figure 1.10. Various categories of soft molecular crystals.

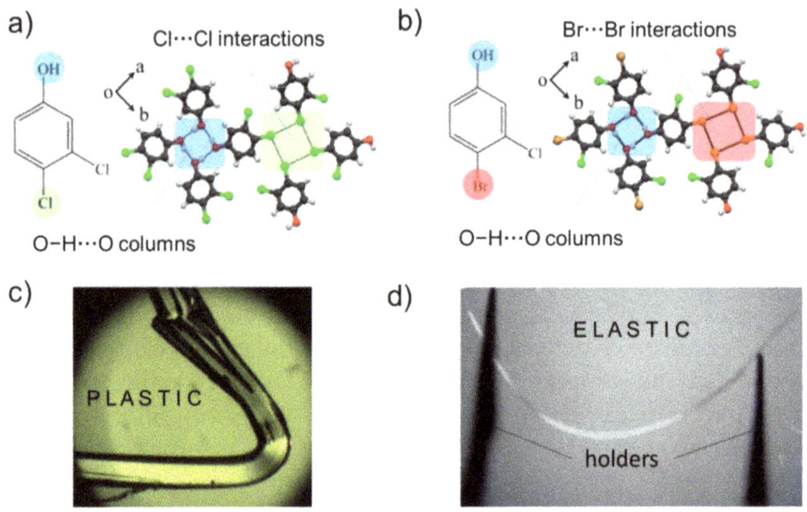

Figure 1.11. (a) Weak Cl···Cl interactions in 3,4-dichlorophenol. (b) Stronger Br···Br interactions in 4-bromo-3-chlorophenol. (c) Plasticity of 3,4-dichlorophenol crystal. (d) Elastic deformation of a single crystal of 4-bromo-3-chlorophenol via increased structural isotropy. Reprinted with permission from [58]. Copyright (2018) American Chemical Society.

dislocations and defects. Conversely, the elastic ones (for example: 3-chloro-4-bromophenol crystal) show reversibility restoring the supramolecular architectures (figure 1.11) [59]. Plastic bending has been observed in 3,4-dichlorophenol [60] and hexachlorobenzene [60] crystals. The molecular single crystal of 1,3,5-trichloro-2,4,6-triiodobenzene shows shearing under mechanical load [61]. *Super elastic crystals* (for example: terephthalamide) are a type of plastic crystals that undergo reversible mechanically induced physical deformation via gross domain transformation involving phase transition [62].

In 2020, my research group at the University of Hyderabad discovered the *pseudo-plasticity* of so-called *elastic* acicular *microscale* crystals on a coverslip while manipulating them using an atomic force microscopy (AFM) cantilever tip [63]. The pseudo-plastic crystals display permanent deformation even after the removal of applied stress due to the adhesion of the crystal's flat surface with the substrate. As a result, we exploited these crystals and performed different types of mechanical manipulations with an AFM cantilever tip to construct many proof-of-principle photonic components and circuits (more details on this will be discussed in chapters 3 and 4).

It is important to note that both rigid and flexible crystals can be considered potential candidates for organic photonic integrated circuit (OPIC) construction via mechanical manipulation [64]. For the circuit construction, both elastically (pseudo-plastic) and plastically bending crystals are used; therefore, it is better to look at the examples and their mechanically-triggered bending mechanism in these two crystal classes at some simple level.

At first, an atomic scale picture can be used to better understand the magnitude of macroscopic elasticity. In molecular crystals, the stress is manifested as small changes in the interatomic spacing and interatomic bond stretching. In other words, **Y** can be understood as a measure of resistance to the separation of adjacent atoms, the interatomic bonding forces (figure 1.8(c)). The magnitude of **Y** is proportional to the slope of the curve at the equilibrium interatomic separation, r_0, as per the following relation.

$$E \propto \left(\frac{dF}{dr}\right)_{r_0} \tag{1.97}$$

where F is force and r interatomic distance. For plastic deformation, due to the movement of many molecules from the equilibrium position, the breaking of weak bonds between molecular neighbors and the reforming of bonds with new neighbors arises. For crystalline materials, plastic deformation arises by a process called '*slip*' (dislocation motion occurs).

To correlate molecular crystal's mechanical flexibility with the structure, understanding their crystal packing, slip, and defects is important. The molecular crystals are stabilized by weak interactions—hydrogen bonds and van der Waals interactions, and possess characteristic structural anisotropy. Thus, their mechanical property is different from extended solids. The deformation of molecular crystals by mechanical force has been reported by crystallographers over the years. However, it is crucial to establish a systematic correlation between crystal bending and crystal packing/intermolecular interactions across various molecular crystals.

Desiraju and his co-workers proposed a structural basis for crystal's flexibility for two isomorphous molecules, namely, 3,4-dichlorophenol and 3-chloro-4-bromophenol, based on halogen bonding (figure 1.11) [58, 59]. The former isomorphic molecular crystals displayed plasticity and possessed 1D helical O–H···O columns and weak Cl···Cl interactions (both type-I and -II halogen bonding) between adjacent columns. However, partial replacement of Cl···Cl interactions by the stronger Br···Br (type-II) altered the mechanical property from plastic to elastic (figure 1.11). The higher interaction strength (Br is more polarizable than Cl) along the direction of bending in 3-chloro-4-bromophenol prevents easy molecular slippage; thereby, crystal plasticity. They noted that the enrichment of structural isotropy provides crystal elasticity.

Reddy *et al* reported a co-crystal of caffeine and 4-chloro-3-nitrobenzoic acid (CAF-CNB) solvated with methanol and its elasticity by mechanically bending with a metal pin and forceps (figure 1.12(a)) [65]. The CAF-CNB single crystals exhibited elasticity when mechanical force was applied to make a loop, but recovered their acicular original shape after the force was removed (figure 1.12(b)). It was reported that the crystal bending is also supported by the methanol solvate. Supporting this finding, the crystals did not exhibit elasticity upon desolvation, even though the overall molecular packing structure was intact.

Clegg *et al* investigated the molecular level factors governing the CAF-CNB elasticity using micro-focused synchrotron radiation at 100 K [66]. In the crystal lattice, both CAF and CNB form hydrogen bond dimers, which form π–π stacks along the

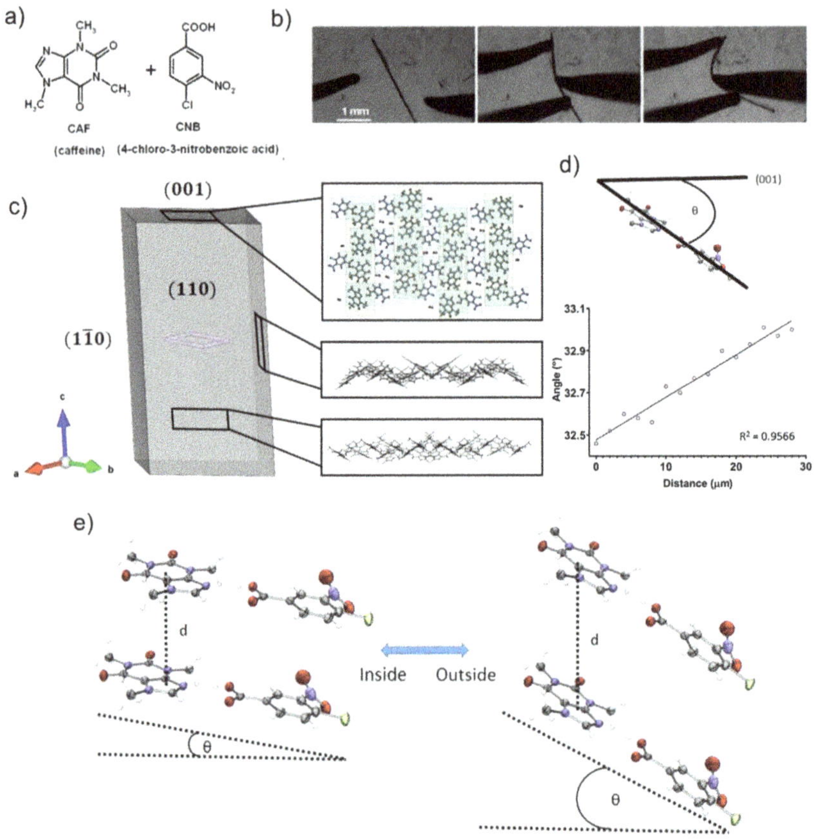

Figure 1.12. (a) Molecular structure of caffeine and 4-chloro-3-nitrobenzoic acid. (b) Elastic bending of a needle-like single crystal. (c) Crystal packing with respect to the face-indexed crystal. The interlocked comb-like structure on the (001) face is shown as green (hydrogen bonding) and blue rectangles. The wireframe view of the (110) and (−110) faces emphasizes the molecules that have their mean planes perpendicular to the face. (d) The change in angle between the mean plane of the hydrogen-bonded dimer and the (001) plane moving from the inside to the outside of the bend. (e) Rotation compared to the (001) face and movement apart within the π-stack (d) facilitates the required cell deformation. $\Delta d = 0.04$Å, $\Delta\theta = 0.54°$. Disorder in the nitro-group is removed for clarity. (a and b) [65] John Wiley & Sons. (2022). (c–e) Reproduced from [66]. CC BY 4.0.

c-axis. The molecules are located perpendicular to (110) or (−110) planes (figure 1.12(c)). Due to the change in the *c*-axis of the bent crystal, the π–π stacks move apart along with the increase of the angle between the mean plane of the H-bonded dimers and the (001) plane (figure 1.12(d)). The molecules perpendicular to (110) and (1–10) planes also undergo deformation. Consequently, the *a,b*-plane expands at the concave side and contracts at the convex side, involving molecular rotation (figure 1.12(e)).

However, the photonic attributes of the CAF and CNB co-crystal have not been explored yet. My group investigated the caffeine acicular crystal along with other pharmaceutically important crystals and confirmed their waveguiding property (see chapter 2; section 2.1.3 for more details) [67]. One of the important aspects of the investigation was the use of Raman microspectroscopy in these crystal waveguides to determine nano/micro-scale defect sites that appeared during the natural growth process.

During the elastic bending of a straight crystal, substantial separation of the molecules from the equilibrium position occurs on the convex side of the crystal. Simultaneously, molecules moderately compress on the concave side of the crystal (figures 1.13(a) and (b)). Examination of the physical morphology of the crystal termini

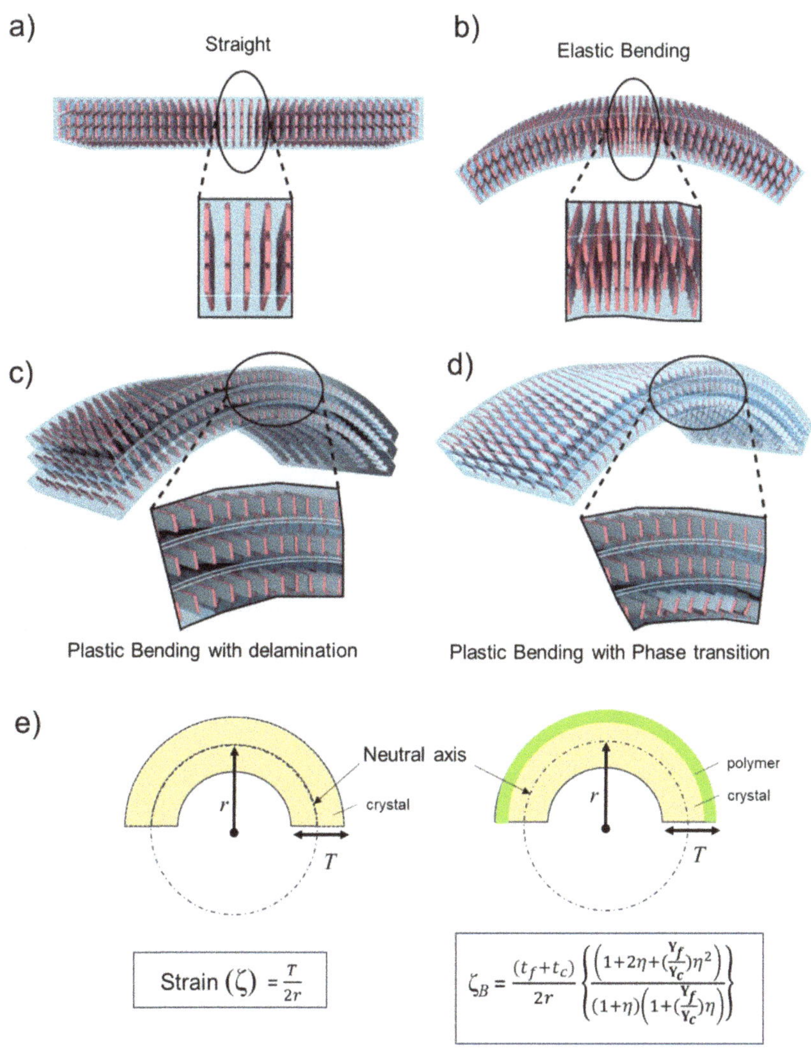

Figure 1.13. The general mechanism for elastic and plastic bending of molecular crystals. (a) Sketch of a straight crystal before bending. The molecules are cartooned as a three-dimensional grid of blocks. (b) Mechanism of elastic bending. (c) Mechanism of plastic bending by delamination. (d) Mechanism of plastic bending by phase transition. (e) Estimation of the strain value of a bend monolithic crystal and (polymer) coated crystal. (a–d) Reproduced from [68]. CC BY 4.0.

is used to identify the crystal type. For example, in the co-crystal of CAF-CNB, the angles between the crystal faces at the crystal termini are ideally well-preserved confirming its elasticity (figures 1.12(b) and 1.13(b)). The *plastic* bending in crystals is facilitated by individual molecular layer sliding, but the layers remain in contact with each other. Plastic crystals are of two types, the one which delaminates during deformation and the other which does not delaminate. Plastic bending of crystals (for example: hexachlorobenzene crystals) [60] usually ensues by partial delamination, due to the sliding of crystal layers over each other (figure 1.13(c)). Relatively short crystals may exhibit changes in the angles between the faces of the crystal at its termini. In long crystals, for example, 2-(methylthio)nicotinic acid crystal, the delamination is not generally anticipated to advance through the entire length of the crystal. In some crystals, such as terephthalic acid, the applied localized stress triggers a partial phase transition on the concave side of the crystal [68]. The two phases cohabit in the bent crystal region (figure 1.13(d)).

Another important directly measurable parameter of a bent crystal is its strain (ζ), which can be calculated using the relation

$$\zeta = \frac{T}{2r} \tag{1.98}$$

where T is the thickness of the crystal and r is the bending radius (figure 1.13(e)). However, it is important to note that the thickness is not uniform in molecular crystals leading to error in the measured ζ.

To avoid optical loss, a crystal coated with low refractive index polymer can be used as a waveguide. In this situation, the bending strain of the waveguide can be calculated using a bilayer model, [69] as given in the following equation:

$$\zeta_{\mathrm{B}} = \frac{(t_{\mathrm{f}} + t_{\mathrm{c}})}{2r}\left\{\frac{\left(1 + 2\eta + (\frac{\mathbf{Y_f}}{\mathbf{Y_c}})\eta^2\right)}{(1+\eta)\left(1 + (\frac{\mathbf{Y_f}}{\mathbf{Y_c}})\eta\right)}\right\} \tag{1.99}$$

where ζ_{B} is the bending strain of the bilayer model, t_{f} is the thickness of the polymer film, t_{c} is the thickness of the crystal, η is the polymer thickness divided by the crystal thickness ($t_{\mathrm{f}}/t_{\mathrm{c}}$). $\mathbf{Y_f}$ and $\mathbf{Y_c}$ denote the Young's modulus of the polymer and crystal, respectively.

Excellent accounts and reviews on the correlation between mechanical compliance with the molecular crystal structure can be found in the literature [57, 60, 70].

1.6.2 Molecules forming mechanically flexible crystals

In *mechanically compliant crystals* the flexibility of the crystal under external mechanical forces is primarily governed by the molecular packing within the crystal lattice, rather than by changes in the chemical skeletal structure (such as bond cleavage and formation). In contrast, for a specific class of molecular solids, intrinsic flexibility can arise without the application of external mechanical force, induced by changes in the chemical skeletal structure due to factors like heat, light, or chemical vapors.

The mechanical flexibility of molecular crystals primarily arises from the peculiar arrangement of molecules within the crystal lattice. This molecular packing is intricately governed by a variety of intermolecular interactions, including hydrogen bonds, halogen bonds, π–π stacking, and van der Waals forces. In the quest to design molecular crystals that exhibit mechanical flexibility, chemists have endeavored to introduce diverse functional groups into the molecular structure, with the expectation of achieving mechanical compliance in the resulting crystals. The molecular crystals which exhibit mechanically-triggered flexibility are provided in table 1.1 along with their mechanical and optical properties.

The initial studies of Desiraju *et al* on the structural basis for mechanically flexible crystals were foreseen on a simple phenyl-substituted molecule like hexachlorobenzene (**81**) [126]. But as more research paced into this attractive field of mechanically flexible crystals, new molecular design strategies and crystal growth processes have been keenly developed to explore new possibilities to achieve mechanical flexibility in organic molecular crystals.

In this direction, several flexible organic crystals have been successfully synthesized using Schiff base (**1–43**) [71–94], azine formation (**44–49**) [95–100], a combination of Schiff base and base condensation (**50–54**) [101, 102], base condensation (**55–61**) [103–109] and other reactions featuring a range of chemical motifs as the backbone such as substituted phenyl derivatives (**1–24**) [71–79], pyridyl derivatives (**25–29,35**) [80–82, 86], chiral amines (**30,31**) [83], benzimidazole (**32**) [84], napthyl (**33–41,43,47,48**) [85–92, 94, 98, 99], pyrene (**40**) [91], anthracene (**42,72,73,77**) [93, 117–119, 122], phenothiazine (**57**) [105], carbazole derivatives (**59,75**) [107, 120], benzil compounds (**62**) [111], indenone derivatives (**63–66**) [110], diaryl beta-diketone compound (**67**) [112], barbituric derivatives (**68**) [113], dimethyl 2,5-diaminoterephthalate (**69**) [114] and its derivatives (**70,71**) [115, 116], dithieno[3,2-a:2′,3′-c]phenazine (**74**) [120], naphthalene diimide derivatives (**76**) [122], difluoroboron avobenzone (**78**) [123], 4-trifluoromethyl phenyl isothiocyanate (**79**) [124], flufenamic acid (**80**), pyridine derivatives (**82**) [126], and bispyrazolylpyridine derivatives (**83**) [127], as illustrated in table 1.1.

Designing crystals that demonstrate mechanical compliance poses challenges and is somewhat unpredictable. This unpredictability is attributed to the distinct energy landscapes involved and the potential formation of multiple polymorphs.

For instance, crystals of compounds **5–7** exhibited remarkable mechanical plasticity [73]. In the case of compound **5**, the crystal formation involved π–π stacking and slip planes facilitated by the presence of two methoxy (OMe) groups at either end. Crystals of compounds **6** and **7** formed slip planes through the close packing of OMe, Cl/Br, and aromatic C–H groups. Notably, when the chloro group in compound **6** or the bromo group in compound **7** was replaced with alkyl substituents (–Me or ^{t}Bu groups on the aniline side), slip planes failed to form, and the crystals did not exhibit plasticity.

Desiraju *et al* studied a series of Schiff base type flexible crystals from **8–14** exhibiting elasticity [74]. Interestingly, Ghosh *et al* found that when the choro group of compound **16** [75] was replaced with the cyano group, the crystal became brittle. They found that the crystal has slip planes (suitable for plasticity) that lack isotropic

Table 1.1. Chemical structures of representative molecules that form elastic and/or plastic crystals, their mechanical and optical attributes [71–128].

1[71]	2[71]	3[71]	4[72]	5[73]	6[73]	7[73]
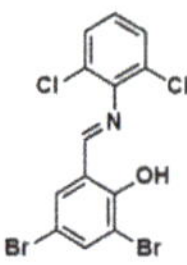	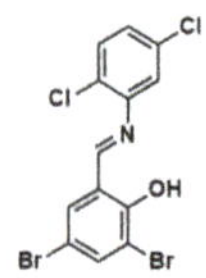	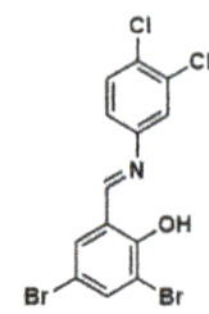	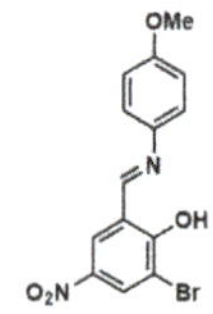	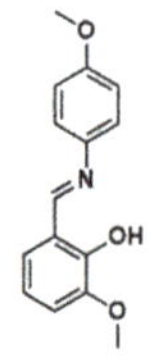	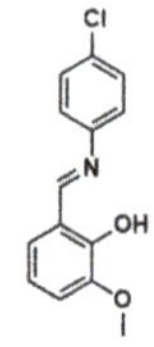	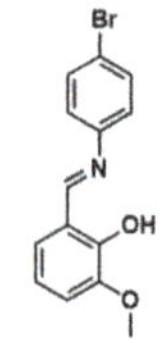
Y = 0.64-0.72 GPa	**Y** = 0.53-0.61 GPa	**Y** = 1.03 GPa	**Y** = -	**Y** = -	**Y** = -	**Y** = -
Hc = -	**Hc** = 0.4208±0.0251 GPa	**Hc** = -	**Hc** =-	**Hc** = -	**Hc** = -	**Hc** = -
FL-BW ≈ 540-775 nm	**FL-BW**= -	**FL-BW** ≈ 500-740 nm	**FL-BW** ≈ 510-750 nm	**FL-BW**= -	**FL-BW**= -	**FL-BW**= -
FL Max ≈ 615 nm	**FL Max** ≈ 615 nm	**FL Max** ≈ 615 nm	**FL Max** ≈ 590 nm	**FL Max**= -	**FL Max**= -	**FL Max**= -
Type = Elastic, Pseudo-plastic	**Type** = Elastic, Pseudo-plastic	**Type** = Elastic, Pseudo-plastic	**Type** = Elastic	**Types** = Plastic	**Types** =Plastic	**Types** = Plastic

8[74]	9[74]	10[74]	11[74]	12[74]	13[74]	14[74]
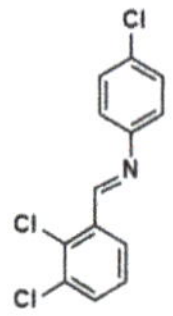	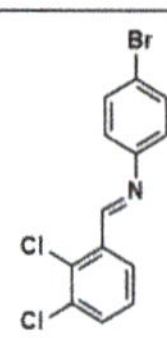	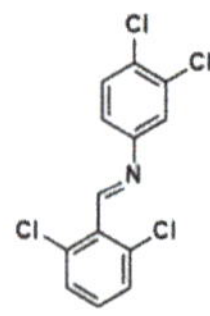	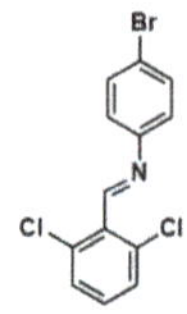	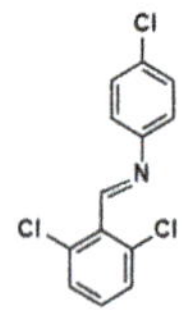	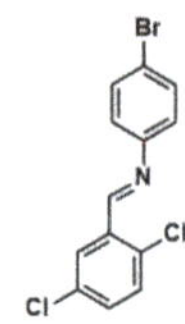	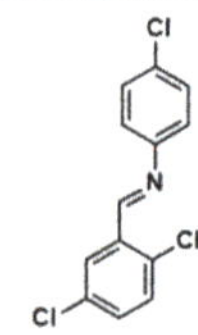
Y = -	**Y** = 0.189 GPa	**Y** = -	**Y** = -	**Y** = -	**Y** = 0.878 GPa	**Y** = -
Hc = -	**Hc** = -	**Hc** = -	**Hc** = -	**Hc** = -	**Hc** = -	**Hc** = -
FL-BW= -	**FL-BW**= -	**FL-BW**= -	**FL-BW**= -	**FL-BW**= -	**FL-BW**= -	**FL-BW**= -
FL Max= -	**FL Max**= -	**FL Max**= -	**FL Max**= -	**FL Max**= -	**FL Max**= -	**FL Max**= -
Type = Elastic	**Type** = Elastic	**Type** = Elastic	**Type** = Elastic	**Type** = Elastic	**Type** = Elastic	**Type** = Elastic

15[75]	16[75]	17[75]	18[76]	19[76]	20[77]	21[77]
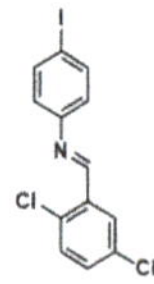	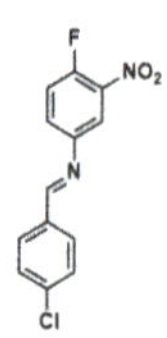	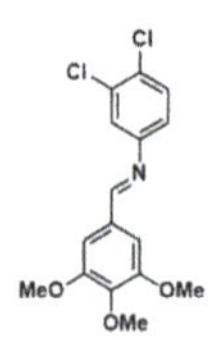	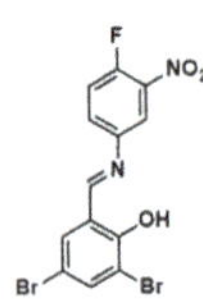	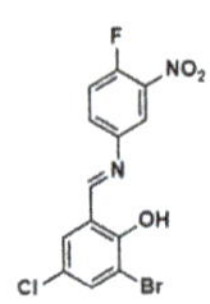	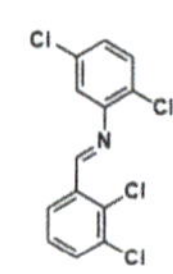	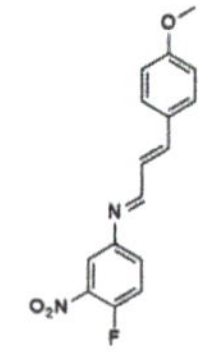
Y = 8.95±1.63 GPa	**Y** = 6.07±0.88 GPa	**Y** =	**Y** = -	**Y** = 0.64-0.72 GPa	**Y** = 3.47±0.6 Gpa	Y_a = 5.59±0.25 GPa; Y_b = 2.28±0.4 Gpa
Hc = 0.226±0.012 Gpa	**Hc** = 0.291±0.03 Gpa	**Hc** = 0.367±0.025 GPa	**Hc** = -	**Hc** = -	**Hc** = 0.156±0.004 GPa	**Hc(a)** = 0.255±0.012 Gpa; **Hc(b)** = 0.106±0.006 Gpa
FL-BW= -	**FL-BW**= -	**FL-BW**= -	**FL-BW**= -	**FL-BW**= -	**FL-BW** ≈ 420-600 nm	**FL-BW** ≈ 500-700 nm
FL Max= -	**FL Max**= -	**FL Max**= -	**FL Max**= -	**FL Max**= -	**FL Max** ≈ 449 nm , 561 nm	**FL Max** ≈ 601 nm
Type = Elastic	**Type** = Elastic	**Type** = Elastic	**Type** = Elastic	**Type** = Elastic	**Type** = Elastic	**Type** = Elastic

22[78]	23[78]	24[79]	25[80]	26[80]	27[80]	28[81]	29[82]
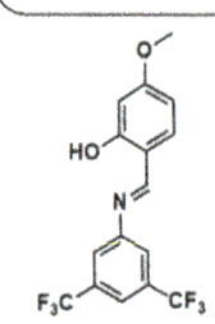	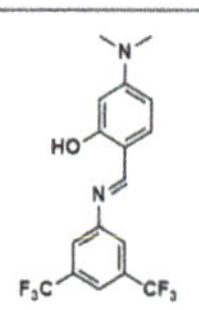	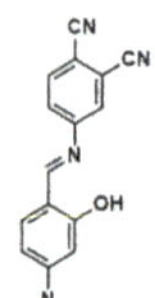	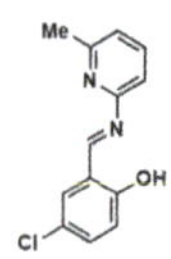	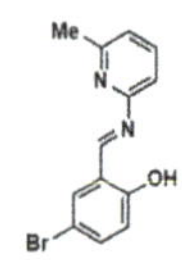	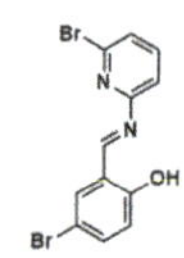	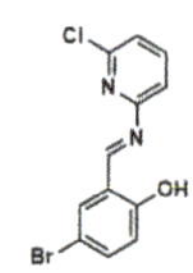	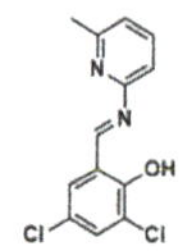
Y =	**Y** =	**Y** = -	**Y** = 8.4 ±1 GPa	**Y** = 4.54±0.45 GPa	**Y** = 7.05±0.25 GPa	**Y** = -	**Y** = 5.7 GPa
Hc =	**Hc** =	**Hc** = -	**Hc** = 0.317±0.033 GPa	**Hc** = 0.204±0.012 GPa	**Hc** = 0.214±0.010 GPa	**Hc** = -	**Hc** = -
FL-BW ≈ 480-800 nm	**FL-BW** ≈ 500-800 nm	**FL-BW** ≈ 475-850 nm 525-850 nm	**FL-BW** ≈ 500-754 nm	**FL-BW** ≈ 500-775 nm	**FL-BW** ≈ 500-747 nm	**FL-BW** ≈ 525-780 nm	**FL-BW**= -
FL Max ≈ 538 nm/-	**FL Max** ≈ 579/590 nm	**FL Max** ≈ 550/625 nm	**FL Max** ≈ 591 nm	**FL Max** ≈ 592 nm	**FL Max** ≈ 579 nm	**FL Max** ≈ 605 nm	**FL Max**= -
Type = Polymorphs I&II Plastic/Brittle	**Type** = Polymorphs I&II Elastic/moderately Elastic	**Type** = Polymorphs I&II Elastic	**Type** = Plastic	**Type** = Plastic	**Type** = Plastic	**Type** = Polymorphs I&II Plastic, Brittle	**Type** = Elasto-plastic

Table 1.1. (*Continued*)

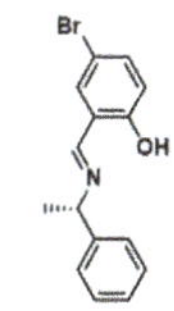 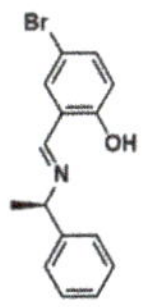 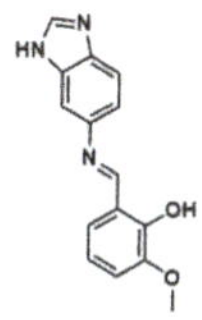 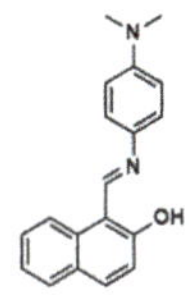 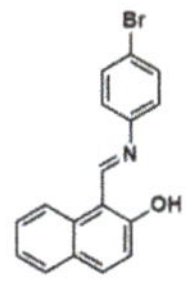 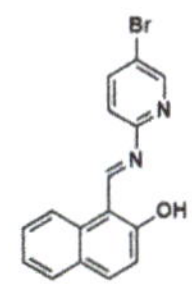 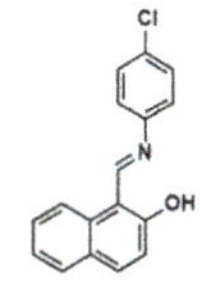

	30[83]	**31**[83]	**32**[84]	**33**[85]	**34**[85]	**35**[86]	**36**[87]
Y	= 0.74-0.79 GPa	= 0.889-0.914 GPa	= 4.51±0.35 GPa, 0.19 ± 0.04 GPa	= 1.02±0.4955 GPa	= 1.02±0.4955 GPa	= -	= 8.22±1.62 GPa
Hc	= -	= -	=12.29±0.54, .27±.01 GPa	= 0.1186±0.0307 GPa	= 0.1186±0.03 GPa	=-	= 0.3532±0.0959 GPa
FL-BW	≈ 445-750 nm	≈ 445-750 nm	≈ 500-750 nm	≈ 530-730 nm	≈ 510-760 nm	≈ 500-650 nm	≈ 470-700 nm
FL Max	≈ 520 nm	≈ 520 nm	≈ 576 nm	≈ 600 nm	≈ 536 nm	≈ 540 nm	≈ 535 nm
Type	= Elastic	= Elastic	= Polymorphs I&II Elastic, Brittle	= Elastic Pseudo-plastic	= Elastic Pseudo-plastic	= Elastic Pseudo-plastic	= Elastic Pseudo-plastic

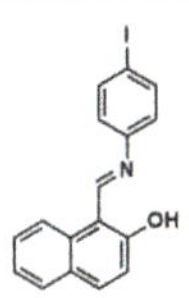 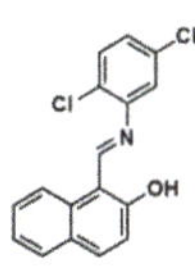 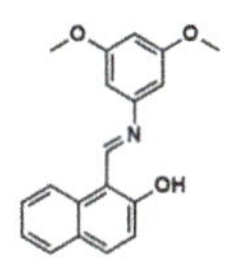 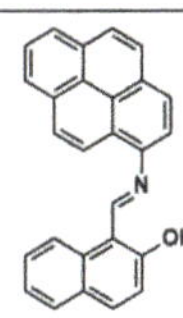 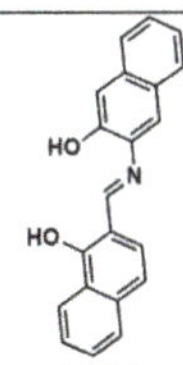 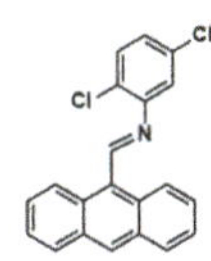 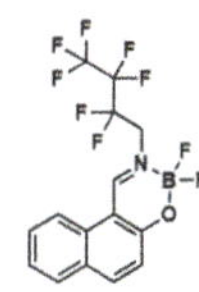

	37[88]	**38**[89]	**39**[90]	**40**[91]	**41**[92]	**42**[93]	**43**[94]
Y	= -	= 11.91 ± 0.2903 GPa	= 6.6 GPa & 8.1 GPa	= -	= -	= -	= 0.27±0.02 GPa (bent region)
Hc	= -	= 0.4208 ± 0.0251 GPa	=	= -	= -	= -	= 19±1.4 GPa (bent region)
FL-BW	≈ 475-700 nm	≈ 490-700 nm	= -	= -	≈ 340-460 nm	= -	≈ 400-630 nm
FL Max	≈ 550 nm	≈ 535 nm	= -	= -	≈ 381 nm	= -	≈ 460 nm
Type	= Elastic and pseudo-plastic	= Elastic	= Polymorphs I&II Elastic, Plastic twisting	= Elastic	**Types** = Elastic	**Types** = Polymorphs I&II Elastic, Brittle	= Plastic, Elastic

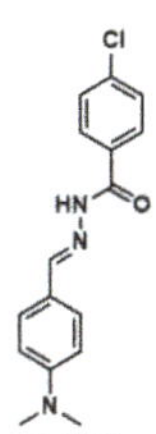 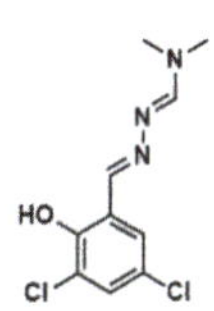 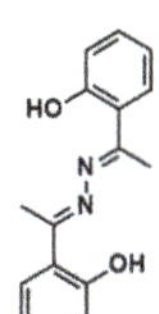 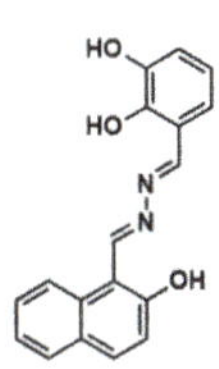 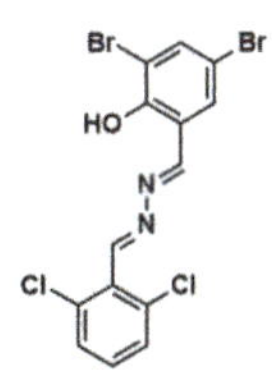 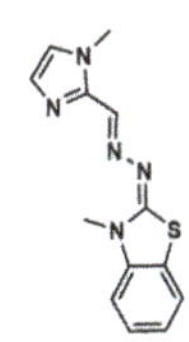

	44[96]	**45**[97]	**46**[98]	**47**[99]	**48**[100]	**49**[101]
Y	= -	= 0.688 GPa	= 0.688 GPa	= -	= 11.91 ± 0.2903 GPa	= -
Hc	= -	= -	= -	= -	= 0.4208 ± 0.0251 GPa	= -
FL-BW	≈ 400-700 nm 400-775 nm	≈ 460-700 nm	≈ 450-700 nm	= -	≈ 525-775 nm	≈ 400-800 nm
FL Max	≈ 450 & 525 nm	≈ 514 nm	≈ 525 nm	= -	≈ 625 nm	≈ 530 nm
Type	= Polymorphs I&II Elastic	= Elastic	= Elastic	= Elastic	= Elastic	= Elastic Photomechanical plasticity

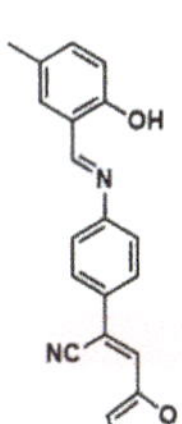 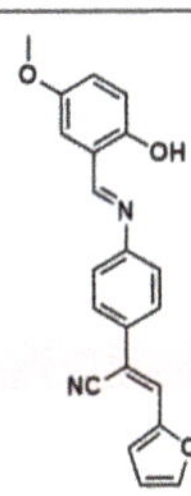 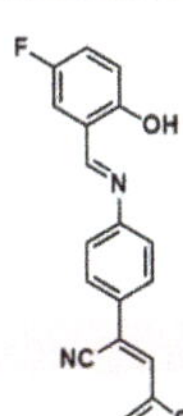 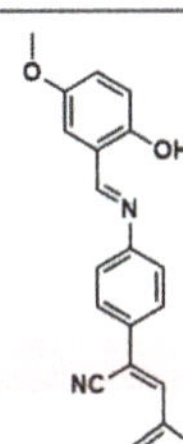 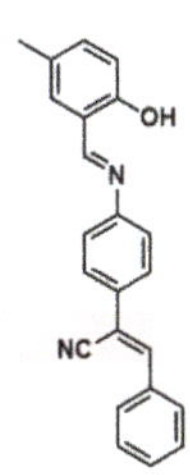 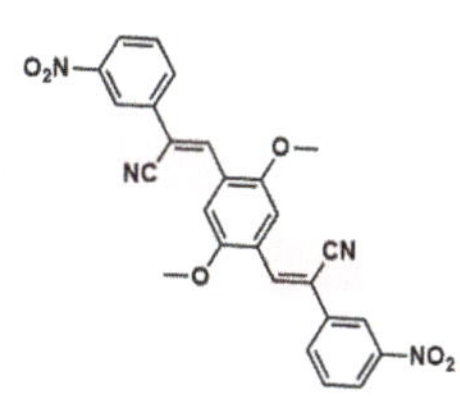

	50[102]	**51**[102]	**52**[102]	**53**[102]	**54**[103]	**55**[104]
Y	= 6.4 GPa	= 2.3 GPa	= 4.2 GPa	= 3.1 GPa	= 4-4.9 GPa	= 0.648 GPa
Hc	= -	= -	= -	= -	= -	= -
FL-BW	≈ 550-800 nm	= -	= -	=	= -	≈ 525-800 nm
FL Max	≈ 610 nm	= -	= -	=	= Orange FL	≈ 637 nm
Type	= Elastic	= Elastic	= Elastic	= Elastic	**Types** = Elastic	= Elastic

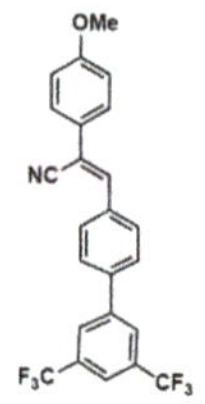

56[104]

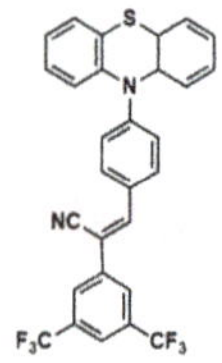

57[105]

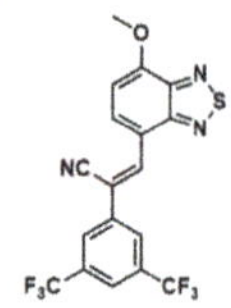

58[106]

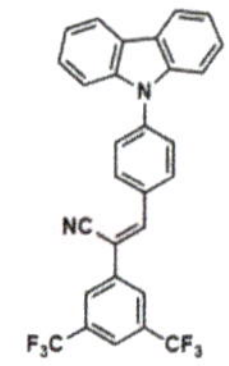

59[107]

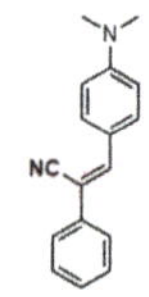

60[108]

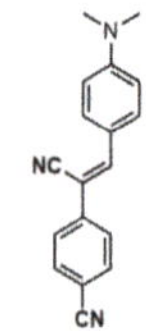

61[109]

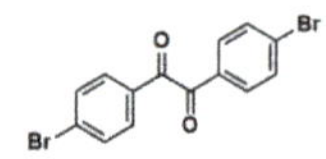

62[111]

56	57	58	59	60	61	62
Y = -	**Y** = -	$\mathbf{Y_a}$ = -	$\mathbf{Y_a}$ = 2.94 GPa	**Y** = -	**Y** = 0.905 GPa	**Y** = -
Hc = -	**Hc** = -	$\mathbf{Y_b}$ = -	$\mathbf{Y_b}$ = 0.115 GPa	**Hc** = -	**Hc** = -	**Hc** = -
FL-BW ≈ 420–660 nm	**FL-BW** ≈ 470-770 nm	**Hc** = -	**Hc** = -	**FL-BW** ≈ 470-700 nm	**FL-BW** ≈ 500-750 nm	**FL-BW** ≈ 450-630 nm
FL Max ≈ 475 nm	**FL Max** ≈ 538 nm	**FL-BW** ≈ 480-755 nm	**FL-BW** ≈ 410-650 nm	**FL Max** ≈ 520 nm	**FL Max** ≈ 556 nm	**FL Max** ≈ 514 nm
Type: Elastic, Pseudo-plastic	**Type:** Elastic Pseudo-plastic	**FL Max** ≈ 576 nm	**FL Max** ≈ 482	**Type:** Elastic	**Type** = Elastic	**Type** = Elastic but plastic on excessive bending
		Type: Elastic, Pseudo-plastic	**Type:** Elastic, Pseudo-plastic			

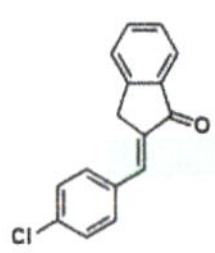

63[110]

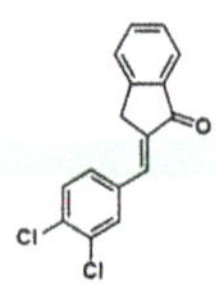

64[110]

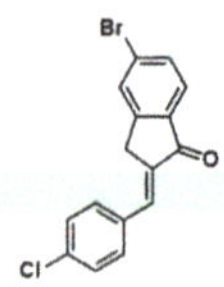

65[110]

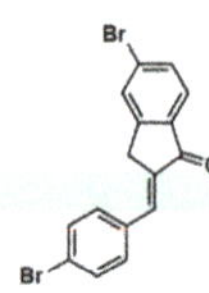

66[110]

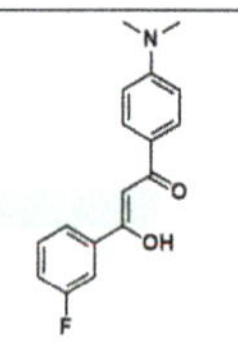

67[112]

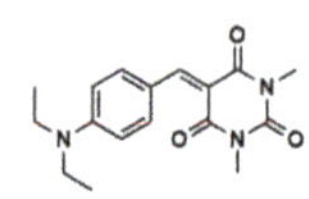

68[113]

63	64	65	66	67	68
Y = 3.45 GPa	**Y** = 9.56 GPa	**Y** = 0.45 GPa	**Y** = 8.98 GPa	**Y** = -	**Y** = 4.56±0.03 G Pa
Hc = 0.16 GPa	**Hc** = 0.382 GPa	**Hc** = 0.035 GPa	**Hc** = 0.204 GPa	**Hc** = -	**Hc** = 0.28±0.01 GPa
FL-BW= -	**FL-BW**= -	**FL-BW** = -	**FL-BW**= -	**FL-BW** = 450-700 nm	**FL-BW** = -
FL Max= -	**FL Max** = -	**FL Max** = -	**FL Max**= -	$\mathbf{FL\ Max_{a,b,c}}$ ≈ 515/, 537/558 nm	**FL Max**= 656 nm
Type = Plastic	**Type** = Slightly Plastic	**Type** = Plastic	**Type** = Plastic	**Type** = Polymorphs I-III Elastic, Plastic and Brittle	**Type** = Elastic

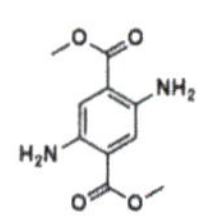

69[114]

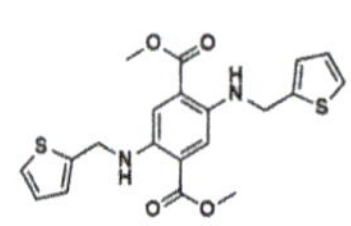

70[115]

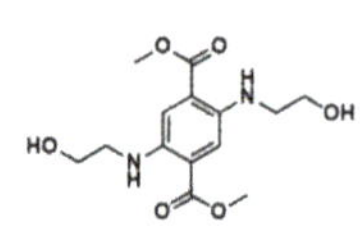

71[116]

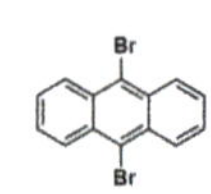

72[117, 118]

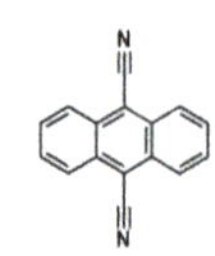

73[119]

69	70	71	72	73
Y = -	**Y** = -	**Y** = -	**Y** = -	**Y** = 0.6±0.2 GPa
Hc = -	**Hc** = -	**Hc** = -	**Hc** = -	**Hc** = -
FL-BW ≈ 500-750 nm	**FL-BW** ≈ 500-800 nm (Orange) 550-800 nm (Red)	**FL-BW** ≈ 550-800 nm 575-750 nm	**FL-BW** ≈ 450-700 nm	**FL-BW** ≈ 480-750 nm
FL Max ≈ 567 nm	**FL Max** ≈ 589/619 nm	**FL Max** ≈ 622/640 nm (both red FL)	**FL Max** ≈ 500 nm	**FL Max** ≈ 516 nm
Type = Elastic (bending/twisting)	**Type** = Polymorphs I&II Elastic, Brittle	**Type** = Polymorphs I&II Elastic, Brittle	**Type** = Elastic Pseudo-plastic	**Type** = Plastic

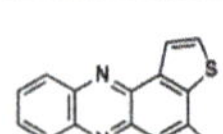

74[120]

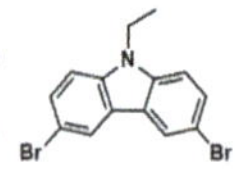

75[121]

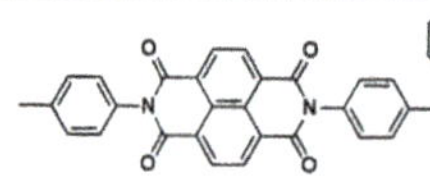

76[122]

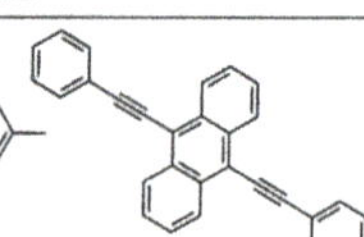

77[122]

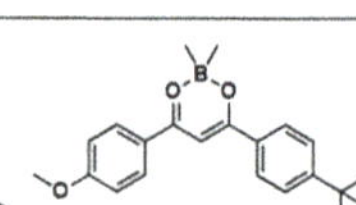

78[123]

74	75	76	77	78
Y = 5.83±0.62 GPa	**Y** = 6.86 ± 0.42 GPa	**Y** = 0.95 ± 0.02 GPa	**Y** = -	**Y** = 5.6 ± 0.46 GPa
Hc = 0.26±0.01 GPa	**Hc** = 0.16 ± 0.013 GPa	**Hc** = 0.30 ± 0.02 GPa	**Hc** = -	**Hc** = 0.206 ± 0.013 GPa
FL-BW ≈ 450-700 nm	**FL-BW** ≈ 400-640 nm	**FL-BW** ≈ 490-760 nm	**FL-BW** ≈ 540-750 nm	**FL-BW** = -
FL Max ≈ 532 nm	**FL Max** ≈ 510/565 nm	**FL Max** ≈ 605 nm	**FL Max** ≈ 595 nm	**FL Max**= -
Type = Elastic Pseudo-plastic	**Type** = Plastic	**Type** = Plastic Pseudo-plastic	**Type** = Elastic	**Type** = Polymorphs I&II Plastic, Brittle

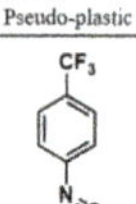

79[124]

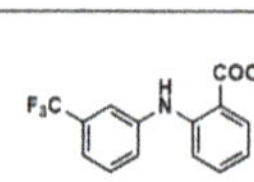

80[125]

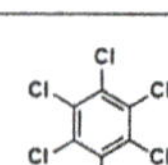

81[126]

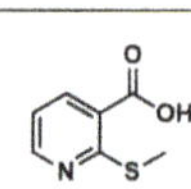

82[126]

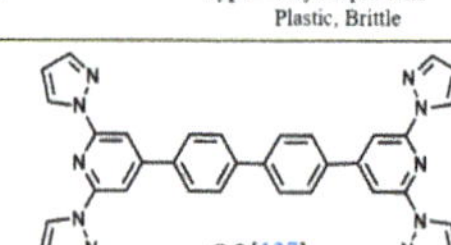

83[127]

79	80	81	82	83
Y = 9.56 ± 1.1 GPa (001)/(00-1)	**Y** = 7.90 ± 0.18 GPa	**Y** = -	**Y** = 15 GPa	**Y** = -
Hc = 0.650 ± 0.014 Gpa (001)/(00-1)	**Hc** = 0.11 ± 0.002 GPa	**Hc** = -	**Hc** = 0.3-0.425 GPa	**Hc** = -
FL-BW = -	**FL-BW** ≈ 390-570 nm	**FL-BW** = -	**FL-BW** = -	**FL-BW** ≈ 380-600 nm
FL Max= -	**FL Max** ≈ 458 nm	**FL Max**= -	**FL Max**= -	**FL Max** ≈ 452 nm
Type = Plastic	**Type** = Plastic	**Type** = Plastic	**Type** = Plastic	**Type** = Elastic Pseudo-plastic

packing; however, halogen bonds and π⋯π stacking, which can act as a structural buffer during bending, are absent. On the other hand, molecules **25–27** all have slip planes and show plasticity [80]. The three-point bending analysis of single crystals of **29** initially demonstrates a pronounced elastic regime, characterized by the crystals reverting to their linear form upon the release of stress, exhibiting an elastic strain (ζ) of approximately 2.5%. Beyond this threshold, when bent further, the crystal displays plastic deformation [82].

Furthermore, polymorphism often has the effect of altering both the optical as well as mechanical properties of the organic crystals. For instance, in the case of molecule **22**, changes in intermolecular packing dictated by molecular conformation in the crystal habit led to two different polymorphs with distinct optical and mechanical properties. In one polymorph, the planar conformation of molecule **22** [78] allowed facile long-range order in the crystal packing with separated layers interacting via π⋯π interactions whose slippage across different layers leads to plastic-like behavior. While, in the other polymorph, molecule **22**'s conformer was highly distorted and prevented the layered arrangement. Thereby the second polymorph exhibited no mechanical flexibility and showcased brittle characteristics under external pressure. In molecule **22**, the polymorph with highly distorted conformation exhibited a photochromic response due to limited stability. However, the other polymorph was stable under photoexcitation and exhibited green emission.

Similarly, molecule **23** (a closely related system to molecule **22** where the methoxy group is replaced by dimethyl amine moiety) formed two polymorphs (orange and yellow emission) and exhibited distinguishable elastic responses [78]. The orange polymorph with strong π⋯π interactions showed a better elastic response compared to the yellow polymorph with nominal mechanical elasticity.

Like in molecule **22**, molecule **28** [81] also exhibited two different polymorphs with plastic (Form-I) and brittle (Form-II) characteristics attributed to anisotropic and 3D interlocked molecular crystal packings, respectively. The variation in crystal packing modified the emission properties of both polymorphs with a slight red shift in the emission of Form-I compared to Form-II. The mechanical flexibility of Form-I allowed the modulation of guided-optical signals' direction via plastic deformation. Compound **32** [84] also forms plate-like and needle-shaped polymorphic crystals with brittle and elastic properties, respectively.

Molecules **42** [93], **70** [115], and **78** [123] also formed two polymorphic crystals, each displaying plastic and brittle characteristics owing to different packing in the crystalline state. The molecular crystal of **71** [116] exhibits elasticity; however, upon heating, it transforms into another polymorph with a brittle character. Notably, molecule **39** [90] was reported to exhibit intriguing properties with two polymorphs —one displaying elasticity attributed to the presence of π–π stacking, and the other demonstrating plasticity owing to the existence of slip planes.

Interestingly, diaryl beta-diketone compound (**67**) [112] crystallizes in three polymorphs I–III exhibiting elasticity, elasticity followed by plasticity and brittleness, respectively. The reason for the peculiar behavior of polymorph II is that the weak C–H⋯F bonds between neighboring (100) planes thwart the molecular sliding.

When the load exceeds the elasticity threshold, irreversible deformation via molecules permanent sliding. Polymorph III of flufenamic acid (**80**) [125] exhibits remarkable plastic bending along the (001) face and demonstrates hand-twistable behavior under applied external force. The polymorph I exhibits isotropic intermolecular interactions along with a crosswise packing structure. The layered structure of polymorph II allows ample space for molecular sliding along the crystal growth direction. In polymorph III, the intermolecular interactions are dispersed homogeneously; without the sliding plane. The molecular arrangement in these three polymorphs exhibits uniformity, while distinctions in optical properties arise from variations in the strengths of $\pi\cdots\pi$ interactions. This difference commonly results in a redshift in FL, with the order observed as I, II, and III.

It is worth considering stable polymorphism as an opportunity rather than a hindrance, as it allows for the creation of diverse crystalline materials with distinct properties and potential applications. Typically, it is highly challenging to design molecules that exhibit elastic as well as plastic mechanical responses. However, two polymorphs of molecule **43** [94] enable the realization of both elastic (thinner crystals) and plastic (thicker crystals) mechanical flexibility depending on the size of the crystal governed by the presence of a smaller or larger number of slip planes. Similar mechanical properties are observed in the polymorphs of molecule **68** [103], where green and yellow emissive polymorphs exhibited elastic and plastic mechanical deformation, respectively. In fact, this molecule showed another polymorph which was brittle, thereby portraying the importance of crystal habitat in controlling various properties.

Apart from crystal packing, the mechanical response also depends on the environment in which the mechanical deformation is performed. The mechanical response of molecular crystals can be significantly different in the micro-domain compared to the macro-scale. My research group at the University of Hyderabad found that crystals of **33–37** [85–88] and **57–59** [105–107] display elastically bendable mechanical characteristics with millimeter-sized crystals while the microcrystals of the same molecules adhered to a glass substrate behave as *pseudo-plastic* due to crystal and substrate strong adhesive interactions (see figure 1.10; a detailed discussion on pseudo-plasticity is provided in chapter 3, section 3.2.3), emphasizing the vital role of environmental factors in miniature molecular crystals. In most cases, waveguiding experiments (as exemplified in case **61** [109]) involving millimeter-sized crystals in a bent geometry were conducted by affixing them to a substrate using glue. From our dozens of experiments on various microcrystals, *pseudo-plastic*ity was observed and this property is key to creating photonic components and circuits. It looks like the aspect ratio plays an important role in observing pseudo-plasticity in crystals.

The optical properties of flexible crystals hold great significance in the development of photonic integrated circuits (PICs). These properties encompass a range of crucial parameters, including solid-state optical absorbance, emissions (both fluorescence (FL) and phosphorescence), optical emission spectral bandwidth, optical energy transfer (arising from the overlap between the emission spectrum of the donor crystal and the absorption spectrum of the acceptor crystal), reabsorption or self-absorption of emissions, chirality, and nonlinear optical (NLO) attributes. These optical characteristics play a pivotal role in the selection of organic crystals for the construction of organic photonic integrated circuits (OPICs). These topics will be extensively discussed in chapter 2, shedding light on the critical considerations for effectively utilizing organic crystals in the construction of integrated photonic devices.

Prior to delving into a detailed discussion on OPICs, it is crucial to gain insight into the historical context. This historical perspective will help elucidate the necessity and motivation for the development of PICs. It will also provide an overview of the different materials platforms that have been employed for the fabrication of PICs, along with an examination of their respective advantages and limitations. Furthermore, it is essential to acknowledge the promising features of organic crystals and their growing prominence as a versatile material platform for photonics applications. Recognizing the unique attributes and potential of organic crystals in the context of photonic technologies is an important foundation for the subsequent discussions on their applications in OPICs.

1.7 Photonic integrated circuits (PICs) [128,129]

1.7.1 Inorganic PICs

In electronic circuits, the flow of electrons through interconnected wires, which link various components like resistors, transistors, capacitors, inductors, and diodes, serves to execute multifaceted operations, including signal amplification, computational tasks, and data transmission. The invention of the field-effect transistor (FET), a fundamental device that harnesses electric fields to regulate the flow of current in a semiconductor, has revolutionized the construction of integrated circuits (ICs) or chips. These ICs are typically manufactured on semiconductor materials, notably doped silicon (Si).

Building upon this foundational development, the groundbreaking invention of the miniature metal–oxide semiconductor field-effect transistor (MOSFET) marked a significant milestone [129–131]. This invention paved the way for the integration of a substantial number of MOSFETs, with sizes scaling down to 5 nm, onto a single chip. This miniaturization of transistors not only made the circuits more compact but also significantly enhanced their speed and cost-efficiency. In alignment with Moore's law, which posits that the number of transistors on an IC doubles every two years, it is foreseen that transistor dimensions will continue to shrink below the 5 nm threshold. ICs have become the backbone of modern computing systems, finding application in computers, mobile devices, controllers, and a myriad of electronic gadgets. More recently, the utilization of the complementary metal–oxide semiconductor (CMOS) fabrication process, a specific variant of MOSFET technology, has gained prominence in IC construction. CMOS offers distinct advantages, such as reduced noise and power consumption, making it an attractive choice for the development of cutting-edge electronic devices [131].

In contrast to electronic circuits, PICs employ photons for their operations [132–135]. This is accomplished by subjecting the light through a series of interconnected optical components, including

- waveguides (with various geometries such as linear, curved, and spiral);
- cavities or resonators;
- amplifiers;
- multiplexers;
- modulators;
- splitters;
- interconnects;
- lasers;
- detectors, and more.

The specific attributes and functions of these components fabricated from molecular crystals will be discussed in greater detail in chapter 2. To start the operation of a PIC, a laser source is used, akin to how an electronic circuit operates with electricity passing through its electronic components. Unlike electrons, light can transport information through various means, including polarization, encompassing both linear and circular polarization states. PICs offer several advantages, including

- miniaturization;
- low optical loss;
- high speed;
- broader bandwidth;
- power efficiency;
- extensive functionality;
- lightweight;
- minimal crosstalk.

In the fabrication of PICs that perform complex functions, achieving high-density integration of various functional optical components is of paramount importance. The functional complexity of PICs is fundamentally determined by the nature and number of optical components that are seamlessly integrated into the circuit (figure 1.14 and table 1.2).

Silicon is an indirect bandgap ($\approx$1.12 eV) material, a widely accepted material for semiconductor and photonic circuit industries, primarily due to its cost-effectiveness. Furthermore, the utilization of silicon in PICs benefits from the well-established infrastructure of CMOS foundries, initially developed for the semiconductor micro-electronics industry. This also makes it easier to integrate photonic components with existing electronic components, allowing for the development of silicon photonics, where both electronic and photonic functions can be integrated on the same chip. This pre-existing infrastructure allows for the relatively cost-effective fabrication of PICs on silicon substrates. In addition to cost advantages, silicon-based PICs leverage the high refractive index contrast between the silicon (n = 3.5) and the silicon dioxide (SiO_2; $n = 1.45$) cladding/wafers (commonly referred to as Silicon-on-Insulator or SOI) [134, 135]. This high refractive index contrast facilitates the

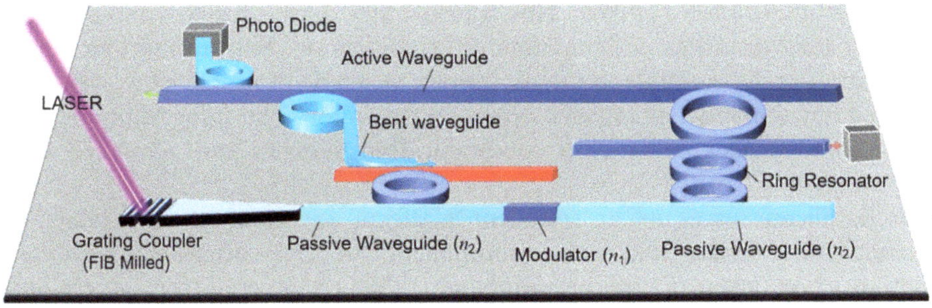

Figure 1.14. Schematic of a photonic integrated circuit (PIC).

Table 1.2. Various optical, mechanical, and crystallographic characteristics of inorganic and organic materials.

Materials	Refractive index (n) at 632 nm	Young's modulus (**Y**, GPa)	Hardness (**H**$_c$, GPa)	Exciton binding energy	Exciton type	Solid-state packing
Si	3.46	130 (100) 187 (110)	12.5	20.6 meV	Wannier–Mott	Crystal (diamond cubic)
SiO_2	1.44	66–75	9.5			Crystal (polymorphic) Amorphous (fused)
Si_3N_4 (α,β,γ phases)	≈2	≈400	50.3	0.6 eV (γ)	Frenkel (γ)	Crystal (trigonal, hexagonal, Spinel type)
InP	3.53	108	7.5	4.8 meV	Wannier–Mott	Crystal
GaAs, CdTe	3.86 2.98	60–115 52	6–10 3	4.2–24.8 meV	Wannier–Mott	Crystal (Zinc blend)
GaN, ZnO	2.37 2.0	210–295 140	11 4.5	25–60 meV	Wannier–Mott	Crystal (Wurzite)
Organic materials	1.56–1.7	4–16	<0.15–0.72	0.5 meV–1 eV	Frenkel	Single crystals

efficient confinement of electromagnetic fields, which means that light travels more slowly in silicon than in air or vacuum, resulting in low optical loss waveguides, even in highly curved geometries. Silicon is transparent in the telecommunications wavelength range (around 1.3 and 1.55 μm), which is crucial for applications in optical communication systems. This transparency allows for the transmission of light over long distances without significant absorption. Silicon photonics enables the integration of multiple photonic components on a single chip, leading to increased functionality and reduced power consumption. This integration is essential for the development of on-chip communication systems and advanced optical interconnects.

While silicon serves as an excellent platform for *passive* components, it is important to consider that *active* components in PICs often require different materials. For instance, lithium niobate ($LiNbO_3$) is used for *electro-optical modulation*[5] (which will be discussed in chapter 2; section 2.2.3) while III/V semiconductors like indium phosphide (InP) and gallium arsenide (GaAs) are chosen for optical amplification and laser sources. However, the fabrication costs associated with III/V semiconductors are notably higher, making them less cost-effective for certain applications. Thus, the choice of material platform in PICs is often dictated by a balance between performance requirements and economic considerations.

Apart from several advantages, there are some negative sides:

- Silicon has a very high thermo–optical coefficient causing unwanted optical band shifts. For example, $\approx$ 1 nm shift for every 10 K rise for the 1150 nm band. As a result, the circuit is very sensitive to temperature variations during long operation and cooling is necessary [135].
- NLO effects such as two-photon absorption and free-carrier absorption can limit the performance of certain devices, particularly at high power levels or in applications where nonlinear effects need to be minimized [136, 137].
- Materials NLO attributes are important for frequency doubling and all-optical signal processing in PICs. Unfortunately, silicon does not show a *Pockels electro-optic effect*[6], which is important to achieve modulation of the signal [136, 137].
- Further, the device functionality is restricted to 1.1–3.8 μm spectral wavelengths in which both Si and SiO_2 are transparent [138].
- Silicon absorbs light at shorter wavelengths, limiting its effectiveness in the visible and ultraviolet spectrum. This can be a disadvantage in applications that require operation in these wavelength ranges.
- Despite the compatibility with CMOS fabrication processes, the fabrication of silicon PICs can still pose challenges, especially when integrating with other materials or components.

Silicon nitride (Si_3N_4) has recently garnered significant attention among researchers in the field of photonic devices due to its broad transparency window spanning from visible to mid-infrared (MIR), coupled with a low thermo-optic coefficient.

[5] Electro-optical modulation: The manipulation of an optical signal by altering its amplitude, phase, or polarization through the application of an external electrical field.
[6] The phenomenon in which the refractive index of a medium changes proportionally to the strength of an applied electric field is known as the linear electro-optic effect.

Photonic crystal waveguides and cavities crafted from silicon nitride with high silicon content have demonstrated enhanced NLO response and reduced NLO losses [139, 140]. In contrast, alternative high-cost materials platforms like Si-on-sapphire [141], Si-on-porous silicon [142], suspended Si [143–148], SiGe-on-silicon [149], and Ge-on-silicon [150] are challenging to fabricate but offer superior performance within the 2–16-μm transparency windows. The incorporation of group III/V periodic elements [151], graphene [152], electro-active polymers [153], $LiNbO_3$ [154], and SiGe [155, 156] introduce heightened modulation capabilities in silicon devices. Notably, CMOS-compatible SiGe stands out by enabling modulation through the *quantum-confined Stark effect* (QCSE)[7] [157] and the *Franz–Keldysh effect*[8] [155].

In contemporary PIC fabrication, a diverse range of technological platforms is employed, including indium phosphide (InP) [158, 159], Si, silica-on-silicon (SiO_2/Si) [160], SOI [161], Si_3N_4 [162], and polymer-based technology. Notably, among these, only direct bandgap InP enables monolithic integration and holds the promise of heterogeneous integration with CMOS electronic chips. The sophistication of modern-day PICs is highlighted by their complexity, with over 1000 optical components seamlessly integrated into a single chip [163].

A recent development involves the utilization of Tantala (Ta_2O_5)-on-insulator for integrated nonlinear and quantum photonics. This choice is motivated by its unique combination, offering low optical absorption, high refractive index contrast, and a high Q-factor (see chapter 2, section 2.2.1.1 for more details about Q-factor) on silicon wafers [164]. This underscores the ongoing efforts to explore innovative materials and integration strategies in the pursuit of advanced and efficient photonic devices.

1.7.2 Organic PICs (OPICs) [64,166,167]

Organic crystals, despite having a relatively low refractive index compared to their main group counterparts, remain an attractive class of materials due to their distinctive characteristics.

- They possess high exciton binding energy, producing room-temperature stable excitons.
- They offer a wide range of optoelectronic properties, the ability to tune optical emission bandwidth, and opportunities for doping and co-crystal growth.
- One of the key features of organic crystals, setting them apart from main group materials, is the presence of non-covalent weak interactions, including hydrogen bonding, halogen bonding, π–π stacking, van der Waals forces, and CT interactions. These interactions play a critical role in determining the optical and mechanical properties of organic crystals.

---

[7] The effect of electric field on the optical absorption and emission spectra of quantum wells.

[8] The effect of electric field on the optical absorption spectra (shifting of the absorption edge) of semiconductors.

- They show physical properties based on factors such as π-conjugation length variation, the electron donor and acceptor characteristics of substituents, chirality, molecular packing, crystal polarity, and crystal symmetry.
- They grow as single crystals in well-defined crystal shapes (1D/2D), and their smooth facets serve as mirror-like light-reflecting surfaces, making them highly effective for guiding and trapping light.
- They display FL or phosphorescence, delayed FL, chiro-optical activity, and NLO activity.
- Many of the CT molecules have large NLO coefficients, which can be advantageous for applications requiring NLO effects, such as frequency conversion and optical modulation. These materials are crucial for developing efficient nonlinear optical devices.
- Some of them are mechanically flexible and lightweight. This flexibility is very useful for the construction of geometrically stable OPICs.
- They possess moderate refractive index (higher than SiO_2).
- Some of them are responsive to external light stimuli and, as a result, show interesting *optomechanical* properties.
- They are generally less expensive than traditional materials used in PICs.
- They can be processed using solution-based techniques, potentially enabling low-cost and scalable fabrication methods such as inkjet printing and spin coating. Additionally, the ease of vapor (sublimation) makes them versatile for various applications.
- Organic crystals can be shaped into various photonically relevant geometries (such as bent waveguides, ring resonators, disc resonators, etc) using focused ion beam lithography. A detailed discussion on this topic is provided in chapter 5.

In the past decade and a half, there has been remarkable growth in the development of laboratory-scale photonic components based on organic single crystals. These components encompass both active and passive elements, including waveguides, cavities or resonators, modulators, and lasers. The intricate details and functioning of these components will be comprehensively discussed in chapter 2.

A recent breakthrough in the field involves the innovative use of plastic and elastic organic crystals to fabricate elementary OPICs on borosilicate glass. This groundbreaking achievement was successfully realized at the University of Hyderabad by developing and employing various micromanipulation techniques assisted by atomic force microscopy (AFM) cantilever tip. This novel approach, known as *Mechanophotonics* and its application to OPICs, is the central focus of this book [64, 166, 167]. Chapters 2–5 will delve into the specifics of this technique, providing insights and practical examples of its application in the field of photonics and integrated circuit construction.

References

[1] O'Brien J L, Furusawa A and Vučković J 2009 Photonic quantum technologies *Nat. Photon* **3** 687–95

[2] Barz S 2015 Quantum computing with photons: introduction to the circuit model, the one-way quantum computer, and the fundamental principles of photonic experiments *J. Phys. B: At. Mol. Opt. Phys.* **48** 083001

[3] Sakurai J J 2003 *Modern Quantum Mechanics* (Reading, MA: Addison-Wesley)

[4] Freedman S J and Clauser J F 1972 Experimental test of local hidden-variable theories *Phys. Rev. Lett.* **28** 938–41

[5] Kwiat P G, Mattle K, Weinfurter H, Zeilinger A, Sergienko A V and Shih Y 1995 New high-intensity source of polarization-entangled photon pairs *Phys. Rev. Lett.* **75** 4337–41

[6] Zellouf D, Jayet Y, Saint-Pierre N, Tatibouët J and Baboux J C 1996 Ultrasonic spectroscopy in polymeric materials. application of the Kramers–Kronig relations *J. Appl. Phys.* **80** 2728–32

[7] Roychowdhury T, Shah D, Hilfiker J and Linford M R 2020 Polymethyl methacrylate: optical properties from 191 to 1688 nm (0.735–6.491 eV) by spectroscopic ellipsometry *Surf. Sci. Spectra* **27** 16002

[8] Nichelatti E 2002 Complex refractive index of a slab from reflectance and transmittance: analytical solution *J. Opt. A: Pure Appl. Opt.* **4** 306400

[9] Nam S B, Reynolds D C, Litton C W, Almassy R J, Collins T C and Wolfe C M 1976 Free-exciton energy spectrum in GaAs *Phys. Rev.* B **13** 761

[10] Christopoulos S *et al* 2007 Room-temperature polariton lasing in semiconductor microcavities *Phys. Rev. Lett.* **98** 126405

[11] Li F *et al* 2013 Fabrication and characterization of a room-temperature ZnO polariton laser *Appl. Phys. Lett.* **102** 191118

[12] Wannier G H 1937 The structure of electronic excitation levels in insulating crystals *Phy. Rev.* **52** 191

[13] Frenkel J 1931 On the transformation of light into heat in solids I *Phy. Rev.* **37** 17

[14] Frenkel J 1931 On the transformation of light into heat in solids II *Phy. Rev.* **37** 1276

[15] Plumhof J D, Stöferle T, Mai L, Scherf U and Mahrt R F 2014 Room-temperature Bose–Einstein condensation of cavity exciton–polaritons in a polymer *Nat. Mater.* **13** 247–52

[16] van Vugt L K, Rühle S, Ravindran P, Gerritsen H C, Kuipers L and Vanmaekelbergh D 2006 Exciton polaritons confined in a ZnO nanowire cavity *Phy. Rev. Lett.* **97** 147401

[17] Kasprzak J *et al* 2006 Bose–Einstein condensation of exciton polaritons *Nature* **443** 409–14

[18] Griffin A, Snoke D W and Stringari S (ed) 1995 *Bose–Einstein Condensation* (Cambridge: Cambridge University Press)

[19] Imamoglu A, Ram R J, Pau S and Yamamoto Y 1996 Nonequilibrium condensates and lasers without inversion: exciton-polariton lasers *Phy. Rev.* A **53** 4250–3

[20] Deng H, Haug H and Yamamoto Y 2010 Exciton-polariton Bose–Einstein condensation *Rev. Mod. Phys.* **82** 1489

[21] Kavokin A 2010 Exciton-polaritons in microcavities: recent discoveries and perspectives *Phys. Status Solidi* B **247** 1898–906

[22] Choi D, Park M, Sung C-Y, Choi H and Cho Y-H 2022 Realization of exciton-polariton condensation in GaAs-based microcavity grown by metalorganic chemical vapor deposition *Phys. Rev. Res.* **4** 033188

[23] Kusano J, Segawa Y, Mihara M, Aoyagi Y and Namba S 1989 Quantum size effect on the exciton polariton in GaAs thin films *Solid State Commun.* **72** 215–8

[24] Cooper D E and Newman P R 1989 Exciton-polariton dynamics and photoluminescence line shapes in cadmium telluride *Phy. Rev.* B **39** 7431–40

[25] Huang R, Yamamoto Y, André R, Bleuse J, Muller M and Ulmer-Tuffigo H 2001 Exciton-polariton amplification based on exciton-exciton scattering in a CdTe quantum well *Quantum Electronics and Laser Science Conference* ed G Agrawal, Y Yamamoto, T Norris and J Thomas (OSA Technical Digest Optica Publishing Group) Paper QTuE31

[26] Franke H, Sturm C, Schmidt-Grund R, Wagner G and Grundmann M 2012 Ballistic propagation of exciton–polariton condensates in a ZnO-based microcavity *New J. Phys.* **14** 013037

[27] Kena-Cohen S and Forrest S R 2010 Room-temperature polariton lasing in an organic single-crystal microcavity *Nature Photon* **4** 371

[28] Fraser M D, Höfling S and Yamamoto Y 2016 Physics and applications of exciton–polariton lasers *Nat. Mater.* **15** 1049–52

[29] Cuevas Á *et al* 2018 First observation of the quantized exciton-polaritonfield and effect of interactions on a single polariton *Sci. Adv.* **4** eaao681

[30] Shen Y R 1984 *The Principles of Nonlinear Optics* (New York: Wiley)

[31] Boyd R W 2008 *Nonlinear Optics* 3rd edn (Cambridge, MA: Academic Press)

[32] Quimby R S 2006 *Photonics and Lasers: An Introduction* (Hoboken, NJ: Wiley)

[33] Göppert-Mayer M 1931 Über elementarakte mit Zwei Quantensprüngen *Ann. Phys.* **401** 273–94

[34] He G S, Tan L-S, Zheng Q and Prasad P N 2008 Multiphoton absorbing materials: molecular designs, characterizations, and applications *Chem. Rev.* **108** 1245–330

[35] Butcher P N and Cotter D 1990 *The Elements of Non-Linear Optics* (Cambridge: Cambridge University Press)

[36] Loudon R 1973 *The Quantum Theory of Light* (Oxford: Oxford University Press)

[37] Kogej T, Beljonne D, Meyers F, Perry J W, Marder S R and Brédas J L 1998 Mechanisms for enhancement of two-photon absorption in donor-acceptor conjugated chromophores *Chem. Phys. Lett.* **298** 1–6

[38] Heflin J R, Wong K Y, Zamani-Khamiri O and Garito A F 1988 Nonlinear optical properties of linear chains and electron-correlation effects *Phys. Rev.* B **38** 1573–6

[39] Dixit S N, Guo D and Mazumdar S 1991 Essential-states mechanism of optical nonlinearity in π-conjugated polymers *Phys. Rev.* B **43** 6781–4

[40] Birge R R and Pierce B M 1986 Semiclassical time-dependent theory of two-photon spectroscopy. The effect of dephasing in the virtual level on the two-photon excitation spectrum of isotachysterol *Int. J. Quantum Chem.* **29** 639–56

[41] Albota M *et al* 1998 Design of organic molecules with large two-photon absorption cross sections *Science* **281** 1653–6

[42] Rumi M *et al* 2000 Structure–property relationships for two-photon absorbing chromophores: bis-donor diphenylpolyene and bis(styryl)benzene derivatives *J. Am. Chem. Soc.* **122** 9500–10

[43] Pawlicki M, Collins H A, Denning R G and Anderson H L 2009 Two-photon absorption and the design of two-photon dyes *Angew. Chem. Int. Ed.* **48** 3244–66

[44] Mak T C W and Trotter J 1964 The crystal and molecular structure of *N*,*N*-dimethyl-*p*-nitroaniline *Acta Crystallogr.* **18** 68–74

[45] Aboulfadl H, Bonin M, Burgner M, Brahimi K and Hulliger J 2013 Molecular recognition at growing crystal faces revealed by phase sensitive second harmonic microscopy *RSC Adv.* **3** 25145

[46] Kamiya M, Sekino H, Tsuneda T and Hirao K 2005 Nonlinear optical property calculations by the long-range-corrected coupled-perturbed Kohn–Sham method *J. Chem. Phys.* **122** 234111

[47] Patra A, Venkatram N, Rao D N and Radhakrishnan T P 2008 Optical limiting in organic molecular nano/microcrystals: nonlinear optical effects dependent on size distribution *J. Phys. Chem.* C **112** 16269–724

[48] Jazbinsek M, Mutter L and Gunter P 2008 Photonic applications with the organic nonlinear optical crystal DAST *IEEE J. Sel. Top. Quantum Electron.* **14** 1298

[49] Brunner F D J, Kwon O P, Kwon S-J, Jazbinšek M, Schneider A and Günter P 2008 A hydrogen-bonded organic nonlinear optical crystal for high-efficiency terahertz generation and detection *Opt. Express* **16** 16496

[50] Mamonov E A, Maydykovskiy A I, Mitetelo N V, Venkatakrishnarao D, Chandrasekar R and Murzina T V 2018 Whispering gallery modes in two-photon fluorescence from spherical DCM dye microresonators *Laser Phys. Lett.* **15** 035401

[51] Jaya Prakash M and Radhakrishnan T P 2006 Second harmonic generation from a homologous series of molecular crystals: impact of supramolecular interactions *Chem. Mater.* **18** 2943

[52] Zheng Y, Guan J, Cheng P, Wang H, Xu J and Bu X 2023 Second-order nonlinear optical organic crystals based on a 'click' compound *J. Mater. Chem.* C **11** 6724

[53] Lu J, Hwang H Y, Li X, Lee S-H, Kwon O P and Nelson K A 2015 Tunable multi-cycle THz generation in organic crystal HMQ-TMS *Opt Express* **23** 22723

[54] Labonte D, Lenz A-K and Oyen M L 2017 On the relationship between indentation hardness and modulus, and the damage resistance of biological materials *Acta Biomater.* **57** 373–83

[55] Wang C and Sun c c 2020 The landscape of mechanical properties of molecular crystals *CrystEngComm* **22** 1149

[56] Azuri I, Meirzadeh E, Ehre D, Cohen S R, Rappe A M, Lahav M, Lubomirsky I and Kronik L 2015 Unusually large Young's moduli of amino acid molecular crystals *Angew. Chem. Int. Ed.* **54** 13566

[57] Karothu D P, Halabi J M, Ahmed E, Ferreira R, Spackman P R, Spackman M A and Naumov P 2022 Global analysis of the mechanical properties of organic crystals *Angew. Chem. Int. Ed.* **61** e202113988

[58] Saha S, Mishra M K, Reddy C M and Desiraju G R 2018 From molecules to interactions to crystal engineering: mechanical properties of organic solids *Acc. Chem. Res.* **51** 2957–67

[59] Mukherjee A and Desiraju G R 2014 Halogen bonds in some dihalogenated phenols: applications to crystal engineering *IUCrJ* **1** 49

[60] Reddy C M, Gundakaram R C, Basavoju S, Kirchner M T, Padmanabhan K A and Desiraju G R 2005 Structural basis for bending of organic crystals *Chem. Commun.* 3945–7

[61] Reddy C M, Kirchner M T, Gundakaram R C, Padmanabhan K A and Desiraju G R 2006 Isostructurality, polymorphism and mechanical properties of some Hexahalogenated Benzenes: the nature of halogen···halogen interactions *Chem. Eur. J.* **12** 2222–34

[62] Takamizawa S and Miyamoto Y 2014 Superelastic organic crystals *Angew. Chem. Int. Ed.* **53** 6970–3

[63] Annadhasan M, Karothu D P, Chinnasamy R, Catalano L, Ahmed E, Ghosh S, Naumov P and Chandrasekar R 2020 Micromanipulation of mechanically compliant organic single-crystal optical microwaveguides *Angew. Chem. Int. Ed.* **59** 13821–30

[64] Chandrasekar R 2023 Advanced all-organic microphotonic components and integrated circuits *Adv. Opt. Mater.* **11** 2301124

[65] Ghosh S and Reddy C M 2012 Elastic and bendable caffeine cocrystals: implications for the design of flexible organic materials *Angew. Chem. Int. Ed.* **51** 10319–23

[66] Thompson A J, Price J R, McMurtrie J C and Clegg J K 2021 The mechanism of bending in co-crystals of caffeine and 4-chloro-3-nitrobenzoic acid *Nat. Commun.* **12** 5983

[67] Chandrasekhar N, Reddy E R, Prasad M D, Rajadurai M S and Chandrasekar R 2014 Passive optical waveguiding tubular pharmaceutical solids and Raman spectroscopy/mapping of nano-/micro-scale defects *CrystEngComm* **16** 4696–700

[68] Ahmed E, Karothu D P, Warren M and Naumov P 2019 Shape-memory effects in molecular crystals *Nat. Commun.* **10** 3723

[69] Kim T-, Lee J-S, Kim Y-C, Joo Y-C and Kim B-J 2019 Bending strain and bending fatigue lifetime of flexible metal electrodes on polymer substrates *Materials* **12** 2490

[70] Thompson A J, Chamorro O A I, Nair J A, Price J R, McMurtrie J and Clegg J K 2021 Elastically flexible molecular crystals *Chem. Soc. Rev.* **50** 11725

[71] Annadhasan M, Karothu D P, Chinnasamy R, Catalano L, Ahmed E, Ghosh S, Naumov P and Chandrasekar R 2020 Micromanipulation of mechanically compliant organic single-crystal optical microwaveguides *Angew. Chem. Int. Ed.* **59** 13821–30

[72] Rohullah M, Pradeep V V, Ravi J, Kumar A V and Chandrasekar R 2022 Micromechanically-powered rolling locomotion of a twisted-crystal optical-waveguide cavity as a mobile light polarization rotor *Angew. Chem. Int. Ed.* **134** e202202114

[73] Krishna G R, Devarapalli R, Lal G and Reddy C M 2016 Mechanically flexible organic crystals achieved by introducing weak interactions in structure: supramolecular shape synthons *J. Am. Chem. Soc.* **138** 13561–7

[74] Ghosh S, Mishra M K, Kadambi S B, Ramamurty U and Desiraju G R 2015 Designing elastic organic crystals: highly flexible polyhalogenated N-benzylideneanilines *Angew. Chem. Int. Ed.* **54** 2674–8

[75] Mishra M K, Kadambi S B, Ramamurty U and Ghosh S 2018 Elastic flexibility tuning via interaction factor modulation in molecular crystals *Chem. Comm.* **54** 9047–50

[76] Chinnasamy R, Munjal B, Suryanarayanan R, Peedikakkal A M P, Mishra M K and Ghosh S 2022 Pressure and temperature induced dual responsive molecular crystals: effect of polymorphism *Cryst. Growth Des.* **22** 615–24

[77] Chinnasamy R, Arul A, AlMousa A, Kiran M S R N, Das P, Jalilov A S, Peedikakkal A M P and Ghosh S 2019 Structure property correlation of a series of halogenated schiff base crystals and understanding of the molecular basis through nanoindentation *Cryst. Growth Des.* **19** 6698–707

[78] Zhao H, Zhang X, Chen K, Wu W, Li S, Wang T, Huang X, Wang N, Zhou L and Hao H 2023 Mechanistic study on the structure–property relationship of flexible organic crystals *CrystEngComm* **25** 2600–6

[79] Wu W, Chen K, Zhang X, Wang T, Li S, Zhao H, Zhou L, Huang X and Hao H 2023 Organic crystals with response to multiple stimuli: mechanical bending, acid-induced bending and heating-induced jumping *Chem. Eur. J.* **29** e202202598

[80] Ravi J, Feiler T, Mondal A, Michalchuk A A L, Reddy C M, Bhattacharya B, Emmerling F and Chandrasekar R 2023 Plastically bendable organic crystals for monolithic and hybrid micro-optical circuits *Adv. Opt. Mater.* **11** 202201518

[81] Feiler T, Bhattacharya B, Michalchuk A A L, Silbernagl D, Yasuda N, Feiler T, Sturm H and Emmerling F 2021 Tuning the mechanical flexibility of organic molecular crystals by polymorphism for flexible optical waveguides *CrystEngComm* **23** 5815–25

[82] Bhattacharya B, Michalchuk A A L, Silbernagl D, Yasuda N, Feiler T, Sturm H and Emmerling F 2023 An atomistic mechanism for elasto-plastic bending in molecular crystals *Chem. Sci.* **14** 3441–50

[83] Pan X, Zheng A, Yu X, Di Q, Li L, Duan P, Ye K, Naumov P and Zhang H 2022 A low-temperature-resistant flexible organic crystal with circularly polarized luminescence *Angew. Chem. Int. Ed.* **61** e202203938

[84] Ye Y, Bai Y, Qi L, Li T, Yan L, Yin Q, Zhang Y and Xie C 2023 Flexible optical waveguides in heterocyclic schiff base self-assembled hydrogen-bonded solvates *Cryst. Growth Des.* **23** 1403–11

[85] Ravi J, Annadhasan M, Kumar A V and Chandrasekar R 2021 Mechanically reconfigurable organic photonic integrated circuits made from two electronically different flexible microcrystals *Adv. Funct. Mater.* **31** 2100642

[86] Kumar A V and Chandrasekar R 2023 A mechanophotonic approach toward an organic, flexible crystal optical interferometer *Adv. Opt. Mater.* **11** 202201009

[87] Ravi J, Vinod Kumar A, Annadhasan M and Chandrasekar R 2022 Realization of mechanically maneuverable circuit ports in organic hybrid photonic chip for 360° steering of bandwidth-engineered signals *Adv. Opt. Mater.* **10** 2102545

[88] Annadhasan M, Pradeep V V, Kumar A V, Ravi J and Chandrasekar R 2022 Integrating triply- and singly-bent highly flexible crystal optical waveguides for organic photonic circuit with a long-pass-filter effect *Small Struct.* **3** 2100163

[89] Chinnasamy R, Ravi J, Vinay Pradeep V, Manoharan D, Emmerling F, Bhattacharya B, Ghosh S and Chandrasekar R 2022 Adaptable optical microwaveguides from mechanically flexible crystalline materials *Chem. Eur. J.* **28** e2022009

[90] Liu H, Lu Z, Tang B, Qu C, Zhang Z and Zhang H 2020 A flexible organic single crystal with plastic-twisting and elastic-bending capabilities and polarization-rotation function *Angew. Chem. Int. Ed.* **59** 12944–50

[91] Marandi P, Naim K, Parida K and Neelakandan P P 2022 Molecular logic operations using multistimuli responsive, mechanically tunable elastic organic single crystals *ACS Mater. Lett.* **4** 2499–505

[92] Kumari P, Verma S K, Natarajan K, Ansari S N, Saini A K and Mobin S M 2019 Design and synthesis of a new facile ligand in a dual role: mechanically elastic crystal and selective mitochondria target *Cryst. Growth Des.* **19** 5483–90

[93] Lakshmipathi M, Tothadi S, Emmerling F, Bhattacharya B and Ghosh S 2022 Different mechanical responses of dimorphic forms of anthracene Schiff base crystal *J. Mol. Struct.* **1252** 132182–9

[94] Naim K, Singh M, Sharma S, Nair R V, Venugopalan P, Chandra Sahoo S and Neelakandan P P 2020 Exceptionally plastic/elastic organic crystals of a naphthalidenimine–boron complex show flexible optical waveguide properties *Chem. Eur. J.* **26** 11979–84

[95] Liu B, Liu H, Zhang H, Di Q and Zhang H 2020 Crystal engineering of a hydrazone molecule toward high elasticity and bright luminescence *J. Phys. Chem. Lett.* **11** 9178–83

[96] Liu B, Lu Z, Tang B, Liu H, Liu H, Zhang Z, Ye K and Zhang H 2020 Self-waveguide single-benzene organic crystal with ultralow-temperature elasticity as a potential flexible material *Angew. Chem. Int. Ed.* **59** 23117–21

[97] Tang S, Ye K, Commins P, Li L, Naumov P and Zhang H 2023 Organic crystalline optical waveguides that remain elastic from –196 to ≈200 °C *Adv. Opt. Mater.* **11** 202200627

[98] Saini A K, Natarajan K and Mobin S M 2017 A new multitalented azine ligand: elastic bending, single-crystal-to-single-crystal transformation and a fluorescence turn-on Al(III) sensor *Chem. Comm.* **53** 9870

[99] Chinnasamy R, Ravi J, Pradeep V, Vinay , Manoharan D, Emmerling F, Bhattacharya B, Ghosh S and Chandrasekar R 2022 Adaptable optical microwaveguides from mechanically flexible crystalline materials *Chem. Eur. J.* **28** e202200905

[100] Zheng X *et al* 2022 Multi-stimuli-induced mechanical bending and reversible fluorescence switching in a single organic crystal *Angew. Chem. Int. Ed.* **61** e202113073

[101] Lan L, Yang X, Tang B, Yu X, Liu X, Li L, Naumov P and Zhang H 2022 Hybrid elastic organic crystals that respond to aerial humidity *Angew. Chem. Int. Ed.* **61** e202200196

[102] Liu H, Ye K, Zhang Z and Zhang H 2019 An organic crystal with high elasticity at an ultra-low temperature (77 K) and shapeability at high temperatures *Angew. Chem. Int. Ed.* **58** 19081–6

[103] Cao L, Tang B, Yu X, Ye K and Zhang H 2021 Intense red emissive organic crystals with elastic bending ability and optical waveguiding behaviour *CrystEngComm* **23** 5758–62

[104] Kumar A V and Chandrasekar R 2023 Mechanophotonics: fabrication of a 2×2 hybrid directional coupler from flexible organic crystals *J. Mater. Chem.* C **11** 7995–8001

[105] Kumar A V, Rohullah M, Ravi J, Godumala M, Annadhasan M and Chandrasekar R 2021 Mechanophotonic aspects of a room temperature phosphorescent flexible organic microcrystal *CrystEngComm* **23** 5774–9

[106] Kumar A V, Godumala M, Ravi J and Chandrasekar R 2022 A broadband, multiplexed-visible-light-transport in composite flexible-organic-crystal waveguide *Angew. Chem. Int. Ed.* **61** e202212382

[107] Kumar A V, Godumala M, Ravi J and Chandrasekar R 2023 A pick-and-place technology enabled mechanically programmable spiral optical waveguides integrated hybrid circuits from four extremely pseudo-plastic organic crystals for visible spectral band tunability *Adv. Opt. Mater.* 2302807

[108] Tang B, Yu X, Ye K and Zhang H 2022 Manifold mechanical deformations of organic crystals with optical waveguiding and polarization rotation functions *Adv. Opt. Mater.* **10** 2101335

[109] Lu Z, Zhang Y, Liu H, Ye K, Liu W and Zhang H 2020 Optical waveguiding organic single crystals exhibiting physical and chemical bending features *Angew. Chem. Int. Ed.* **59** 4299–303

[110] Deepak M, Megha S N, Mangalampalli S R N, Kiran , Emmerling F, Bhattacharya B and Ghosh S 2023 Designing the mechanical plasticity of Benzylidene indanones based molecular crystals by crystal engineering *Cryst. Growth Des.* **23** 657–61

[111] Liu H, Bian Z, Cheng Q, Lan L, Wang Y and Zhang H 2019 Controllably realizing elastic/plastic bending based on a room-temperature phosphorescent waveguiding organic crystal *Chem. Sci.* **10** 227–32

[112] Chu X, Lu Z, Tang B, Liu B, Ye K and Zhang H 2020 Engineering mechanical compliance of an organic compound toward flexible crystal lasing media *J. Phys. Chem. Lett.* **11** 5433–8

[113] Peng J, Bai J, Cao X, He J, Xu W and Jia J 2021 Elastic organic crystals based on barbituric derivative: multi-faceted bending and flexible optical waveguide**27** 16036–42

[114] Huang R, Tang B, Ye K, Wang C and Zhang H 2019 Flexible luminescent organic bulk crystal: 2D elasticity toward 3D optical waveguide *Adv. Opt. Mater.* **7** 201900927

[115] Tang B, Liu B, Liu H and Zhang H 2020 Naturally and elastically bent organic polymorphs for multifunctional optical applications *Adv. Funct. Mater.* **30** 2004116–22
[116] Liu B, Di Q, Liu W, Wang C, Wang Y and Zhang H 2019 Red-emissive organic crystals of a single-benzene molecule: elastically bendable and flexible optical waveguide *J. Phys. Chem. Lett.* **10** 1437–42
[117] Hayashi S, Ishiwari F, Fukushima T, Mikage S, Imamura Y, Tashiro M and Katouda M 2020 Anisotropic Poisson effect and deformation-induced fluorescence change of elastic 9,10-dibromoanthracene single crystals *Angew. Chem. Int. Ed.* **59** 16195–201
[118] Kumar A V, Mamonov E, Murzina T and Chandrasekar R 2023 Race-track type resonator integrated active add-drop filter from flexible organic crystals: experiments and numerical calculations *Adv. Opt. Mater.* **11** 2201507
[119] Catalano L, Karothu D P, Schramm S, Ahmed E, Rezgui R, Barber T J, Famulari A and Naumov P 2018 Dual-mode light transduction through a plastically bendable organic crystal as an optical waveguide *Angew. Chem. Int. Ed.* **57** 17254–8
[120] Annadhasan M, Agrawal A R, Bhunia S, Pradeep V V, Zade S S, Reddy C M and Chandrasekar R 2020 Mechanophotonics: flexible single-crystal organic waveguides and circuits *Angew. Chem. Int. Ed.* **59** 13852–8
[121] Kalita K J, Mondal S, Reddy C M and Vijayaraghavan R K 2023 Temperature-regulated dual phosphorescence and mechanical strain-induced luminescence modulation in a plastically bendable and twistable organic crystal *Chem. Mater.* **35** 709–18
[122] Ravi J, Kumar A V, Karothu D P, Annadhasan M, Naumov P and Chandrasekar R 2021 Geometrically reconfigurable, 2D, all-organic photonic integrated circuits made from two mechanically and optically dissimilar crystals *Adv. Funct. Mater.* **31** 2105415–24
[123] Krishna G R, Kiran M S R N, Fraser C L, Ramamurty U and Reddy C M 2013 The relationship of solid-state plasticity to mechanochromic luminescence in difluoroboron Avobenzone polymorphs *Adv. Funct. Mater.* **23** 1422–30
[124] Hasija A, Thompson A J, Singh L S N M, Mangalampalli K S R N, McMurtrie J C, Bhattacharjee M, Clegg J K and Chopra D 2023 Plastic deformation in a molecular crystal enables a piezoresistive response *Small* **19** 2206169
[125] Liu Y, Yang P, Zhang K, Xu J, Wu S and Gong J 2022 Bendable and twistable crystals of flufenamic acid form III with bending mechanofluorochromism behavior *Cryst. Growth Des.* **22** 1312–8
[126] Reddy C M, Gundakaram R C, Basavoju S, Kirchner M T, Padmanabhan K A and Desiraju G R 2005 Structural basis for bending of organic crystals *Chem. Commun.* 3945–7
[127] Kumar A, Vinod , Rohullah M, Chosenyah M, Ravi J, Venkataramudu U and Chandrasekar R 2023 Amphibian-like flexible organic crystal optical fibers for under-water/air micro-precision lighting and sensing *Angew. Chem. Int. Ed.* **62** e202300046
[128] Coldren L and Corzine S 2012 *Diode Lasers and Photonic Integrated Circuits* 2nd edn (New York: Wiley)
[129] Sze S M and Ng K K 2006 *Physics of Semiconductor Devices* 3rd edn (New York: Wiley)
[130] Taur Y and Ning T H 2009 *Fundamentals of Modern VLSI Devices* 2nd edn (Cambridge: Cambridge University Press)
[131] Razavi B 2000 *Design of Analog CMOS Integrated Circuits* (New York: McGraw-Hill)
[132] Kirchain R and Kimerling L 2007 A roadmap for nanophotonics *Nat. Photonics* **1** 303–5
[133] Bogaerts W, Pérez D, Capmany J, Miller D A B, Poon J, Englund D, Morichetti F and Melloni A 2020 Programmable photonic circuits *Nature* **586** 207–16

[134] Thomson D *et al* 2016 Roadmap on silicon photonics *J. Optics* **18** 073003

[135] Bogaerts W and Chrostowski L 2018 Silicon photonics circuit design: methods, tools and challenges *Laser Photonics Rev.* **12** 1700237

[136] Koos C *et al* 2009 All-optical high-speed signal processing with silicon–organic hybrid slot waveguides *Nat. Photon* **3** 216–9

[137] Tsang H K, Wong C S, Liang T K, Day I E, Roberts S W, Harpin A, Drake J and Asghari M 2002 Optical dispersion, two-photon absorption and self-phase modulation in silicon waveguides at 1.5 μm wavelength *Appl. Phys. Lett.* **80** 416–8

[138] Soref R 2010 Mid-infrared photonics in silicon and germanium *Nat. Photon* **4** 495–7

[139] Khan S, Chiles J, Ma J and Fathpour S 2013 Silicon-on-nitride waveguides for mid- and near-infrared integrated photonics *Appl. Phys. Lett.* **102** 121104

[140] Debnath K, Bucio T D, Al-Attili A, Khokhar A Z, Saito S and Gardes F Y 2017 Photonic crystal waveguides on silicon rich nitride platform *Opt. Exp.* **25** 3214–21

[141] Baehr-Jones T, Spott A, Ilic R, Spott A, Penkov B, Asher W and Hochberg M 2010 Silicon-on-sapphire integrated waveguides for the mid-infrared *Opt. Exp.* **18** 12127–35

[142] Mashanovich G Z, Mashanovich G Z, Milošević M M, Nedeljkovic M, Owens N, Xiong B, Teo E J and Youfang H 2011 Low loss silicon waveguides for the mid-infrared *Opt. Express* **19** 7112–9

[143] Penadés J S, Alonso-Ramos C, Khokhar A Z, Nedeljkovic M, Boodhoo L A, Ortega-Moñux A, Molina-Fernández I, Cheben P and Mashanovich G Z 2014 Suspended SOI waveguide with sub-wavelength grating cladding for mid-infrared *Opt. Lett.* **39** 5661–4

[144] Penades J S *et al* 2018 Suspended silicon waveguides for long-wave infrared wavelengths *Opt. Lett.* **43** 795–8

[145] Zhou W, Cheng Z, Wu X, Zhu B, Sun X and Tsang H 2017 Fully suspended slot waveguides for high refractive index sensitivity *Opt. Lett.* **42** 1245–8

[146] Cheng Z, Chen X, Wong C Y, Xu K and Tsang H K 2012 Mid-infrared suspended membrane waveguide and ring resonator on silicon-on-insulator *IEEE Photon J.* **4** 1510–9

[147] Penades J S *et al* 2016 Suspended silicon mid-infrared waveguide devices with subwavelength grating metamaterial cladding *Opt. Exp.* **24** 22908–16

[148] Soref R 2006 The past, present, and future of silicon photonics *IEEE J. Sel. Top. Quantum Electron.* **12** 1678–87

[149] Brun M, Labeye P, Grand G, Hartmann J M, Boulila F, Carras M and Nicoletti S 2014 Low loss SiGe graded index waveguides for mid-IR applications *Opt. Express* **22** 508–18

[150] Kozak D A, Tyndall N F, Pruessner M W, Rabinovich W S and Stievater T H 2021 Germanium-on-silicon waveguides for long-wave integrated photonics: ring resonance and thermo-optics *Opt. Express* **29** 15443–51

[151] Tang Y, Peters J and Bowers J E 2012 Over 67 GHz bandwidth hybrid silicon electroabsorption modulator with asymmetric segmented electrode for 1.3 μm transmission *Opt. Express* **20** 11529–35

[152] Liu M, Yin X, Ulin-Avila E, Geng B, Zentgraf T, Ju L, Wang F and Zhang X 2011 A graphene-based broadband optical modulator *Nature* **474** 64–7

[153] Koos C *et al* 2014 Terabit/s optical transmission using chip-scale frequency comb sources *Proc. European Conf. on Optical Communication (Cannes, France)* pp 1–3

[154] Chen L, Xu Q, Wood M G and Reano R M 2014 Hybrid silicon and lithium niobate electro-optical ring modulator *Optica* **1** 112–8

[155] Feng D *et al* 2012 High speed GeSi electro-absorption modulator at 1550 nm wavelength on SOI waveguide *Opt. Express* **20** 22224–32
[156] Chaisakul P, Marris-Morini D, Rouifed M-S, Isella G, Chrastina D, Frigerio J, Le Roux X, Edmond S, Coudevylle J-R and Vivien L 2012 23 GHz Ge/SiGe multiple quantum well electro-absorption modulator *Opt. Exp.* **20** 3219–24
[157] Lever L *et al* 2011 Modulation of the absorption coefficient at 1.3 μm in Ge/SiGe multiple quantum well heterostructures on silicon *Opt. Lett.* **36** 4158–60
[158] R *et al* 2010 InP photonic integrated circuits *IEEE J. Sel. Top. Quantum Electron.* **16** 1113–25
[159] Smit M, Baets R and Wale M 2009 InP-based photonic integration: learning from CMOS *Proc. 35th European Conference on Optical Communication* (Vienna,: Austria) 1.7.3–1/2
[160] Wosinski L 2004 Silica-on-silicon technology for photonic integrated devices *IEEE Proc. 6th Int. Conf. on Transparent Optical Networks (Wroclaw, Poland, July 4–8) Proc. IEEE* **2** 274–9
[161] Bogaerts W, Dumon P, van Thourhout D, Taillaert D, Jaenen P, Wouters J, Beckx S, Wiaux V and Baets R G 2006 Compact wavelength-selective functions in silicon-on-insulator photonic wires *IEEE J. Sel. Top. Quantum Electron.* **12** 1394–401
[162] Jalali B 2008 Can silicon change photonics? *Phys. Stat. Sol. (a)* **205** 213–24
[163] Nagarajan R *et al* 2005 Large-scale photonic integrated circuits *IEEE J. Sel. Top. Quant. Electron.* **11** 50–65
[164] Belt M, Davenport M L, Bowers J E and Blumenthal D J 2017 Ultra-low-loss Ta_2O_5-core/SiO_2-clad planar waveguides on Si substrates *Optica* **4** 532–6
[165] Annadhasan M, Basak S, Chandrasekhar N and Chandrasekar R 2020 Next-Generation Orgabic Photonics: The Emergence of Flexible Crystal Optical Waveguides *Adv. Opt. Mater.* **8** 2000959
[166] Chandrasekar R 2022 Mechanophotonics—a guide to integrating microcrystals toward monolithic and hybrid all-organic photonic circuits *Chem. Commun.* **58** 3415–28
[167] Chandrasekar R 2021 Mechanophotonics—mechanical micromanipulation of single-crystals towards organic photonic integrated circuits (OPICs) *Small* **17** 2100277

IOP Publishing

Mechanophotonics for Organic Photonic Integrated Circuits

Rajadurai Chandrasekar

Chapter 2

Properties of organic PIC components

2.1 Waveguides

2.1.1 Electromagnetic field in waveguides

A waveguide is a 1D material structure essential for transporting electromagnetic waves or light from one end to another [1]. The electromagnetic wave propagates through the waveguide with speed, *c*/*n* as per equation (1.35). For the transport of light through the waveguides, the refractive index of the waveguides (n_1) should be always higher than the surrounding medium (n_2). Only in such a situation, the light will be guided through the waveguide material via total internal reflection (TIR), when the light's incident angle, θ_i, is greater than the critical angle θ_c (w.r.t. normal to the interface), i.e., $\theta_c < \theta_i$ (figure 2.1(a)).

Here,

$$\sin \theta_c = n_1/n_2 \tag{2.1}$$

During TIR, only the incident and reflected wave remain within the waveguide, and the transmitted wave is transformed into a kind of surface wave, the so-called *evanescent wave* (or *evanescent field*). The evanescent wave is produced at the interface of the waveguide and it decays exponentially with distance from the interface (figures 2.1(a) and (b)). This means the evanescent wave's amplitude is largest at the interface and decreases exponentially with distance into the n_2 medium. The evanescent wave is a decaying field, and it does not propagate through n_2 medium; however, it holds all the standard characteristics of a propagating electromagnetic wave [1, 2].

The z-dependence of the light's **E** in the low contrast medium (n_2) can be written as [3]:

$$\mathbf{E}(z) = E_o e^{-\xi z} \tag{2.2}$$

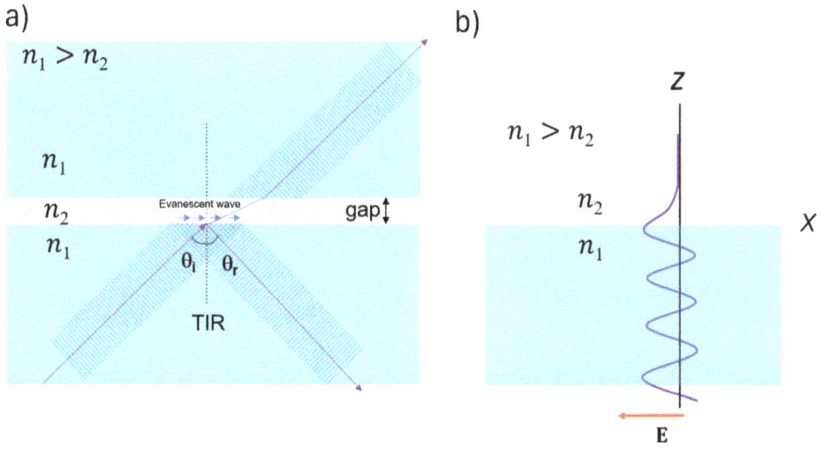

Figure 2.1. (a) Frustrated TIR between two high refractive index materials separated by a thin layer of air. The gap is in the order of wavelength, λ. (b) The exponential decaying of evanescent waves in the lower refractive index medium during TIR.

where,

$$\xi = k_o\sqrt{n_1^2 \sin^2 \theta_1 - n_2^2} \tag{2.3}$$

Here, k_o is the free space wavenumber.

The value of **z** is measured from the boundary of the media, which is in the order of wavelength. From equations (2.2) and (2.3), we can see that the $\mathbf{E}(z)$ is decaying by a factor of e^{-1} at a distance of $\Omega = 1/\xi$ from the interface.

Hence,

$$\Omega = \frac{\lambda_o}{2\pi n_1\sqrt{\sin^2 \theta_1 - \sin^2 \theta_c}} \quad \left(\because k_o = \frac{2\pi}{\lambda_o}\right) \tag{2.4}$$

The typical penetration depth of a visible light evanescent wave into n_2 medium is 50–100 nm.

In PICs, silicon and its derivatives are mainly used as waveguiding material. Usually, these waveguides with nanometer cross-sections are fabricated on a SiO_2 substrate (figure 2.2(a)). The value of n controls the speed of light in a medium and it is an important factor that governs the waveguide's light confinement ability. The higher the n, the better the confinement and slower the light travels.

From equation (1.35), the phase velocity (v_P) of propagating light in a medium is:

$$v_P = c/n \tag{2.5}$$

$$\lambda_p = v_p/\nu = c/n\nu \tag{2.6}$$

Therefore,

$$\lambda_p = \lambda/n \tag{2.7}$$

where $\lambda = c/\nu$ is the wavelength in free space ($n = 1$).

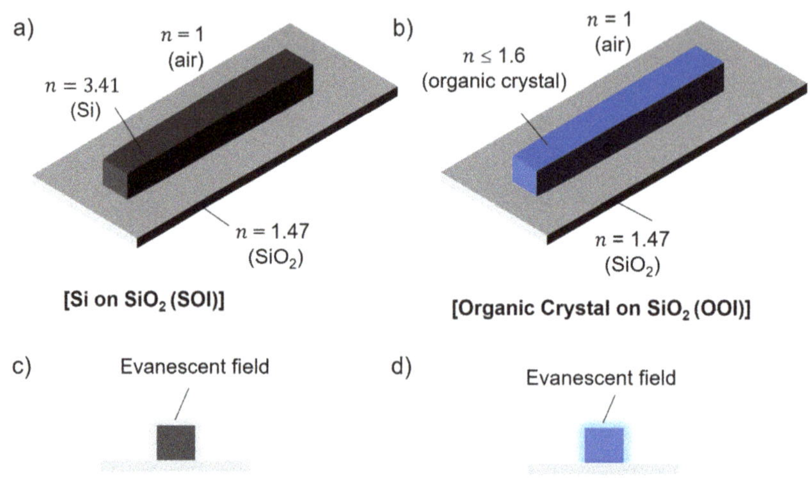

Figure 2.2. (a, b) Schematic of Si waveguide on silica insulator (SOI) and organic crystal waveguide on silica (OOI). (c, d) Schematic view of an *evanescent wave* at the interface and its exponential decay with distance into the rarer medium.

The above relations show that for a given ν, the wavelength of light in the medium (λ_p) is reduced relative to that in free space by a factor of n. Consequently, relative to free space $k = 2\pi/\lambda$, the wavenumber $k_p = 2\pi/\lambda_p$ is also increased by the factor of n.

A Si ($n = 3.4$) waveguide placed on the SiO_2 surface ($n = 1.47$) facilitates effective confinement of propagating light within the Si structure with minimum optical loss (figure 2.2(a)). For example, a 1550 nm (1.55 μm) infrared light used in fiber-optic telecommunication systems propagates in Si waveguides as 455.9 nm. In a nutshell, when a monochromatic wave propagates in media of different n values, except for frequency, its velocity, wavelength, and wavenumber are all altered relative to that in free space medium.

In contrast to free space, light propagates in waveguides as guided *modes*, characterized by the propagation constant. The number of allowed modes for a waveguide and the cut-off wavelength (λ_c) can be calculated through analytical or numerical simulations. Beyond λ_c the waveguide does not allow light transmission. The modes are labeled as transverse electric (TE) and transverse magnetic (TM) depending on the value of the transverse electric field ($\mathbf{E}_z = 0$; $\mathbf{B}_z \neq 0$) or transverse magnetic field ($\mathbf{E}_z \neq 0$; $\mathbf{B}_z = 0$) at the boundary (surface) of the waveguide. Here, z is the direction of light propagation. It should be noted that for transverse electromagnetic (TEM) waves, $\mathbf{E}_z = 0$ and $\mathbf{B}_z = 0$. The TE or TM mode is also expressed as $TE_{m,n}$ or $TM_{m,n}$, respectively, where subscripts m and n stand for the number of half sinusoidal cycle variations along the x-axis and y-axis, respectively (figure 2.3(a)).

Qualitative understanding of light propagation in a nanometer scale (see figure 2.3(b)) matter needs analytical treatment. By taking an approximation that light propagates within optical waveguides of rectangular cross-sections as TM modes, the λ_c for a mode that is guided through the waveguide is given as [3],

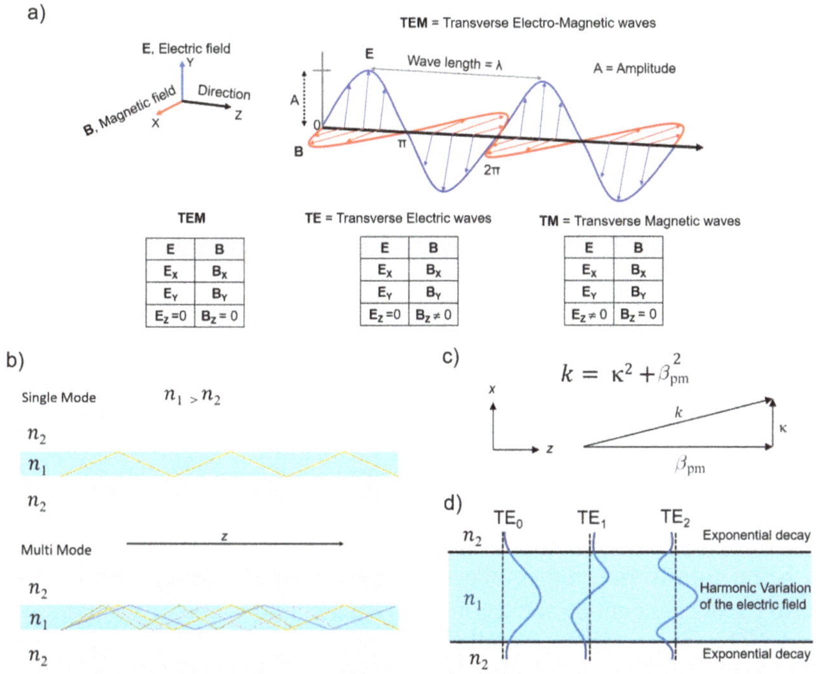

Figure 2.3. (a) The characteristics of TEM, TE, and TM waves. (b) Propagation of single and multimodal light through high refractive index core in optical fibers (left). (c) The relationship between propagation constant β, longitudinal component of the waveguide's wavenumber k, and transverse component of the waveguide's wavenumber κ (right). (d) Propagation of TE polarized modes in a circular waveguide.

$$\lambda_c = \frac{2n_{\perp} w}{m} \tag{2.8}$$

The number of probable modes ($m = 1, 2, 3, 4, \ldots$) is constrained by the following relation,

$$m < \frac{2\,w}{\lambda} \cdot \frac{n_{\perp}}{n_{\parallel}} \sqrt{n_{\parallel}^2 - n_S^2} \tag{2.9}$$

where w is the width of the waveguide, λ is the light wavelength, $n_{\parallel}$ and $n_{\perp}$ are the refractive indices along and perpendicular to the waveguide, respectively, and n_S is the substrate refractive index, which is $\approx$1.5 for silica substrate. For example, according to equation (2.9), the w essential for the propagation of 380 nm light is about 300 nm when $m = 1$ and $n_{\perp}$ and $n_{\parallel}$ are about 1.65.

The effective refractive index, n_{eff} is the ratio of the velocity of light in vacuum to the velocity of a mode (m) for a given polarization in the direction of propagation in a waveguide along the z-direction.

$$n_{eff} = \frac{c}{\vartheta_{z\text{pm}}} = \frac{\nu\lambda_o}{\nu\lambda_{z\text{pm}}} = \frac{\lambda_o}{\lambda_{z\text{pm}}} = \frac{\frac{2\pi}{k_o}}{\frac{2\pi}{k_{z\text{pm}}}} = \frac{k_{z\text{pm}}}{k_o} = \frac{\beta_{\text{pm}}}{k_o} \tag{2.10}$$

Here, $k_{z\text{pm}} = \beta_{\text{pm}}$, and the latter term is called *propagation constant* for a given mode with a particular polarization (TE or TM) and it is pointed towards light propagation direction in a waveguide that is the longitudinal component of the waveguide's wavenumber, k. The transverse component is perpendicular to the propagation direction of the waveguide's wavenumber (figure 2.3(c)).

The order of the TE modes defines the number of zero field regions in the waveguide. One can also see that the field harmonically varies in the waveguide core for different modes. For the lower order mode (TE_0), the electric field is concentrated at the center of the waveguide. However, the modes continue to be guided through the core of the waveguide as long as $n_2k < \beta_{\text{pm}} < n_1k$. When $n_2k > \beta_{\text{pm}}$, optical power leaks out the core into the cladding region or lower refractive index region (if no cladding) (figure 2.3(d)).

2.1.2 Evanescent wave coupling

Let us assume a thin layer of air separates two waveguides of refractive index, n_1, and the air gap (n_2) between them is in the order of light wavelength (figure 2.1(a)). Besides being totally internally reflected, the input light penetrates as evanescent waves into the air gap, enters the second waveguide, and weakens the light in the first waveguide. This phenomenon is called *frustrated* TIR (F-TIR), and it is an optical analog of the quantum mechanical barrier penetrating tunneling effect called *optical tunneling* (see also chapter 1, figure 1.2).

Minimizing the gap between waveguides is a critical requirement to facilitate efficient optical energy transfer through evanescent coupling. In 1966, Coon *et al* [5] employed a configuration with double right-angle prisms and a gap, which ranged from 3.5 to 8.5 times the wavelength of visible light (specifically, the 546.1 nm Hg line). Their observations revealed that the intensity of transmitted light through the gap exponentially decreased as the gap width increased. Subsequently, in 1986, Zhu *et al* [6] successfully reduced the separation between the double prisms to a mere 0.2 times the wavelength of a He–Ne laser. This F-TIR concept has found application in silicon photonics for splitting light signals into two by utilizing evanescently coupled waveguides, known as directional couplers.

The F-TIR technique is widely applied in spectroscopy, nonlinear optics, and particularly, integrated optics. In 2013, my group demonstrated the F-TIR concept to split light signals by using two X-crossed organic crystal *passive* waveguides [7]. Subsequently, in 2020, by using two C-shaped bent *active* crystal waveguides coupled in their convex regions, a directional coupler was fabricated (see chapter 4, section 4.2) [8]. Further details regarding active and passive organic waveguides will be provided in the following section 2.1.3.

This book presents several examples of leveraging the concept of evanescent waves to optically couple monolithic and hybrid organic crystal waveguides, as well as waveguides with resonators for the construction of directional couplers (DCs), interferometers, add-drop-filters (ADFs), wavelength division multiplexers (WDMs), and integrated circuits (ICs).

2.1.3 Types of waveguides

Waveguides can be categorized into *passive* and *active* types, contingent on the nature of the light they guide [9]. Passive waveguides directly transmit input light from one end to another, because the wavelength of the incident light lies outside the optical absorption window of the waveguide material. In contrast, active waveguides are designed to absorb incoming light, particularly in the ultraviolet (UV) or visible spectrum, and subsequently emit photoluminescence (PL), which can include fluorescence (FL) or phosphorescence. This emission is characterized by a Stokes shift resulting from non-radiative processes associated with molecular vibrations. Ultimately, this emitted PL will be self-guided through the 1D waveguide towards the output terminal. These distinct waveguide types, passive and active, hold considerable significance in the development of OPICs.

In 2003, Rubahn *et al* conducted a study in which they grew para-hexaphenyl (*p*-6P) nanofibers emitting blue light, measuring 400 nm in width, on freshly-cleaved mica and NaCl substrates through vacuum sublimation (figure 2.4(a)) [4, 10]. The alignment of substrate dipoles interacted with *p*-6P molecules, guiding the parallel growth of multiple fibers that extended up to 1 mm in length. On mica, each *p*-6P molecule exhibited an orientation at an angle of 76° with respect to the long axis of the fiber. Notably, the characteristic fiber widths were 400 nm on mica and 300 nm

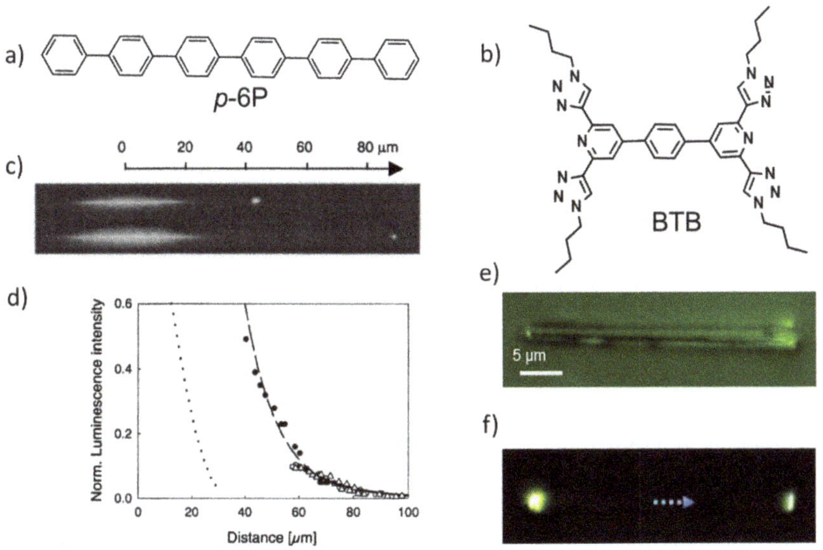

Figure 2.4. (a, b) Structure of para-hexaphenyl (*p*-6P) and back-to-back coupled 1,4-bis(1,2′:6′,1′-bis(3-butyl-1H-3,4,5-triazolyl)pyridin-4′-yl)benzene (BTB), respectively. (c) FL images along the fiber on mica for two distances between the excitation and outcoupling region. (d) Guided FL intensity of a 400 nm wide *p*-6P nanofiber waveguide as a function of distance (open symbols) and for a 300 nm wide fiber on NaCl (closed circles). The spatial excitation profile is given in a dotted line. Theoretical fittings are given in dashed and continuous solid lines. (c, d) Reprinted from [10], with the permission of AIP Publishing. (e) Confocal microscopy images of BTB tube passively guiding the 488 nm light. (f) The image was recorded with a 488 nm long-pass edge filter. (e, f) [11] John Wiley & Sons. (2012).

on NaCl. Due to the refractive index disparity between the substrate and the fiber (higher n), the latter functioned as an active waveguide, facilitating the guidance of blue FL, as depicted in figure 2.4(c). The dependencies of the outcoupled luminescence intensity at varying distances were measured for fibers on mica (indicated by open symbols) and on NaCl (illustrated as closed circles) (see figure 2.4(d)). Based on both experimental data and theoretical calculations, the estimated value of the imaginary part of the dielectric tensor component $\varepsilon_{\parallel}''$ along the long axis of the fiber was determined to be 0.012.

In 2012, my research group reported the fabrication of the first passive organic microtubular optical waveguides. These 1D organic crystalline microtubes were formed through self-assembly from an acetonitrile solution of 1,4-bis(1,2′:6′,1′-bis(3-butyl-1H-3,4,5-triazolyl)pyridin-4′-yl)benzene, abbreviated as **BTB** (depicted in figure 2.4(b)). When excited with a 488 nm laser, as the wavelength is situated outside the optical absorbance range of **BTB** molecules, these 1D microcrystals functioned as passive waveguides. Analysis of an optical image recorded with a 400 nm long-pass edge filter revealed the occurrence of inelastic interactions between the guided 488 nm light and **BTB** molecules, resulting in Raman scattering [11].

2.1.4 Bent (mechanically flexible) waveguides

The construction of OPIC mandates the utilization of waveguides capable of guiding light along different trajectories in a two-dimensional substrate plane. This requirement underscores the need for mechanically flexible or bendable waveguides that can regulate the direction of transducing light. The organic waveguides discussed in section 2.1.3, by contrast, lack the requisite flexibility and cannot be bent to manipulate the trajectory of light. Consequently, bent optical waveguides emerge as an indispensable photonic component for circuit construction. Furthermore, the inherent flexibility of such optical waveguides enables the creation of diverse elements, including ring resonators, modulators, interferometers, and more, within the circuit architecture.

In 2005, Takazawa *et al* [12] reported, for the first time, thiocyanine (TC) dye-based flexible nanofibers acting as an active type of organic waveguide (figure 2.5(a)). These nanofibers are *H*-aggregates[1] of TC that grew up to 300 μm long during self-assembly and are mechanically bendable using a glass fiber tip. However, the crystalline nature of these fibers and the structural underpinnings of their mechanical flexibility warrant further investigation. Notably, these nanofibers exhibited distinct emission colors depending on the solvents and concentrations used in their preparation. Upon excitation with a 405 nm laser in three different positions of the bent fiber, the PL images showed guided emission of PL from both ends of the fiber (figure 2.5(a)). Further, the PL spectra showed that the guided PL intensity decreases with an increase in the guided distance. Analysis of PL images and spatially-resolved spectra disclosed that these fibers functioned as single-mode

[1] In *H*-aggregation the molecules stack face-to-face, with their transition dipoles oriented side-by-side. This type of aggregation leads to a blue shift of the absorption spectrum. In contrast, in *J*-aggregation, the molecules are stacked head-to-tail, and the dipoles aligned transversely. This leads to a red shift of the absorption spectrum.

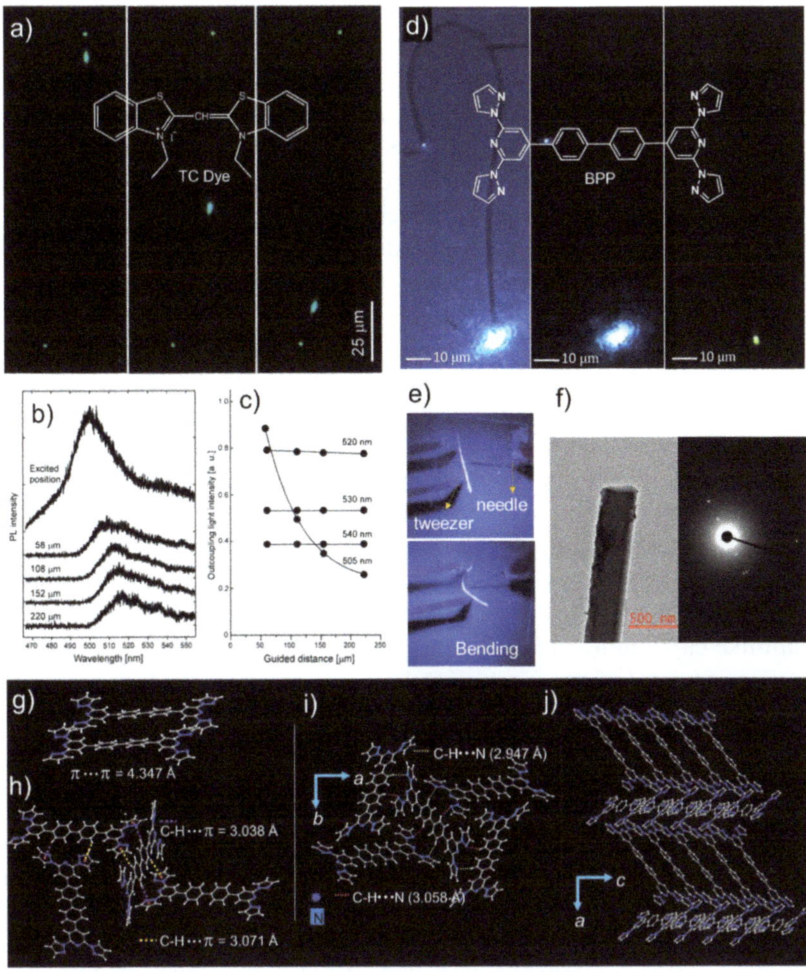

Figure 2.5. (a) PL images of TC fiber on a silica substrate obtained by exciting identical fibers at three different positions. The inset shows the molecular structure of the TC dye molecule. (b) Spatially-resolved PL spectra of the excited position and outcoupling light by excitation at a distance of 58–220 μm from the TC fiber end. (c) The intensity of the outcoupling light at several wavelengths as a function of the guided distance. The intensity at 505 nm was fitted using a first-order exponential function. (a–c) Reprinted with permission from [12]. Copyright (2005) American Chemical Society. (d) Optical images of a flexible passive optical waveguide obtained from BPP on a silica substrate passively guiding 488 nm light. The inset shows the molecular structure of the BPP molecule. [7] John Wiley & Sons. (2013). (e) Three-point bending test on a BPP flexible crystal performed using tweezers and needle under UV light. (f) Transmission electron microscope image and selected-area electron diffraction of a BPP microcrystal. X-ray structure displaying various non-covalent (g) π···π, (h) C–H···π, (i) H-bond interactions in a BPP crystal. Arrangement of molecules in the BPP crystals viewed along, (i) *c* and (j) *b*-crystallographic axes, respectively. (e–j) [8] John Wiley & Sons. (2023).

optical waveguides with no discernible loss, achieving 100% propagation efficiency for distances of up to 250 μm within the 520–560 nm range. However, the exponential decay in the intensity of 505 nm outcoupled light suggested reabsorption within the waveguide, as illustrated in (figure 2.5(c)).

In 2013, my group reported [7] the first flexible crystal passive optical waveguides prepared from back-to-back coupled 4,4′-bis(2,6-di(1H-pyrazol-1-yl)pyridin-4-yl) biphenyl (BPP) molecule (figure 2.5(d)). Notably, BPP exhibits no optical absorption of visible photons with wavelengths greater than 420 nm, enabling it to serve as a passive waveguide effectively guiding 488/535/633 nm laser light without exciting any FL, even when subjected to extreme bending angles of up to 180°. These BPP crystals also demonstrated remarkable mechanical flexibility, as evidenced by their bendability using forceps and a needle, and even an AFM cantilever tip (as seen in figure 2.5(e)) [13]. Transmission electron microscopy (TEM) and selected-area electron diffraction (SAED) pattern analysis confirmed the crystalline nature of these microstructures by revealing distinct diffraction spots (figure 2.5(f)). Furthermore, the single-crystal x-ray structure of BPP provided insights into the planar conformation of the BPP molecule (as illustrated in figure 2.5(g)). In the crystalline state, weak $\pi\cdots\pi$ (4.345 Å) and C–H$\cdots\pi$ (3.075 Å and 3.039 Å) interactions, and C–H$\cdots$N hydrogen bonding interactions (2.947 Å and 3.058 Å), played a pivotal role in stabilizing the molecular packing and imparting mechanical flexibility (as denoted in figures 2.5(h)–(j)).

These BPP optical waveguides displayed efficient functionality both in and outside of water, highlighting their utility in microscale sensing applications with amphibian-like capabilities [13].

2.1.5 Optical loss in bent waveguides

PICs are typically designed to be compact, necessitating the intentional bending of long waveguides to accommodate within the limited space available for carrying optical signals across the chip. Silicon waveguides are usually for passive light transport, while organic/polymer waveguides are employed for the transduction of active and passive light. According to equation (1.35), the n of a waveguiding material affects the speed of light propagation within it, meaning that higher n values result in slower light propagation. This characteristic allows light to be guided through highly curved bends with minimal optical loss. Nevertheless, optical loss due to bending in waveguides can be minimized through the use of materials with high n values. However, the optical loss tends to increase as the bending angle of the waveguide becomes more pronounced.

Within a waveguide, the microscopic variations in material smoothness, structural inhomogeneities, or nanoscale defects contribute to optical loss through the scattering of propagating modes, commonly referred to as *scattering loss*. In the case of organic single-crystal waveguides, these structures generally possess smooth surfaces and are largely devoid of microscopic surface irregularities, resulting in minimal scattering loss.

On the other hand, *absorption loss* primarily arises from the material composition and absorption coefficient. This type of loss occurs when the input light interacts with the waveguide material, leading to its conversion into electromagnetic energy at different wavelengths, such as FL or phosphorescence. Typically, semiconductors are more susceptible to absorption loss. However, passive organic waveguides, which operate outside the optical absorption region, exhibit minimal absorption loss. In this context, the loss primarily stems from molecular scattering (Raman Scattering) [7].

Radiation or *bending loss* occurs when the n_{eff} of a specific propagating mode falls below the n of the waveguide or cladding (as observed in the case of silica optical fibers). When light transduces through a sharply bent waveguide, the phase-matching criteria are not met, leading to the leakage of some light from the waveguide (as depicted in figure 2.6). PICs necessitate numerous sharp bends in waveguides to manipulate the path of light, making bending loss an inevitable occurrence. However, this loss can be mitigated by implementing micro bends instead of macro-scale bends, as the latter is associated with significantly higher optical loss. Macro-scale bending loss arises when the inner radius of curvature (r_c) of the waveguide bend exceeds the waveguide's diameter or cross-section.

Numerical calculations employing the finite difference in time-domain (FDTD) method were carried out to calculate the bending-induced optical loss in a rectangular organic passive waveguide with $n = 1.6$. This study considered waveguides in straight form, featuring negligible optical loss, as well as L-shaped and U-shaped geometries. These waveguides were constructed on a glass substrate with an air-cladding layer ($n = 1$). The waveguide possessed dimensions of 0.5 μm in height, 0.5 μm in width, and 6.5 μm in length. The inner radius of curvature, r_c, was set at 250 nm for both the L- and U-shaped waveguides. The results of the calculations are depicted in figure 2.6 [14].

The calculated optical transmission efficiencies of straight, L-shaped, and U-shaped passive waveguides are 99.51%, 21.51%, and 7.17%, respectively. Notably, the primary source of optical loss in these waveguides is attributed to bending-induced losses, as no material loss ($k = 0$) was considered in this analysis. It is evident that the bending-induced optical loss increases as the inner r_c decreases. The decrease in the r_c causes an increase in the optical power loss through the waveguide, and a decrease in transmission efficiency (figures 2.6(m)–(o)). Specifically, when examining L-shaped bent waveguides with different r_c values (i.e., 200, 500, and 750 nm), the transmission efficiencies are found to be 21.51%, 38.45%, and 52.54%, respectively (as illustrated in figures 2.6(m)–(u)). It's important to note that these calculations consider only the fundamental optical mode. In practice, the total transmission efficiency can vary depending on factors such as the excitation mode, polarization state of the excitation beam, the n of the waveguide and surrounding cladding, while keeping the bent angle, waveguide material, and substrate constant.

The light propagation loss of a passive waveguide can be estimated from the power transmission ratio (T) at the output, which is a ratio of output power (P_{out}) over input power (P_{in}). T is related to attenuation coefficient (α), and waveguide length (L) as per Beer's law

$$\mathrm{T} = \frac{P_{out}}{P_{in}} = e^{-\alpha L} \tag{2.11}$$

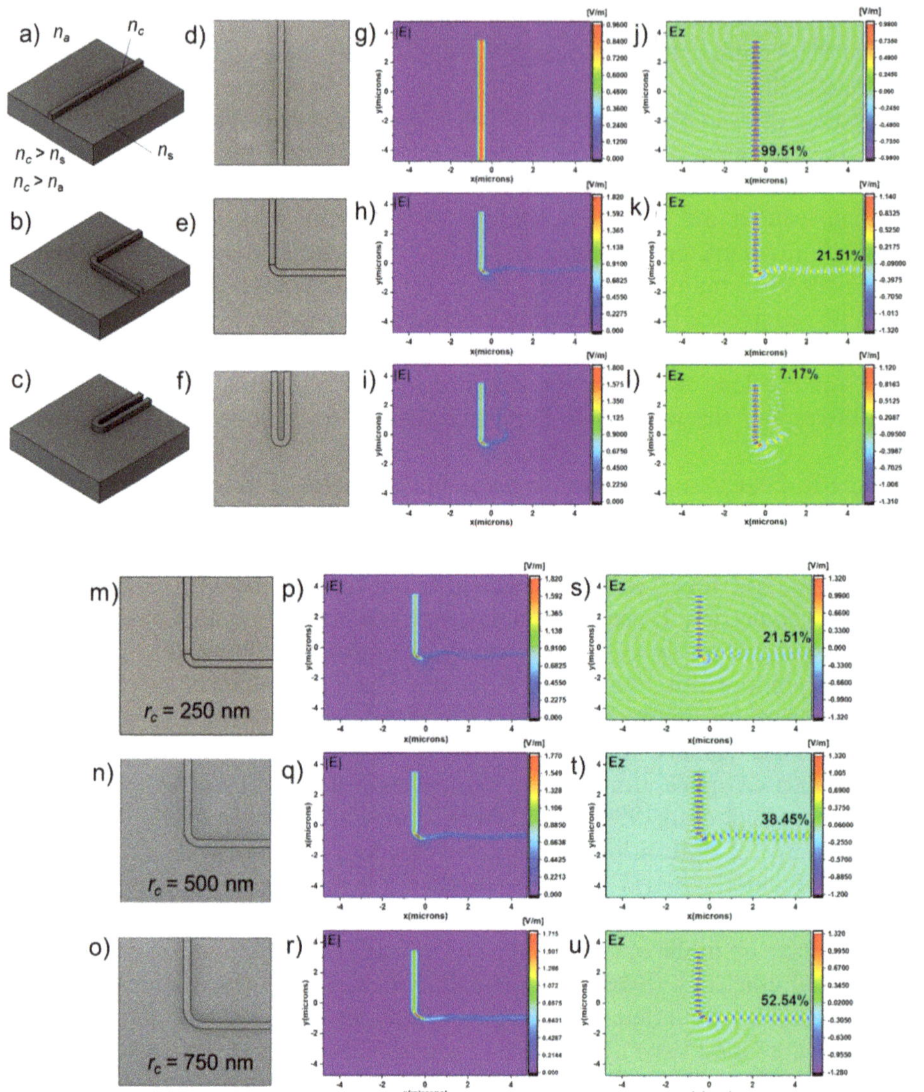

Figure 2.6. (a–c) Three-dimensional (3D) and (d–f) top-view of straight, L- and U-shaped waveguides, respectively, used for the FDTD analysis. (g–i) Total electric field distribution of the waveguides through a longitudinal cross-section. (j–l) Transverse electric field of the waveguides through a longitudinal cross-section. The transmission efficiencies of straight, L-shaped and U-shaped waveguides are 99.51%, 21.51%, and 7.17%, respectively. n_c and n_s are refractive indices of organic crystal and substrate, respectively. (m–o) Top-view of L-shaped waveguides with the $Rc{:}h{:}d$ = 1:1:1; 2:1:1; 3:1:1, respectively, used for the FDTD analysis. (p–r) Total electric field distribution of the waveguides through a longitudinal cross-section. (s–u) Transmission efficiency of the transverse electric field through a longitudinal cross-section for a L-shaped waveguide. [14] John Wiley & Sons. (2020).

Equation (2.11) shows that the power decreases exponentially with L. In decibel (dB) units, α can also be expressed using the following relation,

$$\alpha'(\text{dB mm}^{-1}) = \frac{10}{L}\log_{10}\left(\frac{P_{\text{in}}}{P_{\text{out}}}\right) \tag{2.12}$$

In the case of an ideal optical fiber, the input power is approximately equal to the output power, $P_{\text{in}} \approx P_{\text{out}}$. When expressed in decibels (dB), a lower dB value indicates higher output transmission efficiency. For example, for α' of 1 dB, there is approximately a 20% power loss, while with $\alpha' = 3$ dB, a 50% loss occurs. It is important to note that the optical loss in passive optical waveguides or fibers predominantly results from absorption, scattering, and bending-induced losses.

Unlike commercial optical fibers (made of flexible glass or plastic fibers, surrounded by a low-index dielectric cladding layer), most of the waveguides constructed from organic materials belong to an active type. Therefore, the power (P) terms in equation (2.12) are generously replaced with PL intensity (I). However, the roughness and geometry of the same organic crystal termini (which is a growing facet) differ; as a result, the varied light scattering from these terminals is not usually accounted for. Therefore, a reasonable estimate of a guided PL can be obtained when the excitation/detection direction (orthogonal or nearly parallel) with respect to the waveguide termini should be alike. The correctness of the L value, which is usually estimated using the optical microscopy image, is also crucial. Notably, organic crystal's geometry, dimension, and number of defects differ from crystal to crystal during their growth. Consequently, distinct from commercial optical fibers with consistent core diameters and quality, the parameter α' in organic crystal waveguides is uniquely associated with each crystal and cannot be generalized across all crystal waveguides produced during the crystallization process.

In passive organic waveguides, the absence of optical absorbance precludes PL generation. Optical losses primarily stem from Raman scattering, which occurs as input optical energy is transferred to molecular vibrations. As in most of the demonstrated experiments, the coupling of a laser beam orthogonal to the crystal tip minimizes the light coupling to the crystal waveguide. However, parallel light coupling using a fiber-optic taper increases the light coupling efficiency. To enable direct laser light coupling to the waveguide terminal, my group employed two BPP crystal waveguides naturally positioned in a near V-shaped configuration, although an axial orientation would have been preferable. These waveguides were separated by a terminal-to-terminal gap of approximately 1 μm (as depicted in figure 2.7(a)) [7]. This geometry enabled the direct coupling of output light from waveguide-1 to the terminal of waveguide-2 at junction *B* when laser excitation is applied to point *A* of waveguide-1 (figure 2.7(b)). The passive light transmission efficiency of waveguide-2 was estimated to be 58% (figure 2.7(c)). Moreover, Raman spectroscopy mapping established the light output at points *B* and *C* (figure 2.7(d)).

Another approach to assessing bending-induced optical losses involves comparing the losses in both straight and bent geometries for the same microcrystal waveguide. In this study, the mechanically flexible crystals described in section 1.4 serve as ideal

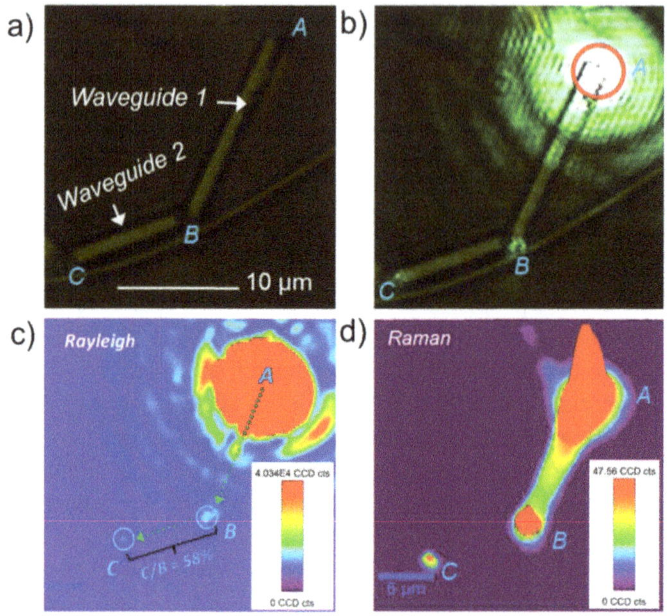

Figure 2.7. (a) Optical image of two waveguides of BPP in tip-to-tip contact. This waveguide configuration was used for the calculation of % optical waveguiding efficiency. (b) Optical image of two tubes in tip-to-tip contact when irradiated at one of the tips (A) with a 488 nm laser. (c) 2D projection of the propagating laser light from point *A* to *B* and from *B* to *C*. The blue circle denotes the passive light output point. The red circle denotes the input point. (d) 3D map of Raman scattered photons. [7] John Wiley & Sons. (2013).

candidates for this purpose. Zhang *et al* [15] investigated mechanically-induced bending loss using a 15–25 mm long flexible acicular single crystal of (*E*)-1-(4-(dimethylamino)phenyl)iminomethyl-2-hydroxyl-naphthalene (DPIN) (figure 2.8(a)). The crystal widths (height not measured) ranged from 40 to 200 μm. These straight crystals could be bent into a half loop by applying force (stress) at the two termini of the crystal using a pair of tweezers (figures 2.8(b) and (c)). After the release of stress, the bent crystal regains its original straight shape. The crystal elasticity is such that it allows screwing of the crystal onto a glass tube (figure 2.8(d)).

They investigated the structural basis for crystal elasticity by probing the molecular packing obtained from single-crystal x-ray diffraction. The analysis shows that the O–H···N intramolecular hydrogen bond (H···N: 1.77 Å, O–H···N: 148.26°) provided near planarity to DPIN; thereby facilitating efficient π-conjugation between the naphthalene and phenyl units, and intermolecular π···π stacking (figures 2.8(e) and (f)). Through the intermolecular C–H···π interactions (dashed green line; H···π: 2.88 Å), the π···π stacks formed a molecular chain along the crystallographic *a*-axis. Furthermore, the two molecular chains are linked by C–H···O hydrogen bonds, forming a dimeric chain that extends along the crystallographic *c*-axis (figure 2.8(f)). The mechanical bending of a straight crystal occurs along the (002) plane.

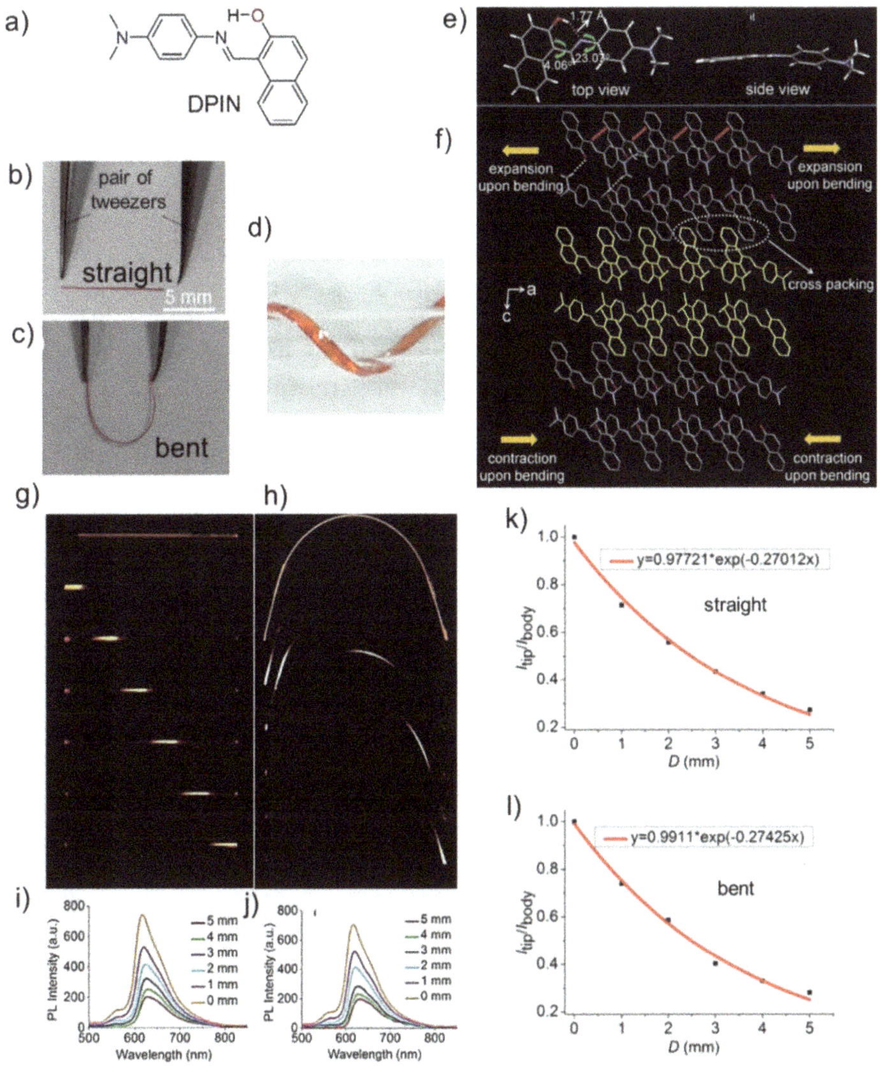

Figure 2.8. (a) Molecular structure of DPIN. (b and c) Elastic bending of a straight DPIN single crystal with a pair of tweezers. (d) A crystal screwed on a glass tube. (e) Crystal structure of DPIN. (f) Molecular packing viewed down the (010) plane along the crystallographic *b*-axis. The red arrows and green dashed lines represent the π···π and C–H···π interactions within a molecular chain, respectively. The dashed yellow lines show the C–H···O hydrogen bonds between molecular chains. FL images of straight (g) and bent (h) crystals excited with a UV lamp (uppermost line) and with 355 nm laser-focused at different positions, and FL spectra collected at a terminal of the crystal in either the straight (i) or the bent (j) state, with the distance between this tip and the excitation site of the laser changed gradually. The I_{tip}/I_{body} decays of a single DPIN crystal in the straight (k) and bent (l) states used for the estimation of the optical loss coefficients by a single exponential fitting of the function. [8] John Wiley & Sons. (2023).

Here, the expansion and contraction of the convex and concave areas, respectively, of a bent crystal are due to the increase and decrease of the π···π stacking distances (figure 2.8(f)). The calculation of the extension/contraction (ϵ) of the convex/concave arc of a bent crystal was accomplished using the equation $\epsilon = T/(d_i + T)$, where T represents the thickness of the crystal and d_i denotes the diameter of the inner arc. The resulting ratio was 2.26%. This calculation enabled the determination of the variation in the π···π distance, which ranged from 3.44–3.52 Å at the outer arc and 3.44–3.36 Å at the inner arc.

Optical excitation of a single crystal of DPIN with a 355 nm laser displayed an orange–red FL ($\lambda_{max} \approx 607$ nm) from it. The crystal showed an active waveguiding tendency both in straight and bent geometries during optical excitation by guiding the light along its long axis towards the crystal termini (figures 2.8(g) and (h)). The waveguide was optically excited at different positions along their growth axis in straight and bent geometries with the laser (figures 2.8(g) and (h)). By examining the ratio of input to output intensity and the propagation distance (D), the optical losses before and after bending the same crystal waveguide were estimated to be 0.270 dB mm^{-1} and 0.274 dB mm^{-1}, respectively (figures 2.8(i)–(l)) by fitting the following single exponential function.

$$\frac{I_{tip}}{I_{body}} = A' \, e^{-\alpha' D} \tag{2.13}$$

where I_{tip} and I_{body} are the FL intensities outcoupled and incidence regions, respectively, D is the distance between the excited site and the tip for collecting emission or more precisely, the light propagation distance and A' is a coefficient. The minimal difference of just 0.004 dB mm^{-1} confirmed that the optical losses induced by bending the crystal are exceedingly minimal.

My group employed a *mechanophotonics* approach to transform straight micro-crystal waveguides into several bent geometries. The details of their fabrication process and comparative analysis of optical losses are given in chapters 3 and 4. Additionally, chapter 4, section 4.1.2., delves into the estimation of strain-dependent optical losses of bent organic crystal waveguides.

2.1.6 NLO (chiral) waveguides

In the field of organic photonics, numerous articles have explored one-photon pumped organic waveguides. However, NLO optical waveguides are very important for achieving all-optical switching. Only a few publications are accessible in the literature on technologically important multi-photon pumped organic crystal waveguides. Importantly, the chirality effect on the guided optical signal has not been actively implemented in organic nanophotonics. The investigation of chirality's effect, such as circular dichroism (CD),[2] on NLO signals like two-photon

[2] The Circular Dichroism (CD) effect arises from the differential optical absorption of left and right circularly polarized light due to the structural asymmetry or chiral nature of the sample. The two main types of circular dichroism spectra are positive and negative CD signals.

luminescence (TPL or 2PL), three-photon luminescence (3PL), and second harmonic generation (SHG), is important to create and guide chiral information-encoded NLO signals pumped by an infrared light source. These chirality-controlled, guided light waves hold great promise for creating miniaturized on-chip information processing and integrated photonic circuits.

In this regard, for the first time, Chandrasekar and co-workers reported [16] the observation of the CD effect in NLO (multi-photon PL and SHG) signals in enantiomeric R- and S-type organic crystal optical waveguides obtained from R- and S-4,4′-(2,2′-diethoxy-1,1′-binaphthyl-6,6′-diyl)-dibenzaldehyde (1-R and 1-S) molecules (figure 2.9(a)). These axially chiral donor–acceptor-type enantiomeric molecules crystallize in a non-centrosymmetric space group ($P2_12_12_1$) and, through self-assembly, give rise to the formation of crystalline microrods (figures 2.9(b) and

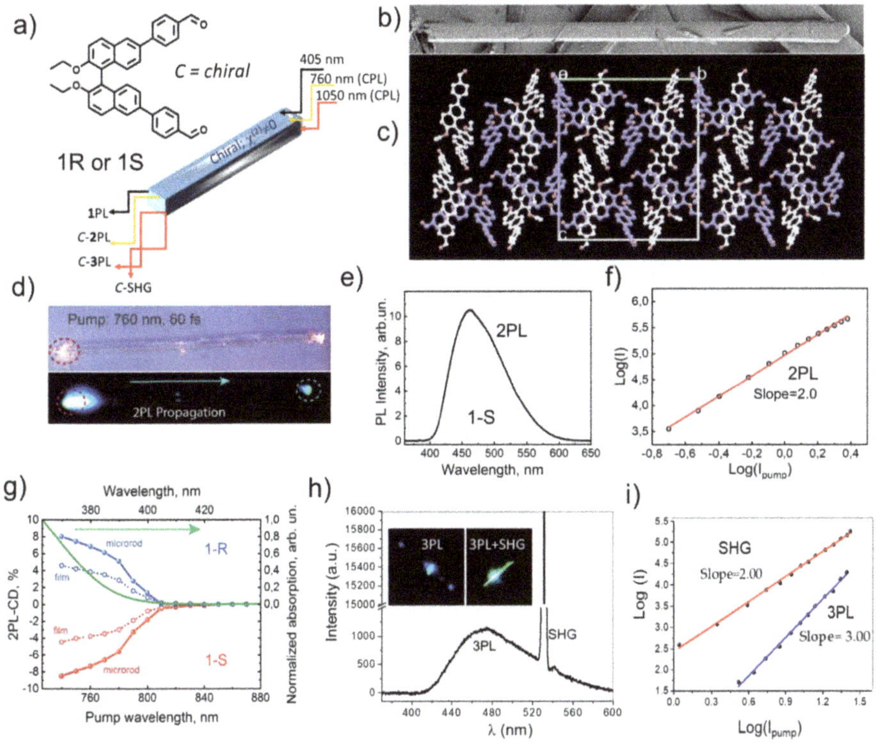

Figure 2.9. (a) Chemical structure of chiral molecules 1-R and 1-S, and cartoon representation of a crystalline chiral microrod producing multi-photon luminescence and SHG at different wavelengths of CPL. CPL stands for circularly polarized light, and *C* for chirality-sensitive processes. (b) FESEM image of a microrod. (c) Packing of chiral molecules along the crystallographic *a*-axis. (d) Optical micrograph of a 1-S microcrystal excited at the left end with the pulsed fundamental radiation at 760 nm wavelength, along with the corresponding 2PL image. (e) 2PL spectrum of a microcrystal waveguide. (f) Logarithmic plot of the 2PL intensity versus the laser power. (g) Plot of 2PL-CD signal versus pump wavelength for 1-S and 1-R microrod and the corresponding homogeneous films. (h) The optical spectrum of a microcrystal revealing the 3PL and SHG signals from the light-guiding crystal. The insets show the corresponding 3PL and 3PL+SHG images. (i) Plots of SHG (in green) and 3PL (in blue) signal intensities vs. the pump power in the log–log scale. [16] John Wiley & Sons. (2019).

(c)). The electron charge transfer mediated by the donor–acceptor groups enables 2PL and 3PL from both R- and S-type microcrystals.

Upon femtosecond (fs) optical pumping with a 760 nm light, these enantiomeric microcrystals generated a 2PL signal and efficiently guided this NLO signal along the waveguide with a low optical loss, $\alpha'(2PL) \approx 0.12 \pm 0.1$ dB μm^{-1} (figures 2.9(d) and (e)). The slope values of signal intensities plotted against the pump power on a log–log scale directly illustrate the two-photon absorption process (figure 2.9(f)). The absence of SHG for shorter fundamental wavelengths ($<$780 nm) is attributed to the presence of strong two-photon absorption and SHG reabsorbance. Interestingly, when the microcrystal waveguide body was excited with a 1050 nm (200 fs, 80 MHz) laser, it displayed 3PL along with SHG (at 525 nm) propagating along the long axis of the crystal waveguide towards the termini (figure 2.9(h) and insets). The $\alpha'(SHG) \approx 0.14 \pm 0.1$ dB μm^{-1} is found to be nearly similar to 2PL, whereas $\alpha'(3\ PL) \approx 0.22 \pm 0.2$ dB μm^{-1} is much larger. Notably, the plots of the intensity of the 3PL and SHG versus the pump laser power in the log–log scale elegantly demonstrate the third-order and second-order dependencies, respectively (figure 2.9(i)).

Significantly, the chiral characteristics inherent in these crystals result in the generation of chiral information-encoded multi-photon PL, manifesting as varying multi-photon PL intensities contingent upon the chiral nature of the waveguides. The spectral dependence of the multi-photon PL circular dichroism (PL-CD) was systematically measured for both 1-R and 1-S crystal waveguides and subsequently compared with their corresponding thin films (refer to figure 2.9(g)). Notably, enantiomeric crystal waveguides exhibit nearly equivalent absolute values of the nonlinear optical circular dichroism (NLO-CD) effect, in addition to an observable enhancement in signal intensity within the multi-photon absorbance spectral region.

2.2 Optical cavities

An optical cavity is a microstructure that traps and stores (see the photon lifetime described in section 2.2.1.1) the totally internally reflected light via repeated light reflection within the structure. Depending upon the geometry of the resonators, they are known as Fabry–Pérot (F–P) and Whispering-Gallery-Mode (WGM) cavities. It is important to note that a cavity can, sometimes, support both F–P and WGMs when the cavity geometry and/or the optical excitation position is conducive enough to support these two resonances.

2.2.1 F–P microcavities

An F–P cavity usually consists of two highly light-reflecting mirrors (M1 and M2) facing each other, separated by a distance (also known as cavity length, L). The light that enters through the back of M1 travels to M2 and gets reflected and is again reflected by the first mirror (figure 2.10(b)). This repeated reflection through the cavity is allowed only when certain conditions for the wavelength and the geometry of the resonator cavity are fulfilled, producing standing electromagnetic waves or optical modes. Otherwise, the light intensity is completely suppressed because of destructive interference, producing free spectral range (FSR), as described in section 2.2.1.1.

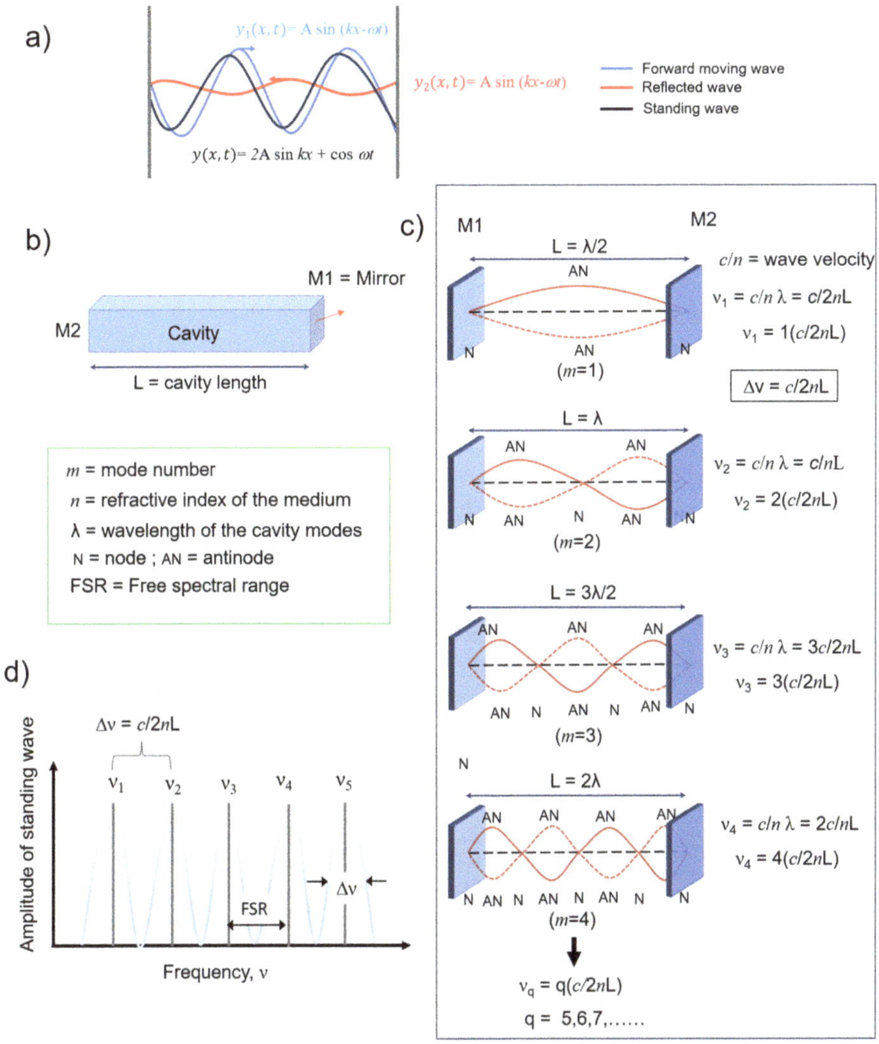

Figure 2.10. (a) Representative standing wave formation due to the interference of forward and reflected waves. (b) A rectangular optical cavity crystal with light-reflecting mirror-like facets. (c) Schematic illustration of optical modes formation in an F–P cavity with a rectangular geometry. (d) Appearance of modes in the FL spectrum.

A standing wave is formed due to the superposition of two light waves of the same frequency propagating in opposite directions (figure 2.10(a)). These waves oscillate in time, but the oscillating wave's peak amplitude in space does not change and remains constant.

The standing wave can be mathematically described as

$$y(x,\ t) = 2\ \mathrm{A}\sin kx + \cos \omega t \tag{2.14}$$

Here, the first term $\sin kx$, ($k = \frac{2\pi}{\lambda}$) denotes the sinusoidal shape of the wave pattern and $\cos \omega t$ signifies the oscillatory up-down motion of the wave at a frequency comparable to a vibrating spring. The symbol A denotes the wave amplitude. The confinement of light within the cavity *quantizes* the frequency (figure 2.10(c)). The simplest mode with one loop ($m = 1$) is called fundamental mode or first harmonic. In the second mode ($m = 2$), the vibration of two loops is called the second harmonic. The qth harmonic consists of m vibrating loops. The higher the number of loops in a mode the faster the vibration. For example, a mode with four loops vibrates four times faster than one loop. The spacing between two adjacent modes is $c/2nL$ (figure 2.10(d)).

2.2.1.1 Parameters of interest in F–P microcavities

Free spectral range (FSR): FSR of an F–P cavity is also inversely related to cavity length L and n by the following equation (figures 2.10(d) and 2.11(c)):

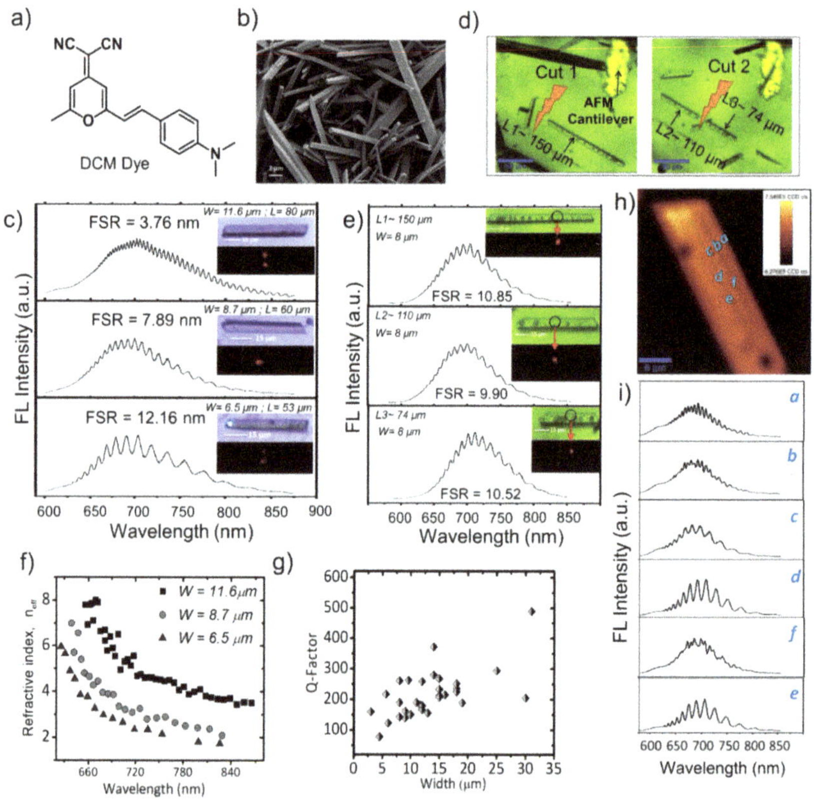

Figure 2.11. (a) Molecular structure of DCM dye. (b) FESEM image of DCM dye crystals. (c) FL spectra of their representative crystals displaying size-dependent FSR values. (d) Mechanophotonic cutting of a long crystal cavity into three segments and (e) their corresponding FSR values. (f) A plot of varying effective refractive index with respect to wavelength. (g) A plot of Q-factor versus the width of the crystals. (h) 2D FL intensity map of a microrod acting as a cavity. The inset shows the CCD counts. (i) The corresponding selected FL spectra of a rod shown in (h). Representative spectra collected at different positions are labeled from (a) to (e). [31] John Wiley & Sons. Copyright (2016).

$$\text{FSR} = \frac{\lambda^2}{2\,\text{L}\,n_{eff}} \tag{2.15}$$

Experimentally, the FSR value can be extracted from the separation between two adjacent optical modes with mode numbers m and $m + 1$, that is $\Delta\lambda_m = \lambda_m - \lambda_{m+1}$.

Note that n is a complex number, $n' + in''$. As n is frequency dependent, $n(\omega)$, the FSR value steadily decreases at lower wavelengths (figure 2.11(c)) and does not follow the $c/2n\text{L}$ trend described in figure 2.10(d).

Takazawa *et al* estimated the real and imaginary parts of refractive indices using a thiocyanine dye-based nanofiber F–P cavity (see figure 2.5) [17]. The frequency-dependent real part of n' values were estimated using equation (2.15) for cavities of different lengths. The frequency-dependent imaginary part of n'', Im(n) was estimated from the PL intensity of the guided signal spots (I) using the relation:

$$I = I_o \exp\left[-\frac{2\,\omega \text{Im}(n)\text{D}}{c}\right] \tag{2.16}$$

where I_o is the PL intensity at the excited spot, ω is the frequency of the light, D is the propagation distance, and c is the speed of light.

Q-factor: This is an important cavity parameter that describes the light-trapping competence or optical gain of a microcavity of any type (F–P and WGM described in section 2.2.2). It can be estimated from the ratio of the light energy stored in the cavity to the light energy dissipated per cycle [18].

$$Q = \frac{\lambda}{\Delta\lambda} = \frac{\text{Stored energy}}{\text{Power loss}} = \frac{2\pi\lambda T_{rt}}{P_{\text{loss}}} \tag{2.17}$$

The value of Q can be estimated easily for a resonance mode of interest as per equation (2.17), where λ is the wavelength at the resonant mode maximum and $\Delta\lambda$ is the line width (full width at half maximum, FWHM) of that mode. A cavity's Q-factor is dependent upon λ, round trip time (T_{rt}) and power loss (P_{loss}) per round trip (assuming $P_{\text{loss}} \ll 1$).

The higher the Q-factor the better the light confinement in the cavity. The magnitude of the Q-factor is diminished by the competing unwanted optical losses that occur in a cavity. The nature of losses is of various types, and the sum of the possible losses that determine the Q-factor of a resonant cavity is given by the following relation [19, 20]:

$$\frac{1}{Q} = \frac{1}{Q_{\text{mat}}} + \frac{1}{Q_{\text{ss}}} + \frac{1}{Q_{\text{rad}}} + \frac{1}{Q_{\text{ex}}} \tag{2.18}$$

where the first three terms are intrinsic losses of a cavity, namely, material absorption loss, (Q_{mat}), scattering loss (Q_{ss}), and radiation loss (Q_{rad}). The fourth term of equation (2.18) is the external coupling loss (Q_{ex}).

Normally, for a passive resonant cavity excited by a white light, a slight light absorption will lead to Q_{mat}. For an active type cavity (for example, organic/inorganic) producing and trapping PL, Q_{mat} is not applicable. However, for materials with self-absorption or reabsorbance of PL this loss (Q_{mat}) can be considered as the re-emission of reabsorbed photons again involving undesirable non-radiative (heat) pathways. Loss due to Q_{ss} is the usual problem encountered by the cavity as slight surface inhomogeneities or roughness and nano/micro contaminants can significantly lower the Q-factor. This type of loss can be eliminated by improving the crystallization or fabrication process. The Q_{rad} is related to the light loss in cavities with curved geometries. Finally, Q_{ex} is related to the loss brought by the coupling of a waveguide or another cavity with a similar or higher refractive index.

Commonly, Q-factors in the range of 10^3–10^6 are considered high and those above 10^7 are ultra-high. High Q-factors are important to achieve microlasers.

The Q-factor of a cavity increases with increasing size and light reflectivity (R) of the cavity. For example, for an F–P cavity, it is the size and percentage reflection of light by M1 and M2 (see figures 2.10 and 2.11(g)) that determine the line width of the optical modes and subsequently, the Q-factor. Therefore, high refractive index crystals with clean and smooth facets reflect the light effectively; thereby, decreasing the mode's line width. Resonators with low-Q values (in hundreds) [21–23] characteristically display the resonance modes spectrum with large FL backgrounds unlike those with high Q-factors (in thousands) [24, 25]. The decay time of internally stored optical energy is higher for high Q cavities. The Q-factor of a cavity usually increases with cavity size and is expected to reach a saturation threshold beyond a certain dimension (depending on the cavity geometry) due to inhibition of TIR at large curvature [21].

Photon lifetime (τ_P): Q-factor is a parameter that determines the time taken for the stored optical energy to exponentially decay from the cavity. High Q-factor cavities allow a higher number of round trips for photons within the cavity. Thus, trapping the photons for a longer time thereby increasing the photon lifetime (τ_P) or cavity ring down lifetime before escaping the cavity via the loss mechanisms [26, 27]. In other words, τ_P defines the $1/e$ decay of trapped electromagnetic energy from its original value. It can be assessed from the following equation:

$$\tau_P = \frac{Q}{2\pi\nu} \tag{2.19}$$

where ν is the linear frequency (angular frequency, $\omega = 2\pi\nu$).

A vital application of higher τ_P (provided by high-Q resonators) is the surge of the light–matter interaction. Therefore, high-Q resonators are used for ultra-high sensing of a single molecule, atom, virus, etc [28, 29].

It is important to note here that the τ_P should not be confused with the FL lifetime. The latter provides the average time the molecule remains in its optically excited state before emitting the absorbed photon. Importantly, during FL lifetime imaging (FLIM) experiments using the TCSPC (time correlated single photon counting) technique on a single luminescent particle or a tiny crystal cavity or a particle loaded with luminescent entities, caution must be exercised as the τ_P can increase the actual FL lifetime value [30].

*Finesse, **F***: This is another important parameter of interest in cavities, which can be calculated from the ratio of FSR to FWHM of a resonance mode [19].

$$\boldsymbol{F} = \frac{\text{Mode spacing}}{\text{Mode width}} = \frac{\text{FSR}}{\text{FWHM}} \tag{2.20}$$

A large $\boldsymbol{F}$ value is usually desirable as it corresponds to a sharper mode. This parameter is entirely dependent upon the loss factor and is independent of the cavity size. Cavities with high reflecting mirror-like surface(s) increase the $\boldsymbol{F}$ value.

Effective mode volume, V_{eff}: This is the ratio of the total optical energy stored inside the resonator in that mode and the maximum energy density of that mode. Smaller V_{eff} is always desirable and it means higher FSR. Further, the lasing threshold also is directly proportional to V_{eff}. The spontaneous and stimulated emission rates are inversely proportional to V_{eff} as well [19].

2.2.1.2 Organic single-crystal F–P microcavities

An F–P cavity can be realized in FL organic single crystals with smooth light-reflecting facets. For example, my group reported acicular DCM dye crystals with rectangular cross-sections to prove their light-trapping ability. As discussed earlier in chapter 1, DCM dye is a strong donor and acceptor molecule (electron push–pull molecule) emitting red FL and possesses NLO characteristics.

The optical excitation (488 nm, CW laser) of self-assembled DCM crystals of length (L)/width (W), 11.6 μm/80 μm, 8.7 μm/60 μm, 6.5 μm/53 μm, revealed crystals acting as F–P cavities with FSR values of 3.76, 7.89, and 12.16 nm, respectively (figures 2.11(a) and (c)) [31]. The FESEM studies confirmed the nearly rectangular cross-section of the microcrystal (figure 2.11(b)). Here, the determination of the height of the crystal is not possible with AFM due to the larger size of the crystal. As the crystals have different W and L, to probe the direction of light-reflecting crystal facets, a crystal cavity of $W = 8$ μm was cut into three segments of lengths 150, 110, and 74 μm using an AFM cantilever tip via the mechanophotonic approach (figure 2.11(d); discussed in detail in chapter 3). All the segments with varying lengths and identical widths showed the same FSR values of about 10 μm within the experimental error limits (figure 2.11(e)). This result confirmed the F–P cavity is not length-dependent and mirror-like reflecting facets are not at the crystal termini, but two opposite side facets. The n_{eff} values estimated using equation (2.15) increased up to 8 at lower wavelengths (figure 2.11(f)), as n is a function of frequency $n(\omega)$. As expected, the crystal cavities exhibited enhanced FL compared to normal crystals with a Q-factor of up to 500 (figure 2.11(g)). The Q-factor also increased with increasing width of the cavities.

Furthermore, FL mapping was performed on a selected rod cavity to understand the electric field distribution within the F–P cavity (figure 2.11(h)). At three representative points *a*, *b*, and *c*, FL spectra showed variation of the FSR in the lateral direction (figures 2.11(h) and (i)). The number of modes and FSR ($\Delta\lambda \approx 16$ nm) persist similarly along the long axis of the rod (see panels (c), (d), and (e) in figures 2.11(h) and (i)) further establishing the role of lateral edges in the F–P cavity resonance.

Yao *et al* reported [32] a cyan-emitting organic nanowire single-crystal waveguide obtained from a 2-(*N*,*N*-diethylanilin-4-yl)-4,6-bis(3,5-dimethylpyrazol-1-yl)-1,3,5-triazine (DPBT) molecule (inset of figures 2.12(a) and (b)). These nanowires were synthesized by injecting a tetrahydrofuran solution of DPBT into aqueous solutions of cetyltrimethylammonium bromide (CTAB) at various temperatures. The x-ray crystal structure exhibited strong π–π stacking and primary growth along the crystal *c*-axis. The formation of 1D nanostructures is a result of molecular stacking along the *c*-axis of the lattice. The AFM investigations revealed the crystal's flat surface and the rectangular cross-section. The TEM image and SAED pattern revealed that the wire is a single crystal growing along the [001] direction, featuring flat end facets. This preferential growth along the [001] direction can be attributed to the CTAB micelles, resulting in the formation of one-dimensional nanowires.

DPBT is a NLO molecule with $\sigma_2(\lambda)$ of 160 GM (see equation (1.90)). Due to the smooth surface morphology of the crystals, upon optical excitation with a 750 nm fs laser, the nanowire propagates a blue TPL as exciton–polaritons (EP) along the wire's long axis from the right to the left terminal (inset of figure 2.12(b)).

Further, the crystal acts as an F–P microcavity by shuttling the photons between the two opposite facets of termini. The flat end facets of the crystal play a crucial role in efficiently reflecting the guided PL, a key factor for achieving optical gain and

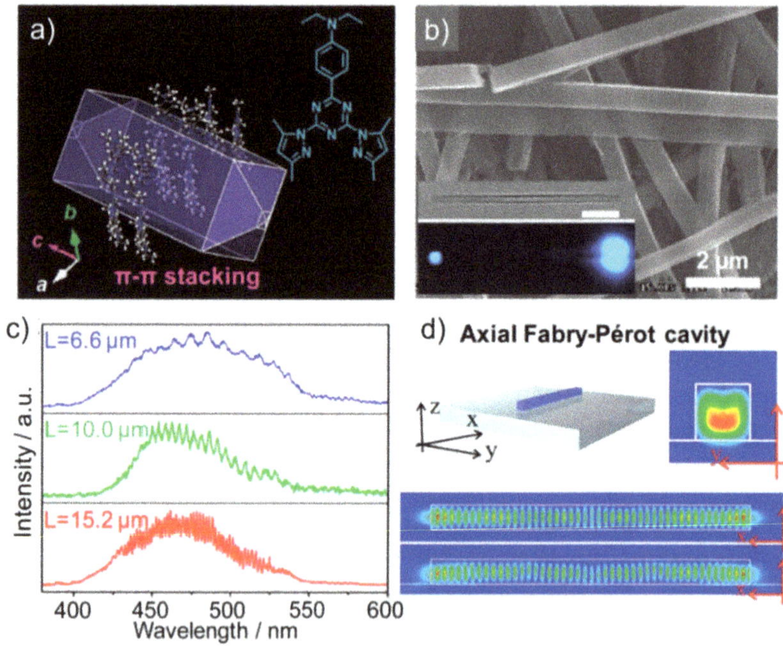

Figure 2.12. (a) 1D growth morphology of DBPT molecules along the *c*-axis through π–π interactions. (b) Microscopy image of the DBPT nanowires. Inset: waveguide images of a single wire. (c) Modulated waveguide spectra with different wire lengths. (d) Numerically calculated electric field distribution of a DPBT nanowire FP-type cavity. Reprinted with permission from [32]. Copyright (2011) American Chemical Society.

amplification of stimulated emission. The nanowires of varying lengths, $L = 6.6$, 10.0, 15.2 μm, exhibited a decrease of FSR between the F–P modes as per equation (2.15) confirming the direction of the mirror-like light-reflecting crystal termini. The FDTD numerical calculations confirmed the axial electric field reflection by the crystal's termini (figure 2.12(d)).

Chandrasekar and Murzina *et al* [33] reported a two-photon pumped NLO organic microcrystal F–P cavity obtained from trapezium-shaped single crystals of a 9,9,9′,9′-tetramethyl-9H,9′H,9′H-[2,2′:7,2′-terfluoren]-9′-one (TTO) molecule (figures 2.13(a) and (b)). Excitation with a Ti:Sapphire 760 nm pulse laser (pulse

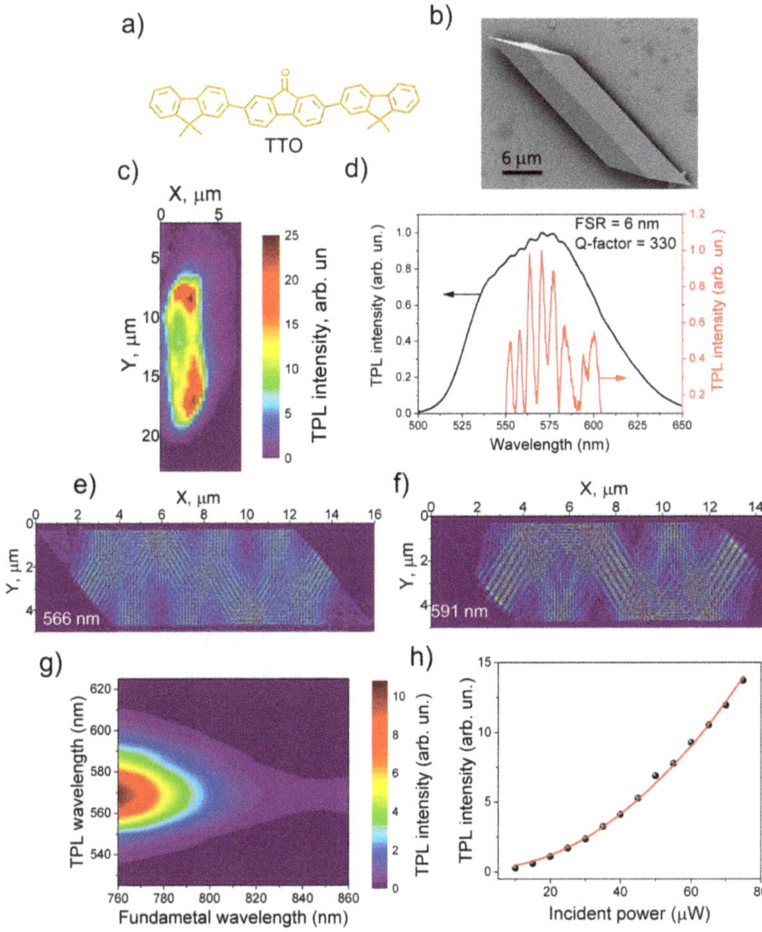

Figure 2.13. (a) Molecular structure of TTO. (b) FESEM image of a single TTO microcrystal. (c) TPL image of a TTO microcrystal F–P cavity. (d) Single-particle spectrum (black line) of a single microcrystal showing F–P modes in the TPL envelope. The background-extracted TPL spectrum with F–P modes is shown in the red line. (e) and (f) FDTD calculations of electric field distribution within the structure in the cross-sections parallel ($\lambda = 566$ and 591 nm) to the substrate. (g) TPL intensity map of a millimeter-sized TTO crystal as a function of fundamental wavelengths (760–860 nm). (h) Plot of TPL intensity versus incident pump power. [33] John Wiley & Sons. Copyright (2018).

width 60 fs, repetition rate 800 MHz) produced TPL from the microcrystals with the emission band nearly identical to one-photon pumped FL (figure 2.13(d)). The TPL intensity map revealed a high concentration of electric field at two opposite diagonal facets of the crystal due to light inward reflection (figure 2.13(c)). Interestingly, the TPL spectrum displayed numerous optical modes with an FSR of 6 nm along with a broad TPL background hinting at the cavity's low Q-factor (figure 2.13(d)). The background-subtracted TPL spectrum tangibly disclosed F–P optical modes generated by the TTO microcrystal cavity (figure 2.13(d), red lines) with a Q-factor of 330.

FDTD numerical calculations were performed to see the optical field localization of representative modes (example: $\lambda = 566$ nm and 591 nm) within the microcrystal cavity. The results (figures 2.13(e) and (f)) showed that the crystals support F–P modes and the direction of the mode circulation depends on the crystal's geometrical shape. As one can see, no optical signal was found at the sharp tips of the trapezium-shaped crystal, and modes can be readily spotted on the side planes of the microcrystal that describe the intense optical signal observed on the edges of the cavity during FL imaging.

Qualitative estimation of the TPA absorption cross-section can be obtained by exciting the crystals with different pump wavelengths. The fundamental wavelength (760–860 nm) dependency of TPL intensity demonstrated a steady increase in the TPL intensity when the excitation wavelength decreased from 760 to 870 nm (figure 2.13(g)). The quadratic dependence of the intensity versus fundamental beam power established a two-photon absorption process (figure 2.13(h)). As a result of resonant coupling, the TTO micro-resonators enhanced the TPL four times compared to thin films of comparable thickness [33].

2.2.2 WGM cavities

Acoustic (elastic) waves undergo phenomena like diffraction, reflection, and interference. The amplification of acoustic waves in some man-made structures has been known since the 16th century across the world. The most prominent ones built in the same century are the Gol Gumbaz mausoleum of Bijapur and St. Paul's Cathedral of London (figure 2.14). Lord Rayleigh attempted to explain the reason for the amplification of sound waves inside the dome of the cathedral using his WGMs theory [34]. Raman and Sutherland carried out experiments in the same cathedral and revised Rayleigh's theory of WGMs [35]. In acoustic WGMs, the bound pressure waves created by the sound (audible wavelength = 0. 017–17 m or frequency = 20 Hz–20 kHz) circle around the concave inner surface of the dome (100 feet or 30 m) by repeated reflections; thus, amplifying the sound via constructive interferences.

Analogous resonance-driven amplification effects of light waves can be realized in material objects with axial symmetry. Within these structures, light waves can be confined and circulate along the periphery, drawing a parallel to the propagation of sound waves along the curved walls of whispering galleries. These structures, commonly referred to as WGM cavities or resonators, capitalize on resonance

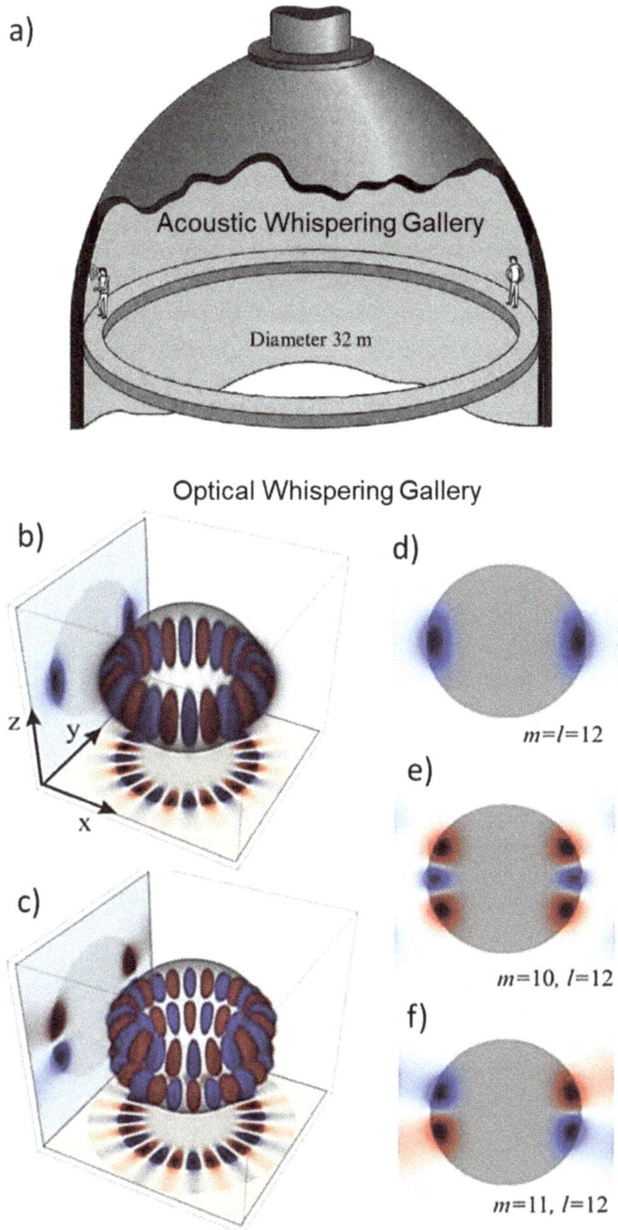

Figure 2.14. (a) Cartoon representation of the whispering gallery under the dome of St. Paul's Cathedral in London. Analytic calculation of eigenmode profiles in a spherical WGMR of 1.5 μm in diameter and for the refractive index of 1.75, for $l = 12$. Reproduced from [45]. © IOP Publishing Ltd. All rights reserved. Panels (b, d) correspond to the case of $m = l$; panels (c, f) to $m = 11$; (e) stands for $m = 10$. (b–f) [25] John Wiley & Sons. Copyright (2018).

phenomena to amplify optical signals. However, the diameter of the object should be in the range of microns.

For example, for visible light of wavelength 400–800 nm, the cavity diameter should be in the range of microns (about 1–20 μm). The shape and size of the cavities are factors that determine the wavelength of the amplified light waves. Apart from dielectric (silica, silicon, polymers, etc) spheres, WGM resonance has been observed in geometrically precise, high-index microscale organic structures such as spheres [22, 24], rings [23], disks [34], tubes [35], inverted frustums [37], inverted hemispheres [31], and ellipsoidal granules [38] (figure 2.15).

My group, in collaboration with Murzina of Moscow State University, reported several organic NLO WGM cavities. My group also reported the fabrication of WGM cavities from single crystals via focused ion beam lithography, which is discussed exclusively in chapter 5 of this book. The Q-factor of WGM microcavities is higher than the F–P cavities offering high energy density and long photon lifetime for light–matter interaction and related processes useful for all-optical devices [39], lasing [40, 41], filters [42], slow light devices [43], and sensing [44] applications.

In a WGM microcavity, light that enters via TIR from the rarer medium is tightly trapped by manifold reflection at the boundary of the circular cavity and air. After one round trip around the cavity, the polychromatic light undergoes interference. Only the light of particular wavelengths that meet the *resonance criterion* given in equation (2.21), is self-reinforced

$$p\,\lambda = 2\pi\, R\, n_{\text{eff}} \tag{2.21}$$

(intensity amplified sharp signals) while others are destroyed. Here, n_{eff} is the effective refractive index of the resonant mode, R is the radius of the cavity, λ is the light wavelength in vacuum. Here p is an integer number related to circulating

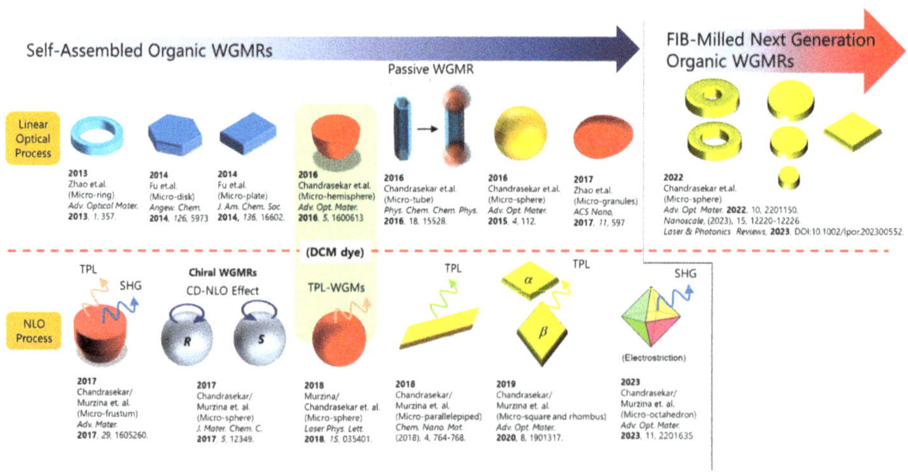

Figure 2.15. Cartoon representation of the development timeline of self-assembled organic WGMRs of various shapes and photonic functions. [25] John Wiley & Sons. Copyright (2018).

photon angular momentum (angular mode number). Typically, a trapped white light (say from 400 to 750 nm) in the cavity will amplify only selected wavelength signals and appear as sharp peaks of very narrow line width, known as optical modes. Here, the circumference of the cavities is important to control of number of modes appearing in the WGM spectrum. A similar WGM effect can be seen in the optical emission spectra of any loss-less single microcavity of axial symmetry self-generating broad-band photoluminescence. These optical modes can be seen by recording the emission spectrum of the WGM cavities using a high-resolution spectrometer (at least a vibrational spectrometer).

The origin of these optical modes in WGM cavities or spherical coordinate systems is well described based on adjoint Legendre polynomials, spherical Bessel and Hankel functions [45]. The optical modes are categorized by integer mode numbers as *azimuthal* (m), *polar* (l), and *radial* (r) that are typically $m \gg 1$, $l \gg 1$, and $r \geqslant 1$, respectively. The numbers l and r are related to the electromagnetic field in the azimuthal (equatorial) and radial directions, respectively. The angular mode number $p = l - m = 0, 1, 2, \ldots$ gives the number of the field oscillations in the polar θ-direction.

For example, the distribution of electromagnetic field within an organic microsphere (diameter of 1.5 μm and refractive index of 1.75) is shown in figure 2.14. The effective and long-living WGMs (long photon lifetime) are confined close to the microsphere's equator. The WGM emission spectrum also contains information about the polarization (TE/TM) state of radiation. This can be seen in the spectrum as a series of pairs of peaks due to TE and TM polarization of a particular mode. The TM-polarized modes resonate at a lower wavelength than the TE for the same mode number (r), which can be understood from the following approximate equations with correction terms. The detailed derivation of these equations (2.22) and (2.23) can be found elsewhere [45–47].

$$\lambda_{TE} \approx \frac{2\pi R n_1}{p + 1.856 p^{\frac{1}{3}} + \left(\frac{1}{2} - \frac{n_1}{\sqrt{n_1^2 - 1}}\right)} \tag{2.22}$$

$$\lambda_{TM} \approx \frac{2\pi R n_1}{p + 1.856 p^{\frac{1}{3}} + \left(\frac{1}{2} - \frac{1}{n_1\sqrt{n_1^2 - 1}}\right)} \tag{2.23}$$

where n_1 is the cavity refractive index. The requirement for a WGM microcavity is $n_1 > 1$ so that light can be confined via TIR.

Here, $\frac{n_1}{\sqrt{n_1^2 - 1}}$ is larger than $\frac{1}{n_1\sqrt{n_1^2 - 1}}$. Therefore, TE mode is resonant at a higher wavelength than TM.

2.2.2.1 Parameters of interest in WGM microcavities

Like F–P cavities, the properties of a WGM cavity can be evaluated from experimental parameters viz. FSR, Q, V_{eff}, τ_P, and $\boldsymbol{F}$. Compared to FP cavities, WGM microcavities show a Q-factor, low mode volume, and a large optical density.

As WGM cavities can show TE/TM modes in the spectra, the FSR value can be measured from the distance between the optical modes of the same polarization (e.g. of the orders m and $m + 1$). Like in the FP cavity, here too, the FSR varies inversely with the cavity size (D, diameter) for a circular cavity (note: square and hexagonal geometries also show WGMs) by

$$\text{FSR} = \lambda^2 \pi D\, n_{\text{eff}} \tag{2.24}$$

The $\boldsymbol{F}$ of a WGM cavity can be calculated from the relation,

$$\boldsymbol{F} = Q\left(\frac{\lambda}{\pi\, n\, D}\right). \tag{2.25}$$

2.2.2.2 High circulating optical power in WGM cavities

Vahala *et al* demonstrated the possibility of generating high circulating optical power (P_{cir}) in comparison to the input power, (P_{in}) in an erbium-doped silicon WGM cavity microlaser [48].

If all light is coupled to the cavity in a steady state, P_{cir} is related to $\boldsymbol{F}$ as follows,

$$P_{\text{cir}} = \left(\frac{\boldsymbol{F}}{2\pi}\right) P_{\text{in}} \tag{2.26}$$

For illustration, when P_{in} of a WGM cavity of $D = 40$ μm is 16 μW. The cavity Q and $\boldsymbol{F}$ values are 4×10^7 and 3×10^5, respectively. The measured P_{cir} built-up is as high as 800 mW. One can see here, that ultra-high $\boldsymbol{F}$ leads to extremely high P_{cir}.

As smaller V_{eff} leads to strong confinement of the circulating power it provides high circulating optical intensity (I_{cir}).

$$O = \frac{V_{\text{eff}}}{L} = \frac{V_{\text{eff}}}{\pi D} \approx (500\ \mu\text{m}^3/125\ \mu\text{m}) = 4\ \mu\text{m}^2 \tag{2.27}$$

$$I_{\text{cir}} = \frac{P_{\text{cir}}}{O} = 20\ \text{MW cm}^{-2} \tag{2.28}$$

As the circulating optical intensity is in the range of megawatts per area, it provides opportunities to achieve NLO signals using CW lasers (see section 2.2.3).

Regrettably, for most of the NLO materials the signal intensities are very weak, since $\chi^{(1)}\mathbf{E} > \chi^{(2)}\mathbf{E}^2 > \chi^{(3)}\mathbf{E}^3$, subsequently, the efficiency of the NLO conversion also decreases. NLO WGM cavities displaying high Q and small V_{eff}, are known to enhance the intrinsically weak NLO signals as Q is proportional to τ_P as per equation (2.19). As a result, the trapped light interacts with matter multiple times, thus enhancing the NLO intensity by several orders of magnitude. The NLO conversion efficiency of the second and third orders is proportional to Q^2/V_{eff} or Q^3/V_{eff}, respectively.

Further, as the observation of NLO effects necessitates high fundamental light intensities; therefore, the damage threshold of cavity molecular materials is important. Typically, polymer for active microlaser media requires a light intensity of ≈ 0.5–2 kW cm^{-2} during CW laser irradiation, which is close to $\approx$4–15 GW cm^{-2} for a nanosecond pulsed laser (10 ns, 10 Hz) [49]. These high values of intensities can easily damage the soft cavities. The use of a fs laser with a pulse duration of tens of fs can avert material damage. Another important issue is the photobleaching effect of organic dyes which is an equally imperative matter like the physical damage of the micro-resonator. However, the photobleaching effect is dominant at lower fundamental beam intensities (irradiated for milliseconds or seconds) as it depends on the number of absorbed photons [50, 51]. Performing NLO studies with higher fundamental beam intensities evades the photobleaching effect as it is far from the organic material's absorption region.

2.2.2.3 Organic single-crystal WGM microcavities

Zhao *et al* [52] conducted a study in which crystalline microrings were fabricated from a 1,5-diphenyl-1,4-pentadien-3-one (DPPDO) molecule, as illustrated in the FL image shown in figure 2.16(a). The distinct circular growth pattern of DPPDO microcrystals was a consequence of the interfacial tension arising from solution droplets (ethanol/water) during the molecular assembly process. The FESEM image

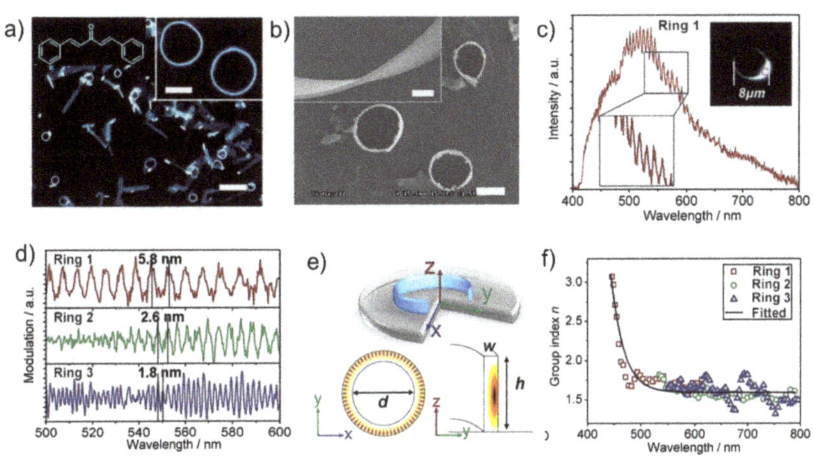

Figure 2.16. (a) FL microscopy image of DPPDO microrings excited with 330–380 nm light. The scale bar is 50 μm. The inset shows the high-magnification image of microrings. The scale bar is 10 μm. (b) FESEM image of microrings. (c) Spatially-resolved FL spectrum outcoupled from a microring of 8 μm diameter. The inset shows the FL microscopy image of the microring locally excited (see violet circle) with a focused 351 nm laser beam. (d) WGM modulation within the range of 500–600 nm in rings 1–3 by removing the FL background. (e) Calculated 3D electric field intensity distribution (λ = 550 nm, n = 1.60) in the microring resonator (d = 4 μm, w = 500 nm, h = 3 μm). (f) The plots of group index versus wavelength in these microrings. [52] John Wiley & Sons. Copyright (2013).

confirmed the belt-like cross-section of the microrings (figure 2.16(b)). These findings suggest that the rings form as a result of the curved growth of belts, induced by the liquid tension within hemispherical-shaped microdroplets during the evaporation process. Notably, the molecular growth exhibited a preferential alignment along the [001], a characteristic elucidated through selected-area electron diffraction patterns.

In the DPPDO microring waveguides, the light emission can circulate and become trapped within the optical modes without outcoupling at the end facets. As a result, the FL spectrum of a microring with a diameter of 8 μm revealed optical modes attributed to WGM resonances (figure 2.16(c)). The WGM resonances in these organic rings were analyzed after removing the background FL in the spectrum. The background-subtracted spectra revealed FSR of 5.8, 2.6, and 1.8 nm at 550 nm for rings 1, 2, and 3, respectively, highlighting the nuanced characteristics of these microstructures (figure 2.16(d)). The calculated group index (n_g) revealed a distinct and pronounced increase below 550 nm. Above 550 nm, it stabilizes at a consistent value (around 1.60), which remains identical across all three microrings of different sizes.

Further insights from FDTD calculations unveiled that the electric field predominantly resides on the outer periphery of the microrings. In the *x*- and *y*-planes, a series of azimuthal modes were identified, with their count diminishing as the cavity diameter increased. The cross-sectional profile (*y*–*z*-plane) indicates that the propagated light can be effectively confined, primarily attributed to the high height/width ratio of the microrings. This confinement occurs although the refractive index of the organic dielectric is close to that of the glass substrate. The estimated *Q*-factor substantially increased from 300 to 400 with the expanding ring diameter, eventually reaching a saturation region at around 450 beyond a ring size of 20 μm.

Fu *et al* reported the fabrication of microscale hexagonal WGM cavities through the self-assembly of p-distyrylbenzene (DSB), as depicted in figure 2.17(a) [53]. The optical pumping of a single-crystalline hexagonal microdisk revealed a captivating phenomenon: alternating bright FL and darkness at six edges, and a distinctive manifestation of D3-WGM type resonances (figures 2.17(b) and (c)). The FL spectrum, however, displayed WGM resonances with a notably weak and low *Q*-factor of 210, spread amidst background vibronic bands (figure 2.17(d)). In addition to hexagonal microdisks, the self-assembly of 1,4-bis(2-methylstyryl)benzene (o-MSB) molecules led to the formation of rectangular microplates, as illustrated in figures 2.17(f) and (g) [54]. Excitingly, optical excitation of these microplates unveiled multi-mode 4-WGM type lasing with an impressive *Q*-factor of up to 1500 (FWHM = 0.3 nm for the 448 nm mode). Particularly, the cavity demonstrated lasing at a threshold of 540 nJ cm^{-2} (figures 2.17(h) and (i)). This comprehensive investigation not only enriches our understanding of WGM cavities but also showcases the diverse optical behaviors exhibited by different microstructures.

Following the exploration of the passive waveguiding property of BPP microtubular waveguides, as detailed in section 2.1.4, my group [36] excited a single BPP microtube with a 355 nm laser, resulting in the emission of a vibrant blue FL that self-guided to the termini (figures 2.18(a) and (b)). The FL spectrum, when collected

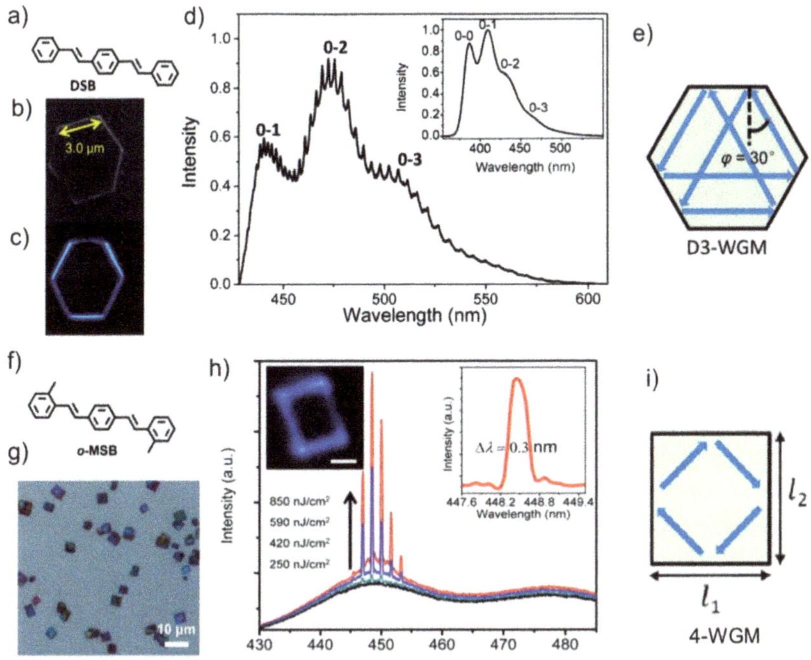

Figure 2.17. (a) The molecular structure of DSB. (b) Scanning electron microscope image of a DSB hexagonal microdisk and (c) the corresponding FL microscopy image. (d) FL spectrum of a hexagonal microdisk with edge length, $d = 3.0$ μm. The insets show the FL spectrum of DSB monomers in THF solution and the molecule structure of DSB. (e) Schematic picture of the light path of the D3-WGM and the angle φ denotes the angle of incidence. (a–e) [53] John Wiley & Sons. Copyright (2014). (f) The molecular structure of *o*-MSB. (g) Optical image of self-assembled rectangular microplates. (h) Pump power dependent FL spectra of microplate excited. The left inset shows the FL microscopy image of the microplate. The scale bar is 5 μm. The right inset shows the FWHM of a peak around 448 nm with $Q \approx 1500$. (i) Schematic representation of the light path of the 4-WGM in the rectangular resonator. (f–i) Reprinted with permission from [54]. Copyright (2014) American Chemical Society.

at the terminal, revealed a distinctive pair of peaks corresponding to TE and TM optical modes, attributed to the breakdown of TE/TM degeneracy (figure 2.18(c)).

The resonances observed in the tube were ascribed to WGMs, arising from the repeated TIR of the trapped FL within the hexagonal tube walls. In accordance with equations (2.22) and (2.23), for the same angular mode number, TM modes resonate at shorter wavelengths than TE modes. Polarization-resolved spectroscopy studies further unveiled that the intensity of the 480 nm (TM) and 485 nm (TE) modes exhibited high sensitivity to the polarizer angle (θ). Specifically, at 90°, the spectrum was dominated by strong TM-polarized WGM resonance modes, while at 0°, equal intensities resulted in a spectrum composed of pure TE modes.

Building upon these findings, the TM/TE polar plots disclosed vertically and horizontally oriented lobes (figures 2.18(d) and (e)), providing a comprehensive visualization of the polarization characteristics of the WGM resonances. This intricate study not only contributes to our understanding of the optical properties

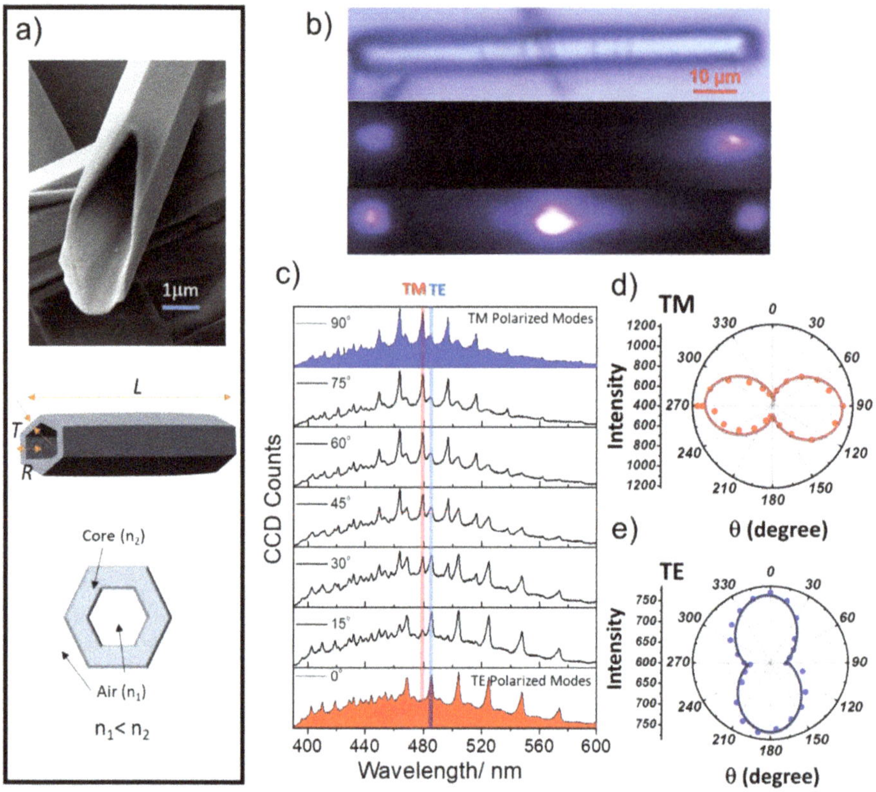

Figure 2.18. (a) FESEM image of hexagonal microtube of BPP (top). The cartoon represents the length (L), radius (R), and thickness (T) of a submicrotube (bottom). (b) FL image and spectrum of an active type submicrotubular resonator excited with a 355 nm UV laser. (c) Polarization-resolved FL spectra of a tube supporting WGM resonances. (d, e) The polar plots of the emission intensity observed for TM (red line) and TE (blue line) polarized modes. Reproduced from [36] with permission from the Royal Society of Chemistry.

of BPP microtubular waveguides but also offers valuable insights into the controllable polarization behavior of WGM modes, opening avenues for advanced applications.

Perylene single crystal is a well-known direct bandgap semiconductor material providing high FL quantum yield. On sublimation, the perylene molecule crystallizes in two polymorphs, I square and II rhombus, emitting yellow and green FL, respectively (figures 2.19(a)–(e)) [55]. In the I and II crystals, the molecules pack as dimeric (sandwich-herringbone motif; *H*-aggregates) and monomeric (γ-type herringbone; *J*-aggregate) forms, respectively.

The high refractive index, shape and light-reflecting mirror-like facets of crystals indicate that these crystals are potential WGM cavities. As expected, upon optical excitation, the polymorphic single crystals exhibit FL spectra (bandwidth: 470–735 nm) with sharp WGMs (figures 2.19(f) and (g)). Further, the 0–0 emission band which appeared at 487 nm is much weaker for the form I cavity compared to form II

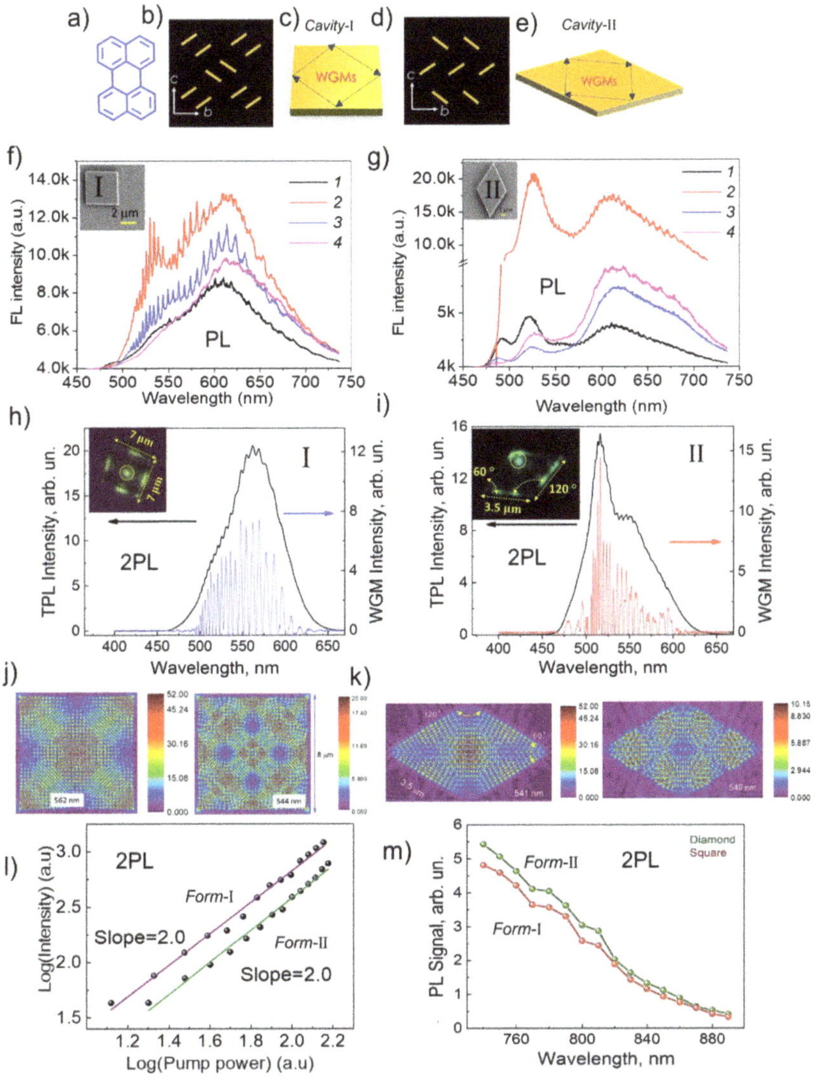

Figure 2.19. (a) Molecular structure of perylene. (c, e) Graphical representation of phase-I and -II WGMs NLO cavities along with their corresponding solid-state packing diagrams (b, d). (f, g) PL spectra of phase-I and -II crystal cavities displaying WGMs, respectively. The insets show the corresponding FESEM image of the crystal cavity. (h, i) TPL or 2-PL spectra of phase-I and -II crystal cavities displaying WGMs, respectively. The insets show the corresponding optical image of the crystal cavity. The blue and red lines show the pure WGM spectra with subtracted non-resonant background luminescence. FDTD calculations: $|\mathbf{E}_z|$ distribution for the wavelengths (see the inset label) that correspond to the WGMs of (j) phase-I with 8 μm edge length and (k) II phase with 3.5 μm edge distance and angles 120° and 60°. (l) 2PL intensity versus pump power for I and II phases of perylene cavities. (m) TPL or 2PL intensity as a function of pump wavelength for both phases, respectively. [55] John Wiley & Sons. Copyright (2019).

and supports their corresponding *H*-type (0–0; forbidden transitions) and *J*-type (0–0; allowed transitions) molecular arrangement, respectively. Due to high crystal anisotropy, they display excitation position-dependent mode intensity and wavelength variations.

Perylene single-crystal polymorphs also show strong two-photon absorption. Upon excitation with a fs ($\lambda = 780$ nm, repetition rate 80 MHz, pulse width 60 fs) laser pulse, both polymorphs show 2-PL in the range of $\approx$460–650 nm. Remarkably, WGM resonances were detected in the 2-PL along with a non-resonant 2-PL background with a Q-factor of $\approx$500. The NLO WGMs spectra extracted from raw 2-PL spectra by subtracting 2-PL background (figures 2.19(h) and (i)) are shown in red and blue lines for form I and II cavities, respectively. FDTD numerical simulations for polymorphic crystals of dimensions [I: L = 8 μm; II: L = 3.5 μm, edge angles 120° and 60°] show that the $|\mathbf{E}_z|$ distribution for λ = 544/562 nm and 541/549 nm, which are the most pronounced WGM peaks of form I and II crystalline cavities, respectively, are shown in figures 2.19(j) and (k). The results show the intricate and distinctive electromagnetic field distribution pattern formed within the cavities for different optical modes. The logarithmic scale plot of TPL (or 2-PL) intensity versus pump power for two polymorphs demonstrated a slope value of 2, confirming the 2-photon excitation process (figure 2.19(l)).

To probe the 2-photon absorption range, the cavities were excited with the fundamental wavelengths in the range 740–890 nm at constant power (figure 2.19(m)). As one can see, the polymorphic cavities exhibited an increase of the 2-PL signal intensity upon decreasing the excitation wavelength. The low signal intensity at 890 nm hints that the 2-PL absorption of perylene crystals is below 900 nm and the maximum absorption is somewhere in the red region. These NLO cavities are useful for infrared laser-pumped lasers and circuits. The geometrical reshaping of these perylene crystal cavities into circular disc, ring, and rectangular cavities using the focused ion beam (FIB) milling technique is discussed in chapter 5.

2.2.3 Electrostriction-mediated mode shifts in an organic polar NLO crystal cavity

The advancement of next-generation organic nanophotonic technologies, crucial for the development of photonic integrated circuits, relies on the utilization of single-crystal NLO microcavities. These microcavities need to be pumped by CW radiation [48, 56] and tuned by an external electric field [57]. The latter stimulus leads to a phenomenon known as the *electrostriction effect*—wherein an electric field force induces elastic (strain) deformation in a dielectric medium [58, 59].

A pioneering study by Chandrasekar and Murzina *et al* [60] reported a novel NLO polar single-crystal optical cavity named University of Hyderabad 1 (UOH1). This crystal, specifically 4-(4-(methylthio)phenyl)-2,6-di(1H-pyrazol-1-yl)pyridine (figure 2.20(a)) [61], exhibited electrostriction when pumped by CW radiation. The FESEM images provided insights into the octahedral shape of the UOH1 microcrystals, showcasing their diverse orientations on the substrate. The crystal

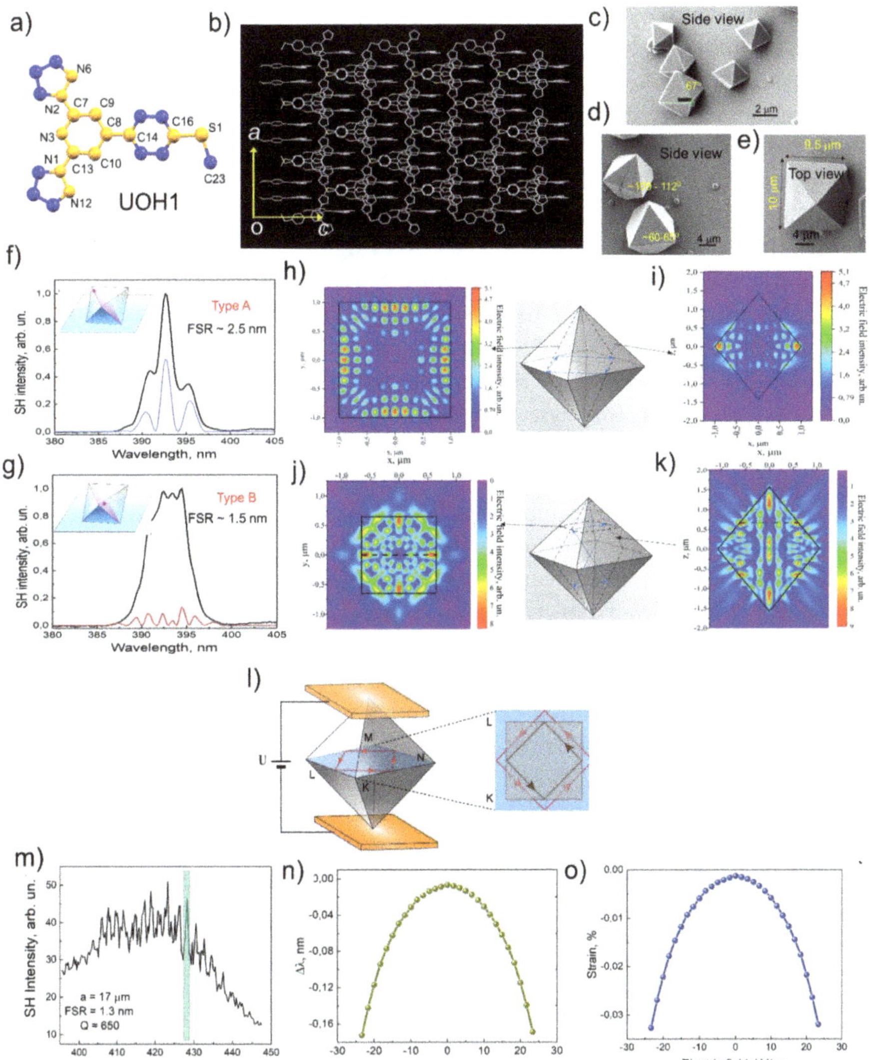

Figure 2.20. (a) Chemical structure of UOH1 and atoms displaying anomalous Born effective charges are shown in an orange color. (b) X-ray crystal structure of UOH1 showing the polar molecular arrangement along the crystallographic *c*-axis. FESEM images of octahedron-shaped crystal with (c) vertices parallel to the substrate, (d) one of the faces facing up and (e) vertex facing up. (f) fs laser-pumped (fundamental $\lambda = 785$ nm) SHG signal with resonant modes with larger FSR (≈2.5 nm) denoting Type A modes. The inset illustrates the excitation spot of an octahedron lying on one of its edges. (g) fs laser-pumped SHG peak with the same wavelength and resonant modes denoting type A + type B modes with smaller FSR (≈1.5 nm). The inset illustrates the excitation spot of an octahedron lying in the geometrical center of the octahedron. (h–k) FDTD numerical calculations of WGMs and bow-tie-like modes in a 2 µm octahedron crystal cavity in two geometrical cross-sections. (l) Schematic of an octahedron crystal cavity under the static uniform externally applied electric field and 2D schematic illustration of the WGM contour compression: the distance between the metal contact plated and the tips of the microcrystal is about 10 µm and the distance between the electrodes is about 60 µm. (m) CW laser-pumped SH excitation spectrum of a 17 µm octahedron with WGMs for electric field measurements. The field-induced effects of (n) the mode and (o) strain. [60] John Wiley & Sons. Copyright (2022).

possesses a non-centrosymmetric, polar space group (figure 2.20(b)), and demonstrated SHG upon fs optical pumping. The unique octahedral shape of these microcrystals, with sizes reaching up to 20 μm, made them technically suitable for in-depth investigations into the NLO cavity effect (figure 2.20(c)). Moreover, the octahedron-shaped crystals of UOH1, characterized by high light-reflecting facets, contribute to the SHG signal. Unusually, this SHG signal also incorporates optical modes with high *Q*-factors, reaching up to approximately 1000.

Remarkably, the microcavity exhibits a SHG signal even when excited with a CW laser at an intensity of 10^7 W cm^{-2}, (about 5×10^4 times lower than in the fs pulse regime). This notable response is attributed to the high second-order nonlinear optical coefficient of 11.187×10^{-9} esu in comparison to urea.

An in-depth exploration into the coupling of the electrostatic field and lattice displacement, analyzed through the Born effective charge[3] (BEC) tensor, has unveiled the polar nature of the UOH1 crystal. The BEC values for atoms C7, C8, C9, C10, C13, C14, C16, nitrogen (N), and sulfur (S) atoms exhibit an abnormal increase in various crystallographic directions, particularly evident for the principal Z^*_{zz} component. This anomaly strongly indicates that the polarization direction aligns with the crystallographic *c*-axis (figure 2.20(a)). Furthermore, this finding suggests a transformation from covalent to ionic character in these atoms, signifying ferroelectric instability (figure 2.20(b)).

Upon exciting different areas, denoted by the red dot in figures 2.20(f) and (g), on the triangle facet facing upwards of the octahedron crystal, pronounced SHG was observed along with optical modes characterized by FSR values of 2.5 nm (figure 2.20(f)) and 1.5 nm (figure 2.20(g)). The distinct FSR values suggest the presence of two resonance types, labeled A and B, originating from varied light circulation geometries within the crystal. This difference is attributed to the varying length of the circulation route with a phase increment divisible by 2π. The type A resonances correspond to WGMs and are located in the main octahedron square section of the cavity. Conversely, the type B resonances represent *bow-tie-like* (BT) modes and are situated in a perpendicular rhombus section, which is also perpendicular to the crystal's facets.

As the crystal is a polar cavity, a DC electric field was applied to manipulate the resonant SHG signal frequency. Using two electrodes, the electric field was applied in the direction perpendicular to the WGM circulation path (figure 2.20(l)). As a result, a particular mode shifts as a second power on the applied electric field as shown in the plot of wavelength shift ($\Delta\lambda$) vs the electric field strength (figures 2.20(m) and (n)). The results show the domination of the electrostriction effect (quadratic in an electric field) over the piezo effect (linear in an electric field) [62, 63]. An alternative stress effect (change in refractive index) can be considered. Nonetheless, in solids, electrostriction generally dominates [57]. The electric field of about 23 kV cm^{-1} (the local field in the microstructure is about 10 times smaller) causes elongation of an octahedron cavity in the field direction and its compression

[3] Born effective charge: In crystalline insulators the change in electric polarization induced by atomic displacements, under zero macroscopic electric field.

in a perpendicular direction of a WGM contour. As a result, the WGM peak wavelength shifts to $\Delta\lambda \approx 0.18$ nm (figure 2.20(n)). This result indicates that the resonator sensitivity is

$$\left(\frac{d\lambda}{d\mathbf{E}}\right)_{\max} \approx 1.79\frac{pm}{V/m} \tag{2.29}$$

to the applied electric field. By using the following equation, for a TE-WGM in an octahedron cavity [64],

$$\lambda_N^{\mathrm{TE,\ WGM}} = \frac{2\sqrt{2}nd}{N + \frac{4}{\pi}\mathrm{arctg}\left(n\sqrt{\frac{n^2-3}{2}}\right)} \tag{2.30}$$

with the value of $n \approx 1.8$ and calculated mode number of $N \approx 140$, a maximum deformation value $\Delta d \approx 5$ nm was estimated. The related dependence of strain on the applied electric field is shown in figure 2.20(o). Notably, these electrostriction-mediated shifts of the resonant modes are useful for ultra-fast narrowband optical switching and filtering applications.

2.2.4 Organic photochromic multi-block crystal as light intensity modulators

Lithium niobate ($LiNbO_3$) is a well-known material that acts as a modulator through the electro-optic effect. By applying an electric field, the refractive index of the material is intentionally increased, effectively slowing down the speed of propagating light pulses. In a notable study reported by my group in 2015 [65], a light-responsive organic crystal was employed to create a modulator-implanted waveguide. This crystal undergoes reversible transformations between different isomeric states, allowing for precise control over the modulation of light propagation. The crystal features back-to-back coupled 2,6-bis(pyrazolyl)pyridine (bpp) fragments bridged by a dithienylethene unit. This structure undergoes photochromic reversible cyclization (closed-state; L-c; blue) and cyclo-reversion (open-state; L-o; colorless) reactions when exposed to UV and visible light, respectively (figures 2.21(a)–(c)). In the L-o state, the 1D crystal passively transduces visible light from one terminal to another [9, 11]. Conversely, in the L-c state, the crystal absorbs visible light, converting the molecules to an L-o state in a time-dependent manner, influenced by the laser power. This results in a delay for the propagating light to reach the output.

Leveraging this phenomenon, an L-c isomer can be selectively created locally through UV light irradiation, effectively serving as a modulator in an L-o isomeric waveguide. In this configuration, the L-c isomer acts as a light shutter, controlling the flow of light over time. Notably, this method allows for the repeated implantation or erasure of modulators within a waveguide, enabling the modification of optical output intensity as a function of the delay time (τ).

In a comprehensive exploration, a series of modulators (L-c domains) were deliberately created at regular intervals along a crystal measuring approximately

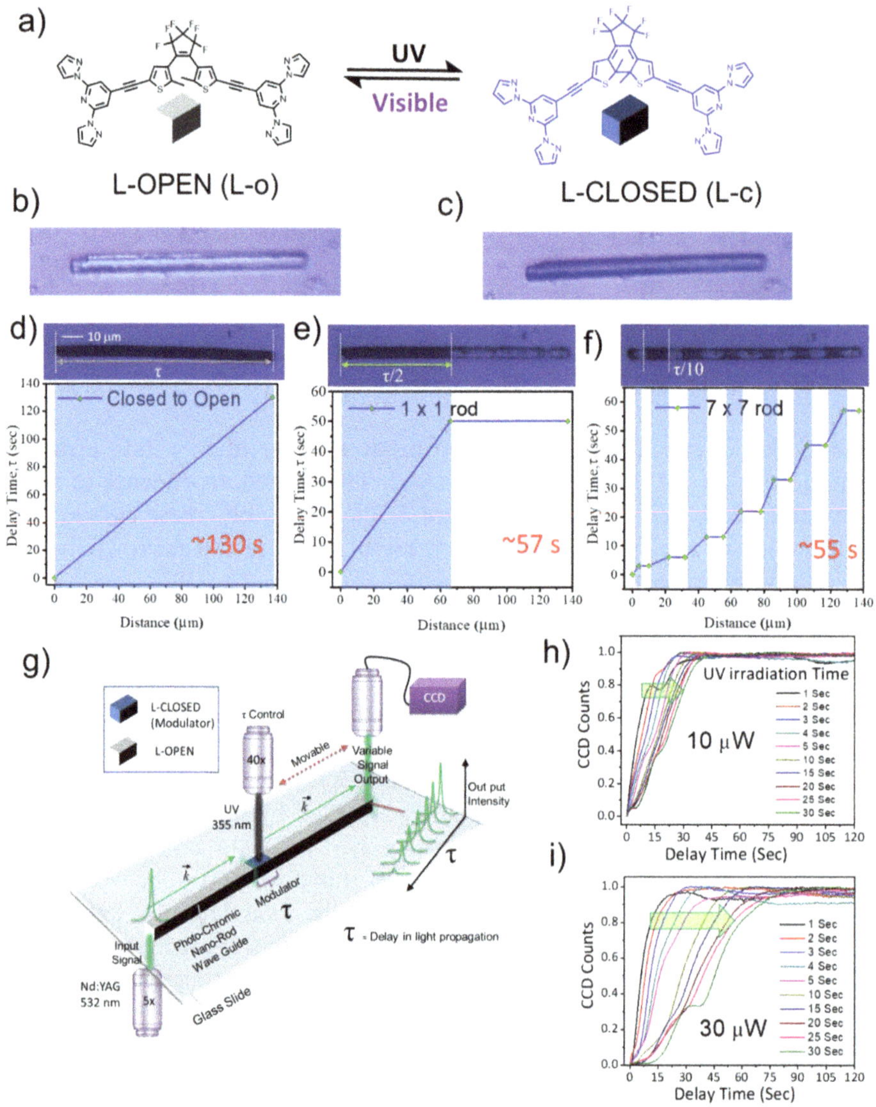

Figure 2.21. Chemical structure of photo-switching molecules (a) in its open (L-o) and closed (L-c) states. (b and c) Confocal micrographs of L-o microrod crystal grown from MeOH/$CHCl_3$ solution and its reversibly interconversion to L-c state microrod under UV/visible light, respectively. Tuning the delaying time using a series of modulators. (d) An L-c microrod modulator (length ~102 μm) delays the output light by ≈ 130 s. (e) and (f) The same rod integrated with modulators of different delay times ≈ $\tau/2$ and $6\times(\tau/10)$. The solid straight lines are guides for the eye. (g) Schematics of the experimental set-up used for tuning the delay time of the output light by creating L-c domains (modulators) of different volumes in the crystal shown as an inset. (h) A plot of normalized 532 nm CCD counts at the output as a function of delay time displays the effect of modulators created by 10 μW UV laser exposure for 1–5, 10, 15, 20, 25, and 30 s. (i) Similarly, plots for irradiation with 30 μW UV laser at various durations. [65] John Wiley & Sons. Copyright (2015).

102 μm in length using a focused UV laser. The irradiation of visible laser light (532 nm; Nd:YAG) at one of the termini of the L-c crystal induced a delay in the outcoupling of passive light at the opposite terminal by approximately $\tau \approx 130$ s (figure 2.21(d)). Subsequently, a 50% L-c (blue) and 50% L-o (colorless) isomeric state configuration was established within the same crystal (figure 2.21(e)). It was anticipated that the delay time would be halved, or $\tau/2$, compared to the previous L-c crystal. The experimental value obtained was approximately 57 s, closely aligning with the estimated value. Similarly, the creation of the 6× ($\tau/10$) domain within the same crystal resulted in a light delay of 55 s (figure 2.21(f)).

Subsequently, to precisely fine-tune the delay time within a single L-c domain, an experimental set-up was devised utilizing a transmission mode confocal microscope (figure 2.21(g)). An L-o crystal measuring approximately 108 μm in length and 3.1 μm in width was secured on the microscope stage within a coverslip. Positioned from the bottom, a 532 nm laser was directed at one of the left terminals of the crystal, while a UV laser was targeted at the top center of the crystal. Notably, the top UV laser objective is mobile, serving a dual purpose: (i) performing point irradiation of the UV laser at the center of the crystal to establish an L-c domain or modulator, and (ii) collecting a 532 nm output signal from the right terminal of the crystal (figure 2.21(g)). This experimental configuration allowed for precise control and manipulation of the crystal's optical properties, enabling the creation and modulation of L-c domains, thereby facilitating the study and optimization of delay times within the experimental framework.

Initially, a 10 μW UV light was applied for just 1 s at the crystal center to create an L-c modulator. Subsequently, the same objective was repositioned to the crystal output terminal. A 532 nm laser (11 μW) was directed at the crystal's left terminal, and the passive output signal was continuously collected for 120 s using the objective focused on the right terminal until the signal intensity reached a maximum and constant value. During this 120 s interval, the signal output was initially minimal, gradually increasing to a maximum value after a delay time of approximately 21 s. This delay time reflects the duration required for the conversion of the L-c domain to L-o.

Similarly, while maintaining the UV laser at 10 μW, the UV exposure time at the crystal center was incrementally extended to 2, 3, 4, 5, 10, 15, 20, 25, and 30 s to augment the L-c domain size. The corresponding delay times versus signal intensity plot exhibited a sigmoidal trend, indicating that an increase in the UV exposure time progressively elevated the τ value from about 21 up to 35 s (figure 2.21(h)). To further extend the τ value, the UV laser power was increased to 30 μW, and the UV exposure time was varied from 2 to 30 s. This adjustment allowed fine-tuning of the τ values to the range of minutes (figure 2.21(i)).

This systematic approach of selectively introducing L-c domains at different positions within the crystal not only demonstrated precise control over the delay times but also showcased the scalability and adaptability of the technique. The ability to modulate delay times by altering the composition of isomeric states opens avenues for tailoring the temporal characteristics of light propagation through these

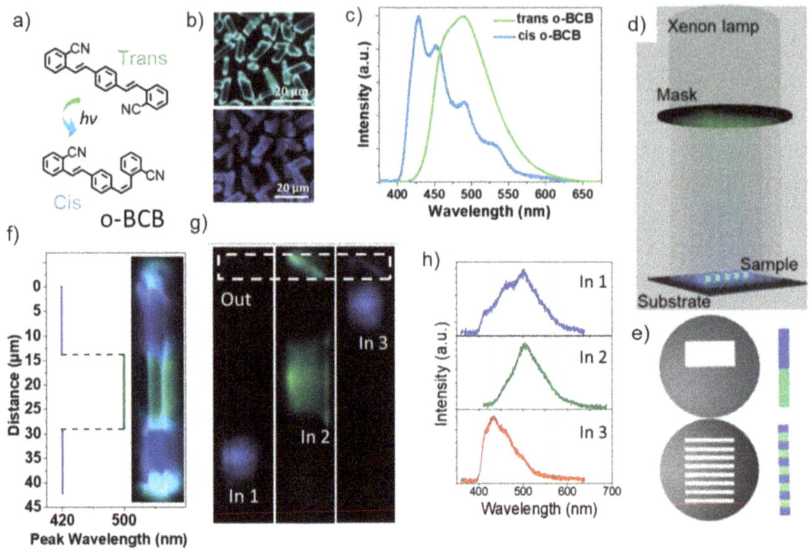

Figure 2.22. (a) The molecular structure of trans-o-BCB and cis-o-BCB during the photochromic effect. (b) FL microscope image of trans-o-BCB and cis-o-BCB crystals; (c) emission spectra of trans-o-BCB microcrystals (green line) and cis-o-BCB structures (blue line). (d) The fabrication system for the heterostructure microcrystals. (e) The mask shapes used for the preparation of the heterostructures. (f) Emission peak wavelength changes along the three-block heterostructure, inset: the corresponding FL microscope image. (e) FL microscope images of the three-block heterostructures excited at different positions. (h) Collected output light spectra when the different input ports in (g) were excited. [66] John Wiley & Sons. Copyright (2019).

crystals, presenting opportunities for applications in advanced optical devices and information processing systems.

In 2019, Liao *et al* [66] employed a similar strategy, utilizing cis-to-trans photoisomerizable organic crystals of 2,2′-((1*E*,1′*E*)-1,4-phenylenebis(ethene-2,1-diyl))-dibenzonitrile (o-BCB) (figure 2.22(a)). When exposed to xenon light, the green-emitting trans-o-BCB microcrystals underwent a photo-isomerization process, transforming into blue-emitting trans-o-BCB. The PL spectra revealed distinct emission bands with an 80 nm blue shift during the cis-o-BCB (λ_{max} = 420 nm) to trans-o-BCB (λ_{max} = 500 nm) isomerization. Remarkably, these crystals also functioned as active optical waveguides. The optical loss coefficient for the cis-o-BCB crystals (0.245 dB μm^{-1}) was found to be higher than that for the trans-o-BCB crystals (0.0355 dB μm^{-1}). This difference in optical loss was attributed to a slight change in the surface smoothness of the microcrystals induced by the photochemical interconversion. This observation highlights the impact of the photo-isomerization process not only on the emission properties of the crystals but also on their performance as active optical waveguides, demonstrating the intricate interplay between the structural changes induced by photochemistry and the resulting optical characteristics.

To create multi-block crystals, they used a mask between the xenon source and the crystals. As a result, the mask partially blocked light irradiation on selected regions of the crystal allowing the photochemical reaction to occur only at certain exposed regions of the crystal, consequently forming the heterostructures (figures 2.22(d) and (e)). They excited a triblock crystal with a 375 nm UV laser at different regions of the crystal to observe blue, blue/green (mixed), and green bands, which were outcoupled from the terminal of the crystal confirming passive/active mixed waveguiding mechanisms (figure 2.22(f)). For input 1, resonance energy transfer occurred, so the output light spectrum expanded from 400 to 650 nm (a mixed blue/green), corresponding to the FL of trans and cis forms of o-BCB crystal (figures 2.22(g) and (h)). For input 2, only the green wavelength band was detected as output via a passive mechanism. When input 3 was excited, only the blue wavelength band was detected at one terminal of the crystal.

2.2.5 Organic single-mode lasers, mode filters, and laser arrays

The distinctive microcavity properties exhibited by organic nano/microcrystals can serve as a high-gain medium for generating stimulated emissions. Zhao *et al* [67] explored a quasi-four-level singlet process in regular organic microlasers. In this gain mechanism, the quasi-four-level process involves absorption between vibronic levels ($|00\rangle$ to $|2A\rangle$ or $|1A\rangle$), with subsequent emission occurring between $|10\rangle$ to $|01\rangle$ states. This process results in population inversion, a large Stokes shift, and consequently, minimal reabsorption loss (figures 2.23(a) and (b)). Leveraging the four-level process within the microcavity effects of organic microcrystals, the authors demonstrated the creation of low-threshold organic microlasers.

They utilized a self-assembled 2,4,5-triphenylimidazole (TPI) molecular nanowire measuring 15 μm in length and 400 nm in width. These nanowires functioned as optical waveguides upon excitation with a pulsed 325 nm laser in the UV range of the electromagnetic spectrum (figure 2.23(c)), evident from the 0–1 ($|10\rangle$ to $|01\rangle$) spectral band. The emission spectrum also featured optical modes attributed to the cavity effect. The emission band exhibited a substantial Stokes shift of approximately 5900 cm^{-1}, a notable contrast to inorganic counterparts (around 700 cm^{-1}). Pump power dependent waveguiding experiments unveiled a laser-like emission peak centered at 374 nm above a certain pump threshold due to the 0–1 band stimulated transition (figure 2.23(d)). This approach not only underscores the unique advantages of the quasi-four-level singlet process in organic microlasers but also highlights the potential for developing efficient, low-threshold organic microlasers with the incorporation of microcavity effects in organic microcrystals.

Achieving clean single-mode lasing in organic crystal cavities poses a significant challenge, prompting the development of various strategies. Techniques such as distributed feedback gratings [69], optical pumping with spatial variation [70], and parity-time symmetry breaking [71, 72] have been employed to manipulate the mode. Additionally, the use of smaller cavities has been explored to increase the FSR, facilitating single-mode lasing. Another effective approach involves a coupled cavity system, where one cavity acts as a spectral filter for the resonance modes of the other, leading to an enlargement of FSR and a reduction in the threshold pump density [68, 73–75].

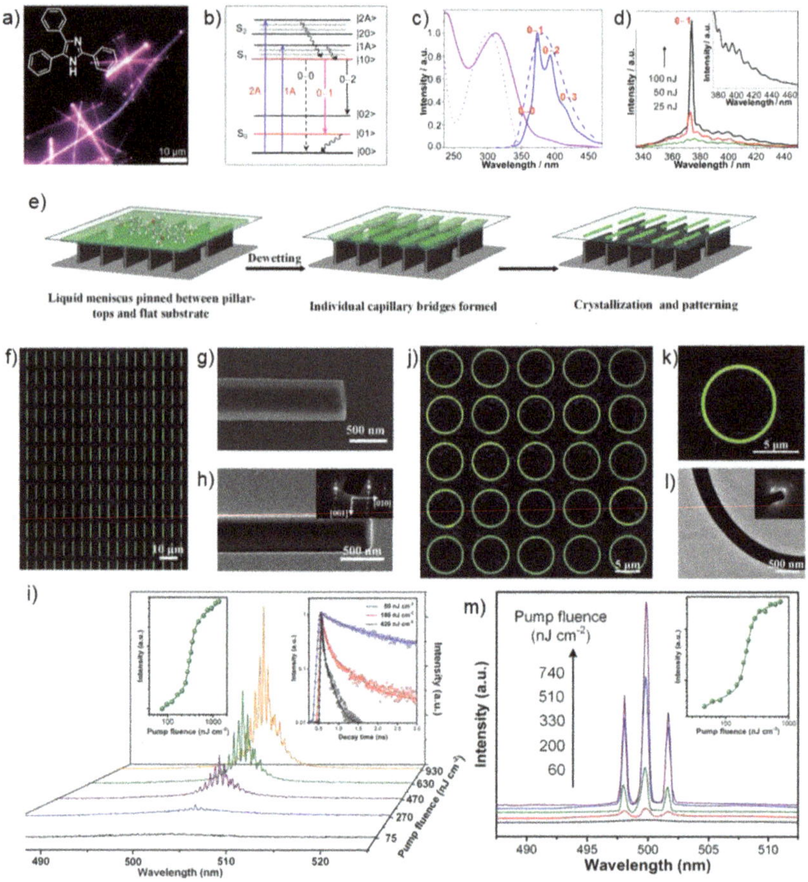

Figure 2.23. (a) PL image of TPI nanowires. (b) Quasi-four-level transition diagram for regular organic microlasers. (c) Emission (blue) and absorption (purple) spectra of TPI nanowires (solid) and monomers (dashed). (d) Lasing spectra of a TPI nanowire. (e) Schematic illustration of the capillary-bridge lithography technique for fabrication of 1D organic single-crystal arrays. (f) FL image presents large-scale TDSB 1D organic arrays. (g) SEM, (h) TEM images of 1D organic single crystal. The inset shows a SAED pattern recorded from the region shown in (h). (i) μ-PL spectra of nanowires under a set of pump fluence. The inset shows a double logarithmic curve of an organic laser with a low threshold of 410 nJ cm^{-2} (left) and time-resolved photoluminescence at different pump fluence (right). The S-shaped curve presents from spontaneous emission via amplified spontaneous emission to full lasing emission. (j) FL image of a TDSB microring array and (k) a single microring. (l) TEM image illustrates a segment of a microring. Inset is the corresponding SAED pattern, indicating that molecules stack into a low-ordering state in the microring. (m) Power-dependent PL spectra of an 8 μm diameter microring. Inset is the integrated PL intensity as a function of pump density of a microring laser with a diameter of 8 μm, showing a lasing threshold around 310 nJ cm^{-2}. (a, c, d) [67] John Wiley & Sons. Copyright (2008). (b) Reprinted with permission from [78]. Copyright (2016) American Chemical Society. (e–m) [76] John Wiley & Sons. Copyright (2017).

Jiang *et al* presented an effective capillary-bridge lithography technique for crafting high-quality arrays of 1,4-dimethoxy-2,5-di[4-(methylthio)styryl]benzene (TDSB) organic single crystals [76] (figure 2.23(e)). The selection of TDSB was based on its *H*-aggregated cofacial molecular packing, characterized by high material gain and a low lasing threshold [77]. The controlled recession of the 'gas–liquid–solid' triphase contact line during the dewetting process is achieved through a template featuring periodically arranged micropillars and asymmetric wettability. This results in the formation of individual capillary-bridge arrays anchored onto the tops of the micropillars. The precise location and geometry of these capillary bridges are strictly regulated by the asymmetric-wettability micropillars, leading to the controlled confinement of nucleation sites and growth orientations of organic crystals. The use of appropriate micropillar-structured templates enables the fabrication of organic single-crystalline 1D structures and microring arrays characterized by high crystallinity, tunable size, and precise positioning. A low-magnification FL microscopy image (figure 2.23(f)) of the produced sample illustrates the formation of extensive organic crystal arrays characterized by uniform size and accurate positioning. Examining a single crystal of TDSB through the SEM reveals its smooth surface and flat end facets (see figure 2.23(g)). The crystallinity is further affirmed by the SAED pattern (figure 2.23(h)).

The emission spectra at various pump fluences[4] near the threshold from a sample of 1D crystals with a cavity length of 20.0 μm are shown in figure 2.23(i). At a low pump fluence of 75 nJ cm^{-2}, the 1D organic crystal manifests a broad peak centered at 500 nm, with an FWHM of 25 nm. As the input power increases, a series of narrow peaks around 500 nm exhibit a rapid rise in intensity. The log–log scale plot of integrated peak intensity versus pump fluence illustrates an 'S'-shaped profile, indicating the transition from spontaneous emission through amplified spontaneous emission to full lasing emission (figure 2.23(i), left inset). The determined lasing threshold stands at 410 nJ cm^{-2} based on the 'S'-shaped curve. Beyond the lasing threshold, the micro-photoluminescence (μ-PL) image reveals two bright green spots at the ends of the 1D crystal, illustrating a pronounced axial waveguiding behavior and characteristic F–P lasing emission.

The dark-field FL images of 8 μm diameter microrings with approximately 0.5 μm width are shown in figures 2.23(j) and (k). The TEM image shows a section of microring with uniform width and a smooth boundary, while the SAED pattern indicates low crystallinity due to continuous changes in molecular packing orientation (figure 2.23(l), inset). Despite the low crystallinity, no visible cracks are observed, enabling effective light confinement and lasing emission from these organic microrings. At a low pump fluence of 60 nJ cm^{-2}, the FL spectrum is dominated by a broad spontaneous emission. As the pump fluence increases, robust lasing emission peaks emerge. Analyzing the integrated PL intensity as a function of pump fluence on a log–log scale reveals an S-shaped curve with a distinct lasing threshold of 310 nJ cm^{-2}, notably lower than that of an F–P laser (see figure 2.23(m), inset). These arrays of 1D structures and microrings produced low-threshold F–P and WGM lasing emissions, respectively.

[4] 'Fluences' refers to the optical energy delivered per unit area; energy/area of the laser spot in the units of J cm^{-2}.

References

[1] Saleh B E A and Teich M C 2018 *Fundamentals of Photonics* 2nd edn (New Delhi: Wiley)
[2] Milosevic M 2013 On the nature of the evanescent wave *Appl. Spectrosc.* **67** 126
[3] Quimby R S 2006 *Photonics and Lasers: An Introduction* (Hoboken, NJ: Wiley)
[4] Balzer F, Bordo V G, Simonsen A C and Rubahn H-G 2003 Optical waveguiding in individual nanometer-scale organic fibers *Phys. Rev.* B **67** 115408
[5] Coon D D 1966 Counting photons in the optical barrier penetration experiment *Am. J. Phys.* **34** 240–3
[6] Zhu S, Yu A W, Hawley D and Roy R 1986 Frustrated total internal reflection: a demonstration and review *Am. J. Phys.* **54** 601–7
[7] Chandrasekhar N, Mohiddon M A and Chandrasekar R 2013 Organic submicro tubular optical waveguides: self-assembly, diverse geometries, efficiency, and remote sensing properties *Adv. Opt. Mater.* **1** 305–11
[8] Annadhasan M, Agrawal A, Bhunia S, Pradeep V V, Zade S S, Reddy C M and Chandrasekar R 2020 Mechanophotonics: flexible single-crystal organic waveguides and circuits *Angew. Chem. Int. Ed.* **59** 13852–8
[9] Chandrasekar R 2014 Organic photonics: prospective nano/micro scale passive organic optical waveguides obtained from π-conjugated ligand molecules *Phys. Chem. Chem. Phys.* **16** 7173
[10] Balzer F, Bordo V G, Simonsen A C and Rubahn H-G 2003 Isolated hexaphenyl nanofibers as optical waveguides *Appl. Phys. Lett.* **82** 10–2
[11] Chandrasekhar N and Chandrasekar R 2012 Reversibly shape-switching organic optical wave guides: formation of nano-rings, nano-tubes, and nano-sheets *Angew. Chem. Int. Ed.* **51** 3556–61
[12] Takazawa K, Kitahama Y, Kimura Y and Kido G 2005 Optical waveguide self-assembled from organic dye molecules in solution *Nano Lett.* **5** 1293
[13] Vinod Kumar A, Rohullah M, Chosenyah M, Ravi J, Venkataramudu U and Chandrasekar R 2023 Amphibian-like flexible organic crystal optical fibers for underwater/air micro-precision lighting and sensing *Angew. Chem. Int. Ed.* **62** e202300046
[14] Annadhasan M, Basak S, Chandrasekhar N and Chandrasekar R 2020 Next-generation organic photonics: the emergence of flexible crystal optical waveguides *Adv. Opt. Mater.* **8** 2000959
[15] Liu H, Lu Z, Zhang Z, Wang Y and Zhang H 2018 Highly elastic organic crystals for flexible optical waveguides *Angew. Chem. Int. Ed.* **57** 8448–52
[16] Mitetelo N, Venkatakrishnarao D, Ravi J, Popov M, Mamonov E, Murzina T V and Chandrasekar R 2019 Chirality-controlled multi-photon luminescence and second harmonic generation from enantiomeric organic micro-optical waveguides *Adv. Opt. Mater.* **7** 1801775
[17] Takazawa K, Inoue J-I, Mitsuishi K and Takamasu T 2011 Micrometer-scale photonic circuit components based on propagation of exciton polaritons in organic dye nanofibers *Adv. Mater.* **23** 3659–63
[18] Gorodetsky M L, Savchenkov A A and Ilchenko V S 1996 Ultimate Q of optical microsphere resonators *Opt. Lett.* **21** 453
[19] Kippenberg T J 2004 Nonlinear optics in ultra-high-Q whispering-gallery optical microcavities *PhD Thesis* (California Institute of Technology) Caltech ETD: etd-06072004-085555
[20] Armani D K, Kippenberg T J, Spillane S M and Vahala K J 2003 Ultra-high-Q toroid microcavity on a chip *Nature* **421** 925

[21] Tabata K, Braam D, Kushida S, Tong L, Kuwabara J, Kanbara T, Beckel A, Lorke A and Yamamoto Y 2014 Self-assembled conjugated polymer spheres as fluorescent microresonators *Sci. Rep.* **4** 5902

[22] Venkatakrishnarao D, Sahoo C, Mamonov E A, Kolmychek I A, Maydykovskiy A I, Mitetelo N V, Novikov V B, Naraharisetty S R G, Murzina T V and Chandrasekar R 2017 A two-photon pumped supramolecular upconversion microresonator *J. Mater. Chem.* C **5** 12349

[23] Zhang C, Zou C-L, Yan Y, Wei C, Cui J-M, Sun F-W, Yao J and Zhao Y S 2013 Self-assembled organic crystalline microrings as active whispering-gallery-mode optical resonators *Adv. Opt. Mater.* **1** 357

[24] Venkatakrishnarao D, Mohiddon M A and Chandrasekar R 2016 The photonic side of curcumin: microsphere resonators self-assembled from curcumin derivatives emitting visible/near-infrared light *Adv. Opt. Mater.* **5** 1600613

[25] Venkatakrishnarao D, Mamonov E A, Murzina T V and Chandrasekar R 2018 Advanced organic and polymer whispering-gallery-mode microresonators for enhanced nonlinear optical light *Adv. Opt. Mater.* **6** 1800343

[26] Sandeep P and Bisht P B 2003 Cavity quantum electrodynamic effects and control of radiative rate of 9-amino acridine hydrochloride hydrate *Chem. Phys. Lett.* **371** 327

[27] Fujiwara H, Sasaki K and Masuhara H 1999 Pyrene fluorescence dynamics within a polymer microspherical cavity *J. Appl. Phys.* **85** 2052

[28] Lu T, Lee H, Chen T, Herchak S, Kim J-H, Fraser S E, Flagan R C and Vahala K 2011 High sensitivity nanoparticle detection using optical microcavities *PNAS* **108** 5976–9

[29] Dantham V R, Holler S, Kolchenko V, Wan Z and Arnold S 2012 Taking whispering gallery-mode single virus detection and sizing to the limit *Appl. Phys. Lett.* **101** 043704

[30] Venkatakrishnarao D, Mohiddon M A and Chandrasekar R 2017 The photonic side of curcumin: self-assembly of curcumin derivatives into microsphere resonators emitting visible/near-infrared light *Adv. Opt. Mater.* **5** 1600613–13

[31] Venkatakrishnarao D and Chandrasekar R 2016 Engineering the self-assembly of DCM dye into whispering-gallery-mode micro-hemispheres and Fabry–Pèrot type micro-rods for Vis–NIR (600–875 nm) range micro-optical cavities *Adv. Opt. Mater.* **4** 112–9

[32] Zhang C, Zou C-L, Yan Y, Hao R, Sun F-W, Han Z-F, Zhao Y S and Yao J 2011 Two-photon pumped lasing in single-crystal organic nanowire exciton polariton resonators *J. Am. Chem. Soc.* **133** 7276–9

[33] Ravi J, Venkatakrishnarao D, Sahoo C, Naraharisetty S R G, Mitetelo N, Ezhov A A, Mamonov E, Murzina T and Chandrasekar R 2018 A two-photon pumped supramolecular up-conversion microresonator *Chem. Nano. Mat.* **4** 764–8

[34] Rayleigh L 1910 CXII The problem of the whispering gallery *Phil. Mag.* **20** 1001

[35] Raman C V and Sutherland G A 1921 Whispering-gallery phenomena at St. Paul's Cathedral *Nature* **108** 42

[36] Venkataramudu U, Venkatakrishnarao D, Chandrasekhar N, Mohiddon M A and Chandrasekar R 2016 Single-particle to single-particle transformation of an active type organic μ-tubular homo-structure photonic resonator into a passive type hetero-structure resonator *Phys. Chem. Chem. Phys.* **18** 15528

[37] Venkatakrishnarao D L V, Narayana Y S, Mohaiddon M A, Mamonov E A, Kolmychek I A, Maydykovskiy A I, Novikov V B, Murzina T V and Chandrasekar R 2017 Two-photon luminescence and second-harmonic generation in organic nonlinear surface comprised of self-assembled frustum shaped organic microlasers *Adv. Mater.* **29** 1605260

[38] Wei Y, Lin X, Wei C, Zhang W, Yan Y and Zhao Y S 2017 Starch-based biological microlasers *ACS Nano* **11** 597
[39] Piccione B, Cho C H, van Vugt L K and Agarwal R 2012 All-optical active switching in individual semiconductor nanowires *Nat. Nanotech.* **7** 640
[40] Wang X, Liao Q, Li H, Bai S, Wu Y, Lu X, Hu H, Shi Q and Fu H 2015 Near-infrared lasing from small-molecule organic hemispheres *J. Am. Chem. Soc.* **137** 9289
[41] Collodo M C, Sedlmeir F, Sprenger B, Svitlov S, Wang L J and Schwefel H G L 2014 Sub-kHz lasing of a CaF_2 whispering gallery mode resonator stabilized fiber ring laser *Opt. Express* **22** 19277
[42] Vyatchanin S P, Gorodetsky M L and Ilchenko V S 1992 Tunable narrow-band optical filters with modes of the whispering gallery type *J. Appl. Spectrosc.* **56** 182
[43] Asano M, Ozdemir S K, Chen W, Ikuta R, Yang L, Imoto N and Yamamoto T 2016 Lossless microwave photonic delay line using a ring resonator with an integrated semiconductor optical amplifier *Appl. Phys. Lett.* **108** 181105
[44] Ta V D, Chen R, Ma L, Ying Y J and Sun H D 2013 Tuning whispering gallery mode lasing from self-assembled polymer droplets *Laser Photonics Rev.* **7** 133
[45] Oraevsky A N 2002 Whispering-gallery waves *Quant. Electron.* **32** 377
[46] Yang S, Wang Y and Sun H 2015 Advances and prospects for whispering gallery mode microcavities *Adv. Opt. Mater.* **3** 1136
[47] Lam C, Leung P T and Young K 1992 Explicit asymptotic formulas for the positions, widths, and strengths of resonances in Mie scattering *J. Opt. Soc. Am.* B **9** 1585
[48] Polman A, Min B, Kalkman J, Kippenberg T J and Vahala K J 2004 Ultralow-threshold erbium-implanted toroidal microlaser on silicon *Appl. Phys. Lett.* **84** 1037
[49] Argyros A, van Eijkelenborg M A, Jackson S D and Mildren R P 2004 Microstructured polymer fiber laser *Opt. Lett.* **29** 1882
[50] Ippen E, Shank C and Dienes A 1971 Rapid photobleaching of organic laser dyes in continuously operated devices *IEEE J. Quantum Electron.* **7** 178
[51] Eggeling C, Widengren J, Rigler R and Seidel C A M 1998 Photobleaching of fluorescent dyes under conditions used for single-molecule detection: evidence of two-step photolysis *Anal. Chem.* **70** 2651
[52] Zhang C, Zou C-L, Yan Y, Wei C, Cui J-M, Sun F-W, Yao J and Zhao Y S 2013 Self-assembled organic crystalline microrings as active whispering-gallery-mode optical resonators *Adv. Optical. Mater.* **1** 357
[53] Wang X, Liao Q, Kong Q, Zhang Y, Xu Z, Lu X and Fu H 2014 Whispering-gallery-mode microlaser based on self-assembled organic single-crystalline hexagonal microdisks *Angew. Chem.* **126** 5973
[54] Wang X, Li H, Wu Y, Xu Z and Fu H 2014 Tunable morphology of the self-assembled organic microcrystals for the efficient laser optical resonator by molecular modulation *J. Am. Chem. Soc.* **136** 16602
[55] Pradeep V V, Mitetelo N, Annadhasan M, Popov M, Mamonov E, Murzina T V and Chandrasekar R 2019 Ambient pressure sublimation technique provides polymorph-selective perylene non-linear optical micro-cavities *Adv. Opt. Mater.* **8** 1901317
[56] Gan X T, Zhao C-Y, Hu S-Q, Wang T, Song Y, Li J, Zhao Q-H, Jie W-Q and Zhao J-L 2018 Microwatts continuous-wave pumped second harmonic generation in few- and mono-layer GaSe *Light: Sci. Appl.* **7** 17126

[57] Ioppolo T, Ayaz U and Ötügen M V 2009 Tuning of whispering gallery modes of spherical resonators using an external electric field *Opt. Exp.* **17** 16465
[58] Yang C, Zhang H, Liu B, Liu H, Wang C and Lin S 2018 Electrically tuned whispering gallery modes microresonator based on microstructured optical fibers infiltrated with dual-frequency liquid crystals *Nanophotonics* **7** 1333
[59] Yang C, Zhang H, Liu B, Li Y and Liu H 2017 Electrically tunable whispering gallery mode microresonator based on a grapefruit-microstructured optical fiber infiltrated with nematic liquid crystals *Opt. Lett.* **42** 2988–91
[60] Mitetelo N, Ravi J, Mondal S, Venkataramudu U, Mamonov E, Popov M, Maydykovskiy A, Vaitheeswaran G, Murzina T and Chandrasekar R 2022 An organic electro-mechanical-cavity emitting efficiently tunable, continuous-wave-pumped non-linear-optical-modes *Adv. Opt. Mater.* **11** 2201635
[61] Venkataramudu U, Sahoo C, Leelashree S, Venkatesh M, Ganesh D, Naraharisetty S R G, Chaudhary A K, Srinath S and Chandrasekar R 2018 Terahertz radiation and second-harmonic-generation from a single-component polar organic ferroelectric crystal *J. Mater. Chem.* C **6** 9330
[62] Chen B, Li T, Dong Q, Mosconi E, Song J, Chen Z, Liu Y, Ducharme S, Angelis F D and Huang J 2018 Large electrostrictive response in lead halide perovskites *Nat. Mater.* **17** 1020–6
[63] Bai W, Wang L, Zheng P, Wen F, Yuan Y, Ding M, Chen D, Zhai J and Ji Z 2018 Large electrostrictive effect in lead-free $(Bi_{.5}Na_{.5})$ TiO_3-based composite piezoceramics *Ceram. Int.* **44** 8628–34
[64] Dong H, Sun L, Sun S, Xie W, Zhou L, Shen X and Chen Z 2010 Indium oxide octahedra optical microcavities *Appl. Phys. Lett.* **97** 223114
[65] Venkatakrishnarao D, Mohiddon M A, Chandrasekhar N and Chandrasekar R 2015 Photonic micro-rods composed of photo-switchable molecules: erasable hetero structure wave guides for tunable optical modulation *Adv. Opt. Mater.* **3** 1035–40
[66] Li Z Z, Wu J J, Wang X D, Wang K L, Zhang S, Xie W F and Liao L S 2019 Controllable fabrication of in-series organic heterostructures for optical waveguide application *Adv. Opt. Mater.* **7** 1900373
[67] Zhao Y S, Peng A, Fu H, Ma Y and Yao J 2008 Nanowire waveguides and ultraviolet lasers based on small organic molecules *Adv. Mater.* **20** 1661–5
[68] Zhang C, Zou C-L, Dong H, Yan Y, Yao J and Zhao Y S 2017 Dual-color single-mode lasing in axially coupled organic nanowire resonators *Sci. Adv.* **3** e1700225
[69] Persano L, Camposeo A, Del Carro P, Fasano V, Moffa M, Manco R, D'Agostino S and Pisignano D 2014 Distributed feedback imprinted electrospun fiber lasers *Adv. Mater.* **26** 6542–7
[70] Liew S F, Redding B, Ge L, Solomon G S and Cao H 2014 Active control of emission directionality of semiconductor microdisk lasers *Appl. Phys. Lett.* **104** 231108
[71] Feng L, Wong Z J, Ma R-M, Wang Y and Zhang X 2014 Single-mode laser by parity-time symmetry breaking *Science* **346** 972–5
[72] Hodaei H, Miri M-A, Heinrich M, Christodoulides D N and Khajavikhan M 2014 Parity-time–symmetric microring lasers *Science* **346** 975–8
[73] Xiao Y, Meng C, Wang P, Ye Y, Yu H, Wang S, Gu F, Dai L and Tong L 2011 Single-nanowire single-mode laser *Nano Lett.* **11** 1122–6

[74] Xu H, Wright J B, Luk T-S, Figiel J J, Cross K, Lester L F, Balakrishnan G, Wang G T, Brener I and Li Q 2012 Single-mode lasing of GaN nanowire-pairs *Appl. Phys. Lett.* **101** 113106

[75] Ku J-F, Chen Q-D, Ma X-W, Yang Y-D, Huang Y-Z, Xu H-L and Sun H-B 2015 Photonic-molecule single-mode laser *IEEE Photon. Technol. Lett.* **27** 1157–60

[76] Feng J, Jiang X, Yan X, Wu Y, Su B, Fu H, Yao J and Jiang L 2017 Capillary-bridge lithography" for patterning organic crystals toward mode-tunable microlaser arrays *Adv. Mater.* **29** 1603652

[77] Xu Z Z, Liao Q, Shi Q, Zhang H L, Yao J N and Fu H B 2012 Low-threshold nanolasers based on slab-nanocrystals of H-aggregated organic semiconductors *Adv. Mater.* **24** OP216–20

[78] Zhang W, Yao J and Zhao Y S 2016 Organic micro/nanoscale lasers *Acc. Chem. Res.* **49** 1691–700

Chapter 3

Mechanophotonics: mechanical micromanipulation of molecular crystals into various photonic components

3.1 The need for mechanophotonics

The fabrication of OPICs from flexible single crystals demands a meticulous process of shaping microcrystals into various curved geometries such as bent waveguides, ring resonators, splitters, delay lines/phase shifters, interferometers, add-drop filters, etc for the construction of photonic modules (figure 3.1) [1]. These components must then be precisely positioned on a silica substrate and seamlessly integrated. Achieving this level of precision and integration necessitates the development of innovative crystal shaping and micromanipulation techniques, facilitating the assembly of various photonic modules.

Therefore, the elemental step in the construction of OPICs is the mechanical micromanipulation of organic crystals. Nevertheless, this task has remained unexplored for *micro-sized crystals*, primarily due to the absence of a suitable and convenient tool. The requirement of geometrically distinct crystal-based OPIC components has compelled the invention of pioneering micromechanical crystal processing methodologies.

The conventional approach of employing a three-point bending technique with macroscale needles and forceps, supplemented by oil as a lubricant, has proven effective for manipulating crystals within the millimeter to centimeter size range [2, 3]. However, this method is notably ill-suited for micro-sized crystals, where achieving the requisite level of precision in crystal integration is paramount. Furthermore, a common practice in crystal bending experiments has been to employ oil as a lubricating agent.

Nevertheless, the coating of high refractive index oil around the crystal waveguides or cavities is undesirable, and the effect of oil coating on the optical loss has not been investigated. Moreover, the behavior of flexible macrocrystals, which are typically elastic and range in scale from millimeters to centimeters, exhibits a

doi:10.1088/978-0-7503-5336-6ch3

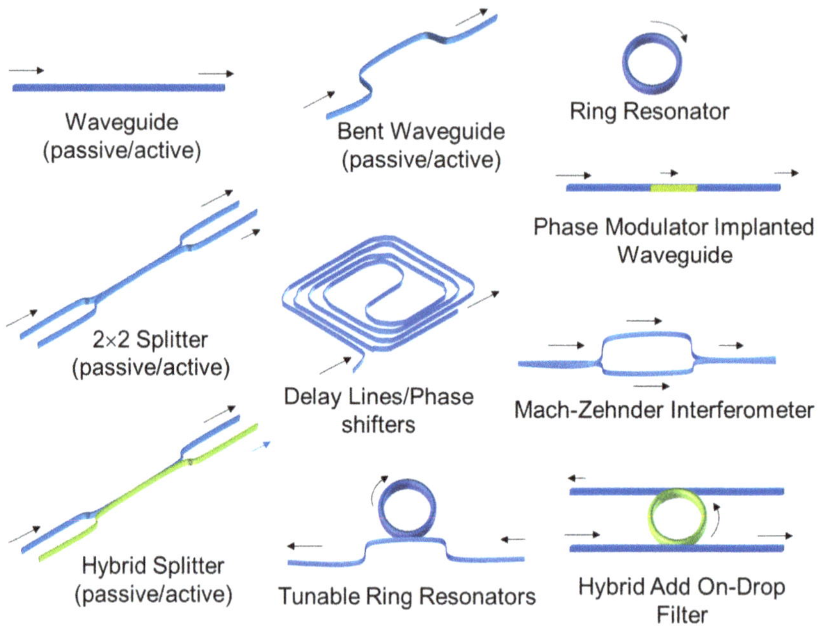

Figure 3.1. Organic crystal photonic modules for the fabrication of PICs. The arrows show the light path. The small arrows show modulation of light intensity or phase or polarization. Reproduced from [1] with permission from the Royal Society of Chemistry.

notable tendency to revert to their original configuration once the mechanical force is removed [2]. This shape reversion renders them unsuitable for the fabrication of permanent crystal-based photonic circuits.

The production of high-quality crystals exhibiting key optical attributes, including a high refractive index, photoluminescence, high quantum yield, NLO attributes, mechanical flexibility, and robust stability, stands as a critical requirement for the construction of OPICs [1]. Notably, miniature PICs are composed of different microscale optical modules. Therefore, the fabrication of PIC components requires micro-sized crystalline materials and a lubricant-free micromanipulation environment, as the removal of lubricants post-manipulation process can prove to be challenging. More importantly, there is a need for the manipulated microcrystals of certain desirable geometry to exhibit shape irreversibility, ensuring that the altered form persists beyond the cessation of mechanical stress. Logically, to address the intricate task of manipulating microcrystals, a micro-sized tool becomes indispensable, such as a sharp tip or needle with controlled operability.

In this context, an innovative approach has materialized in my research laboratory at the University of Hyderabad, capitalizing on contact-mode AFM and confocal optical microscopy, collectively constituting a powerful and precise technique for microcrystal manipulation (figure 3.2) [1, 4, 5]. The AFM cantilever tip serves as the tool for precise micromanipulation, while confocal microscopy facilitates close-range visualization of the crystals undergoing micromanipulation and helps to probe the crystal's photonic characteristics with exceptional detail.

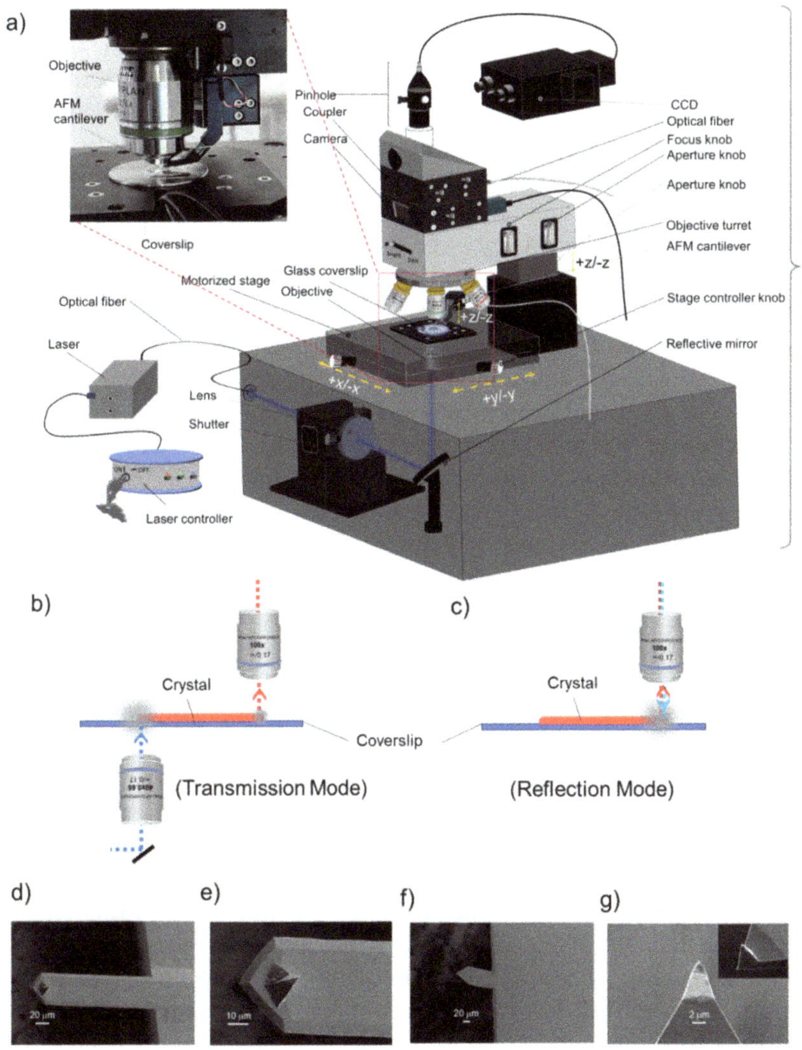

Figure 3.2. (a) Experimental (upright and inverted) optical microscope set-up for the production of laboratory-scale OPICs using the mechanophotonics approach. (b, c) Schematic of transmission and reflection mode geometries. (d, e) Down tip AFM cantilever, (f, g) front tip AFM cantilever used for the mechanical micromanipulation of microcrystals.

3.2 Mechanophotonics - mechanical micromanipulation of organic optical crystals

For several years, my research group has been at the forefront of pioneering the advancement of AFM cantilever tip-based mechanical operations for optical crystal micromanipulation, a technique known as *Mechanophotonics* [1].

Mechanical micromanipulation procedures, encompassing a diverse array of operations such as *lifting* [6–8], *cutting* [9], *moving* [10], *bending* [7, 10–12], *slicing* [7], *transferring* [11, 12], *rolling* [13], *standing* [14], *stacking* [14], *aligning* [10–12], and *integrating* [10–12], stand as pivotal and indispensable in the complex endeavor of constructing organic single-crystal-based monolithic or hybrid OPICs (figure 3.3). These operations are the cornerstone of a multifaceted approach, culminating in the possible realization of large-scale production of OPICs through automation, where each manipulation operation contributes to the precision and intricacy demanded by these minuscule optical circuits. These operations extend beyond crystals to encompass the manipulation of polymer WGM microcavities [15], whether grown or deposited, on low-index silica substrates, thereby advancing the frontiers of microfabrication of coupled resonators (photonic molecules) and coupled-resonator-optical waveguides (CROWs) for light manipulation.

The microscopy set-up employed for the photonics studies requires a transmission mode geometry combining both upright and inverted microscopes (figure 3.2). The judicious selection of low and high numerical aperture (NA)[1] objectives serves the dual purpose of affording a comprehensive panoramic view of the circuit construction site while also offering an intimately detailed examination of the optical attributes intrinsic to the crystals. It is imperative to recognize that the NA of an objective directly influences its capacity to collect and manipulate image-forming optical rays. Consequently, objectives with higher NA values yield resolved images of tiny microcrystals.

In reflection mode (backscattering geometry) experiments (figure 3.2(c)), the sample is optically excited via a careful focusing of laser light employing appropriate objectives, with the ensuing optical emission being adeptly collected using the same objective. For example, a 20× objective can be employed to obtain a broad view of the substrate containing the FL crystal of interest. For detailed photonic studies, higher magnifications such as 100× or 150× objectives are recommended, allowing for a closer examination of the crystals and facilitating precisely exciting and recording of FL. This reflection mode geometry is employed for rapid examination of the photonic attributes of crystals. It is crucial to highlight that the observation of modes and their quality is contingent upon the NA of the objective used for signal detection. This is because different crystals emit radiation in various directions based on their geometry.

In transmission mode experiments, the sample can be excited through an inverted microscope objective and the FL can be recorded using the upright microscope objective and vice versa (figure 3.2(b)). This geometry is crucial for selectively exciting the crystal at one location and recording the FL at other areas of the crystal. It is useful, for instance, in estimating the optical loss of a waveguide, investigating the crystal cavity effect, conducting optical experiments on OPIC, etc.

During the crystal micromanipulation, the coverslip containing the crystals is kept on a piezo-electric stage and can be moved in the *x*- and *y*-axis with

[1] Numerical Aperture (NA): The numerical aperture of a microscope objective is a crucial metric, indicating its capability to collect light and discern fine details in a specimen while operating at a fixed object (or specimen) distance. $NA = n \sin\theta$. Here, θ is one-half of the objective's opening angle and n is the refractive between the objective and coverslip containing the sample.

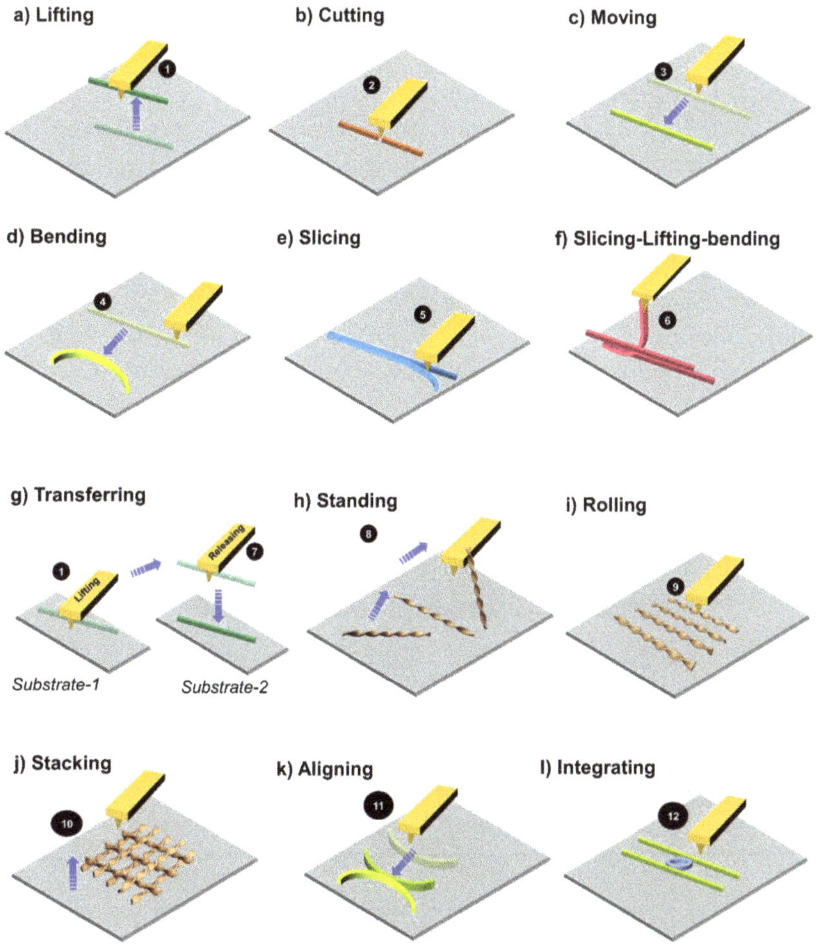

Figure 3.3. Mechanophotonics—various types of operations developed for the mechanical micromanipulation of organic optical crystals with AFM cantilever tip [6–14]. [7] John Wiley & Sons. Copyright (2020).

micro-precision as the AFM tip (positioned close to the objective) can be moved in the *z*-axis (useful for *lifting* and *transfer* operations). Alternatively, the AFM cantilever with the *x*, *y*, and *z* movements can be employed alone or along with the piezo-electric stage moves.

Within the intricate realm of mechanical micromanipulation for microcrystals, as depicted in figures 3.2(d–g), the utilization of both the downward- and forward-tip configurations of AFM cantilevers emerges as a judicious choice. As the micromanipulation involves *contact mode*, the cantilever tip is convenient for executing various mechanophotonic operations. In comparison to downward AFM cantilevers, the forward-tip AFM cantilever enables convenient observation of crystals undergoing micromanipulation.

3.2.1 Lifting of optical waveguiding crystals

In 2014, my group performed the first single-crystal micromanipulation experiment, i.e., demonstrating the possibility of mechanically *lifting* a single tubular crystal of BPP-1 waveguide with an AFM cantilever tip (figure 3.4(a)) [6]. Lifting of crystals involves electrostatic force (attractive) between the microcrystal and cantilever tip. The entire experiment was conducted to validate the intrinsic passive-type waveguiding nature of the crystal when irradiated with 488 nm light, free from any substrate influence. The optical microscope images of an inverted cantilever carrying the crystal waveguide are shown in figures 3.4(b)–(d).

Recently, the lifting of a single flexible microcrystal of *N*-benzylideneanilines derivative, NBA (figure 3.4(e)) from the substrate was demonstrated [7]. The

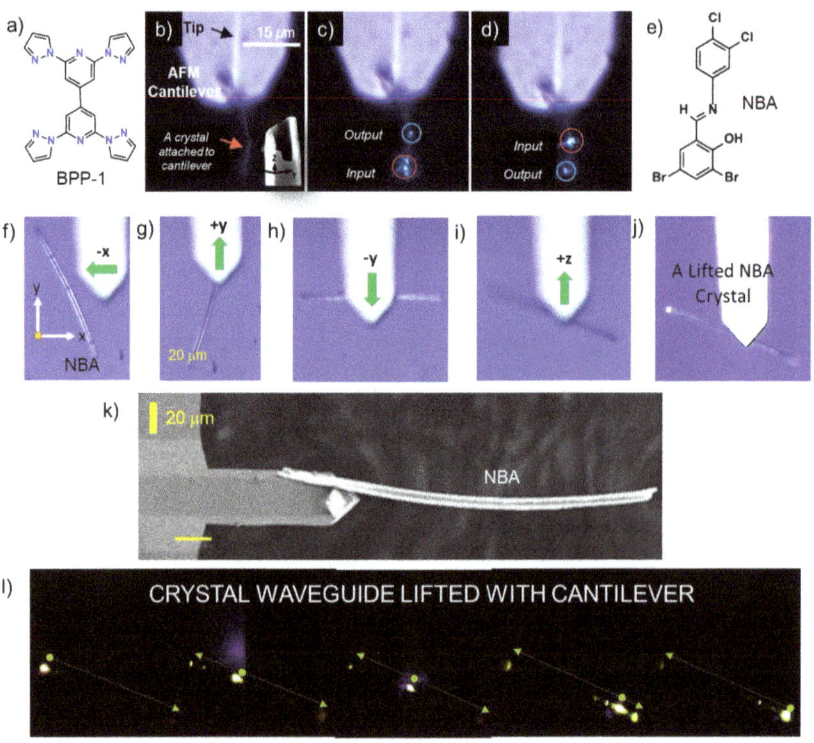

Figure 3.4. (a) Molecular structure of BPP-1. Optical microscope images of (b) a crystal attached to an inverted AFM cantilever (the inset shows the FESEM image of a tubular crystal tip), and (c, d) passive optical waveguiding (488 nm) behavior of the bent tube. The red and blue circles denote laser input and output points, respectively. (a–d) Reproduced from [6] with permission from the Royal Society of Chemistry. (e) Molecular structure of NBA. (f, g) Optical microscope images of an NBA crystal at the substrate and the movement of the cantilever towards the crystal, (h) the crystal being pulled by the cantilever in the forward direction, (i) and then pushed back to connect it to the tip, (j) finally, being lifted. (k) FESEM image of the crystal attached to the cantilever. (l) FL images of a single crystal lifted with an AFM cantilever, displaying active-type waveguiding tendency when excited at different positions. The broken yellow arrows show the direction of light propagation. (e–l) [7] John Wiley & Sons. Copyright (2020).

operation was carried out by primarily maneuvering the crystal backward and forward using the cantilever ($+y$ and $-y$-directions) to facilitate the attachment of the crystal to the tip (figure 3.4(f–i)). The attached crystal is slowly raised from the substrate by lifting the cantilever ($+z$) (figure 3.4(j)). The FESEM image presents a single NBA crystal attached to an inverted cantilever (figure 3.4(k)). The optical waveguiding experiments of the lifted crystals confirmed the propagation of FL (active type) to the crystal terminal or termini (figure 3.4(l)). The essence of the experiment is the demonstration of controlled mechanical manipulation of a crystal in the x, y and z coordinates, which is important for integrating the crystals.

In another experiment, an active-type flexible crystal cavity was transformed into a passive-type cavity with the aid of AFM cantilever-based lifting [8]. For that, a crystalline, hexagonal tubular cavity of BPP was air-lifted using an AFM cantilever tip (figures 3.5(a) and (b)). The cantilever was inverted and optical excitation (355 nm) of the center of the crystal showed blue FL at the excitation area and crystal termini (figure 3.5(c)). The FL spectra recorded at the termini exhibited highly resolved WGMs. Here, the appearance of optical modes is due to repetitive TIR of the confined FL light inside the hexagonal tube walls, consequently producing WGM optical resonances. This same tubular crystal cavity shows no FL when excited with a 488 nm laser, due to the absence of any molecular absorbance, producing only the Raman spectrum. Fascinatingly, increasing the power of the laser burnt the crystal's terminal displaying a bright yellow-orange

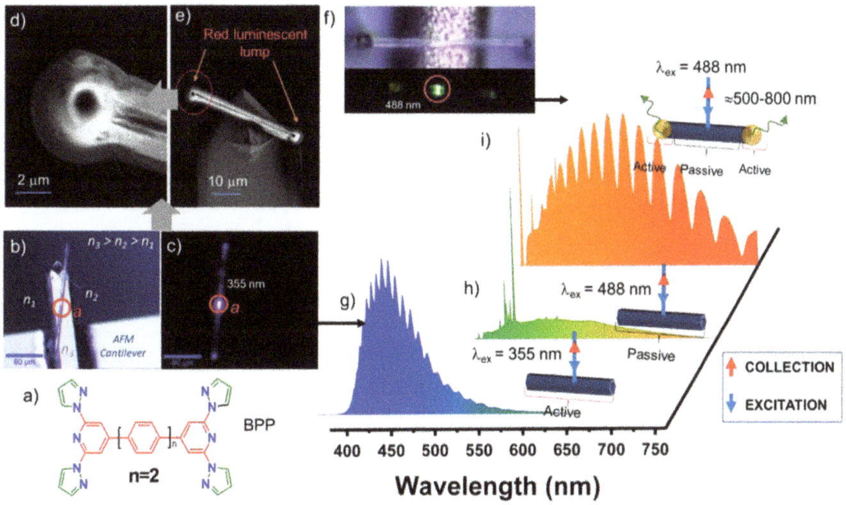

Figure 3.5. (a) Molecular structure of BPP. (b) A crystalline tubular waveguide of BPP attached to an AFM cantilever. (c) FL image displaying laser excitation of tube area (position a), which is in contact with the cantilever. (d, e) FESEM images of the tube with spherical carbonaceous lump termini attached to the cantilever tip. (f) Confocal micrography image of the tube with carbonaceous lumps at both termini. FL images of the same tube excited at the center. (g) FL spectra of the tube supporting active WGMs. (h) Raman spectrum of the tube that is passive to 488 nm light. (i) FL/Raman spectra of tips burnt tube. Reproduced from [8] with permission from the Royal Society of Chemistry.

emission. The burnt crystal termini developed into a black color lump due to the formation of carbonaceous mass (figure 3.5(f)). The FESEM images disclosed the almost spherical shape of the lumps and their smooth surface formed due to the surface tension (figures 3.5(d) and (e)). Interestingly, in the reflection mode experiment, excitation of crystal center with 488 nm light, selectively excited the lumps producing a broad visible-near infrared range (500–800 nm) spectrum with WGMs (figure 3.5(i)). In addition, the emission spectrum also revealed Raman signals (C=C; 1609 cm^{-1}) produced by the BPP molecules (figure 3.5(h)).

Over the years, this crystal *lifting* operation developed and became key to selecting and picking up suitable crystals from any substrate for OPIC assembly, for the fabrication of hybrid OPICs.

3.2.2 Cutting of crystal cavities

The length of naturally grown microcrystals is difficult to control. Gaining control over the length of the microcrystal waveguide and F–P cavity is imperative to fine-tune the light propagation distance and number of optical modes, respectively. As no other technique is available for this crystal tailoring job, my group exploited the sharpness of the AFM cantilever to cut the acicular crystals. For example, in a pivotal experiment conducted in 2016, we showed that a DCM dye-based F–P crystal cavity with a width (W) of 8 μm can be meticulously cut into smaller fragments [9]. An AFM cantilever tip was employed to divide the crystal into three distinct pieces: $L1 = 150$ μm, $L2 = 110$ μm, and $L3 = 74$ μm. This intricate cutting operation resulted in cavities with nearly identical FSR values. Significantly, the observed absence of length dependency on the FSR indicated the presence of mirror-like crystal facets along the lateral sides. This revelation provides valuable insights for the efficient design of optical devices, as discussed in chapter 2 and illustrated in figure 2.11(d). Such precise cutting procedures are instrumental in OPIC construction, allowing the selection of crystals with appropriate lengths for tailored photonic components, especially in cases where crystal length plays a crucial role.

3.2.3 Discovery of pseudo-plasticity: moving/bending of a crystal

3.2.3.1 Curved crystal waveguide/cavity

Organic crystal based bent waveguides and ring resonators are important components for OPIC construction. Flexible single crystals are suitable for the creation of these photonic components. However, the shape reversibility (after the removal of stress) due to a single crystal's elasticity hampers the components geometrical stability. In 2020, my group made the first attempt to bend *N*-benzylideneanilines-based elastic crystals using an AFM cantilever tip on a coverslip without the use of any lubricant oil (figure 3.6(a)) [7]. Earlier, an investigation by Naumov *et al* disclosed that these crystals bend readily in the macroscale during the three-point bending (see figure 3.6(b)) and regain their original straight geometry after the removal of the force, corroborating NBA-2 crystal's elasticity (figure 3.6 (d)) [7]. The Young's modulus of the crystal was in the range of 0.64–0.72 GPa (figure 3.6 (d)). In the x-ray crystal structure of NBA-2 type-II Br···Br, type-II Cl···Cl, and

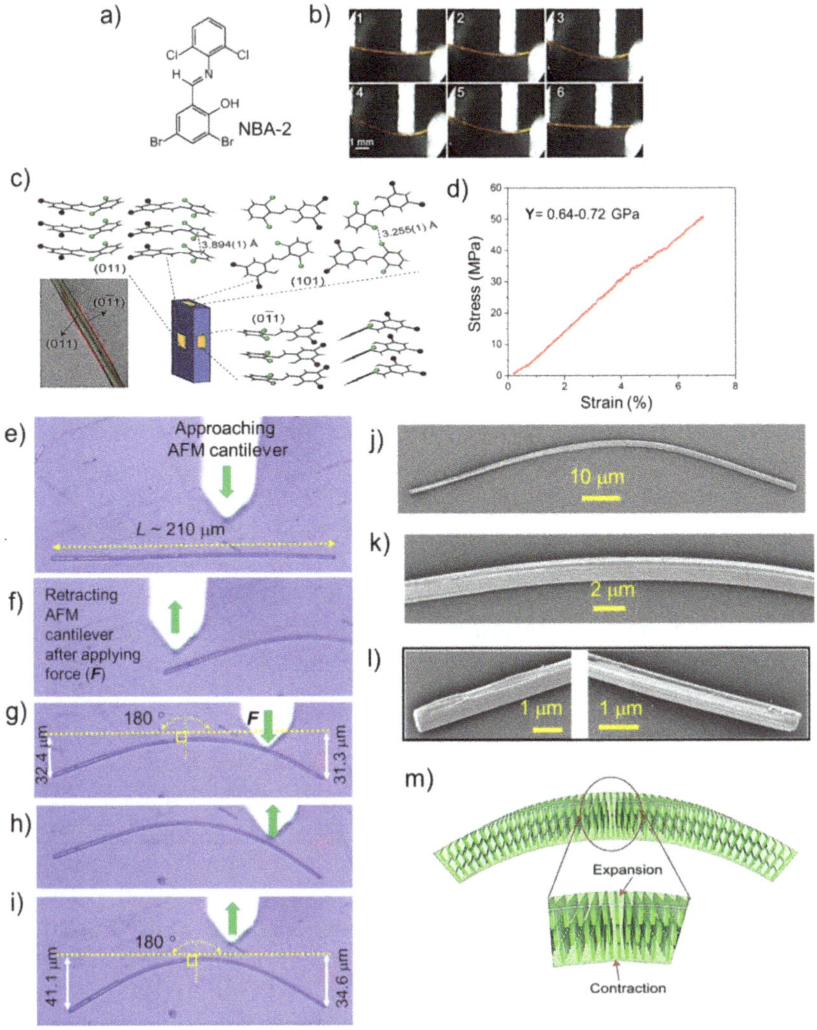

Figure 3.6. (a) Molecular structure of NBA-2. (b) Optical microscopy images showing elasticity of NBA-2 single crystal. (c) X-ray crystal structure depicting the molecular packing of NBA-2. (d) Stress/strain diagram for three-point bending test on an NBA-2 single crystal. Confocal optical microscope images showing the micromanipulation of the crystal of length $L \approx 210$ μm, (e) an approaching AFM cantilever tip, (f) retraction of the tip after applying force at the left side, leaving the crystal bent at the contact point. (g, h) The same procedure was applied to the right side of the crystal. (i) A bent crystal with the shape of an arc. (j–l) FESEM images of the bent crystal. (m) Cartoon depicting molecular motions in the expanded (convex) and contracted (concave) regions of the crystal. [7] John Wiley & Sons. Copyright (2020).

Cl···Br interactions, π···π-stack along the crystallographic *c*-axis were found without slip planes (figure 3.6(c)). It looks like the weak and dispersive interactions between the molecules buffered the developed strain in all directions during crystal bending. During the elastic bending of a crystal, the NBA-2 molecules expand on the convex side and compress on the crystal's concave side (figure 3.6(m)).

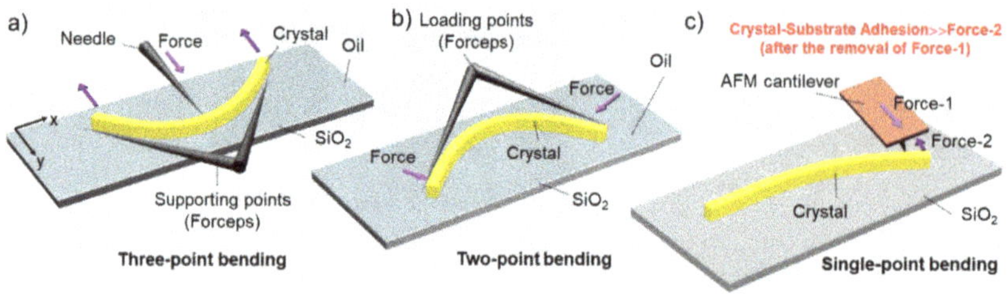

Figure 3.7. Schematic of (a) three-point bending of a crystal kept on an oil-covered SiO_2 substrate using forceps and needle, (b) two-point bending of a crystal kept on an oil-covered SiO_2 substrate using forceps, and (c) single-point bending of a pseudo-plastic crystal with the AFM cantilever kept on the SiO_2 substrate.

The micromechanical investigation was carried out with an AFM cantilever tip at the University of Hyderabad for NBA-2 microcrystals grown on a coverslip. Unlike three-point or two-point bending (figures 3.7(a) and (b)), our method was a single-point and oil-free technique (figure 3.7(c)), wherein cantilever tip force was exerted on the crystal manually at one of its termini, without two crystal supporting points.

For the single-point experiment, a straight NBA-2 microcrystal of length $L \approx 210$ μm grown on a silica substrate was taken (figure 3.6(e)). By exerting force with an AFM cantilever tip, the left side of the microcrystal was pushed to <5 μm distance and the crystal retained its bent geometry even after the withdrawal of tip force (figure 3.6(f)). Similarly, by applying force, the crystal's left side was bent to $\approx$32.4 μm (figure 3.6(g)). This observed unusual behavior showed that the micro-crystal-substrate interaction energy is larger than the crystal shape-retrieval energy (figure 3.7(c)). To confirm this premise, the crystal's right side was further pushed to a distance of $\approx$31.3 μm, producing an arc-shaped crystal. Later, by applying force on the left and right sides of the crystal, a perfect arc shape was realized, wherein the left and the right termini of the crystal were bent to 41.1 μm and 34.6 μm distance, respectively compared to the original straight geometry. The noteworthy conclusion of the experiment is that the microcrystals on the oil-free surface, after mechanical deformation, permanently retain their new geometry unveiling their anomalous *pseudo-plasticity*. As a result of this important discovery, many extremely bent optical waveguiding crystals with a high strain value, including rings and circuits, were created.

The FESEM revealed no noticeable crack or damage due to crystal bending stress and changes in the crystal termini interfacial angles (figures 3.6(h)–(l)). Usually, the defects (of sizes larger than half of the optical wavelength) in crystal optical waveguides reduce the light transmission ability due to unwanted optical loss through scattering.

3.2.3.2 Semi-circular crystal waveguide/cavity

The next logical approach was the precision crafting of a semi-circular crystal geometry. This endeavor began with the selection of a DPIN microcrystal (discussed in figure 2.8) possessing a length of 164 μm (figure 3.8(a, i)) [11].

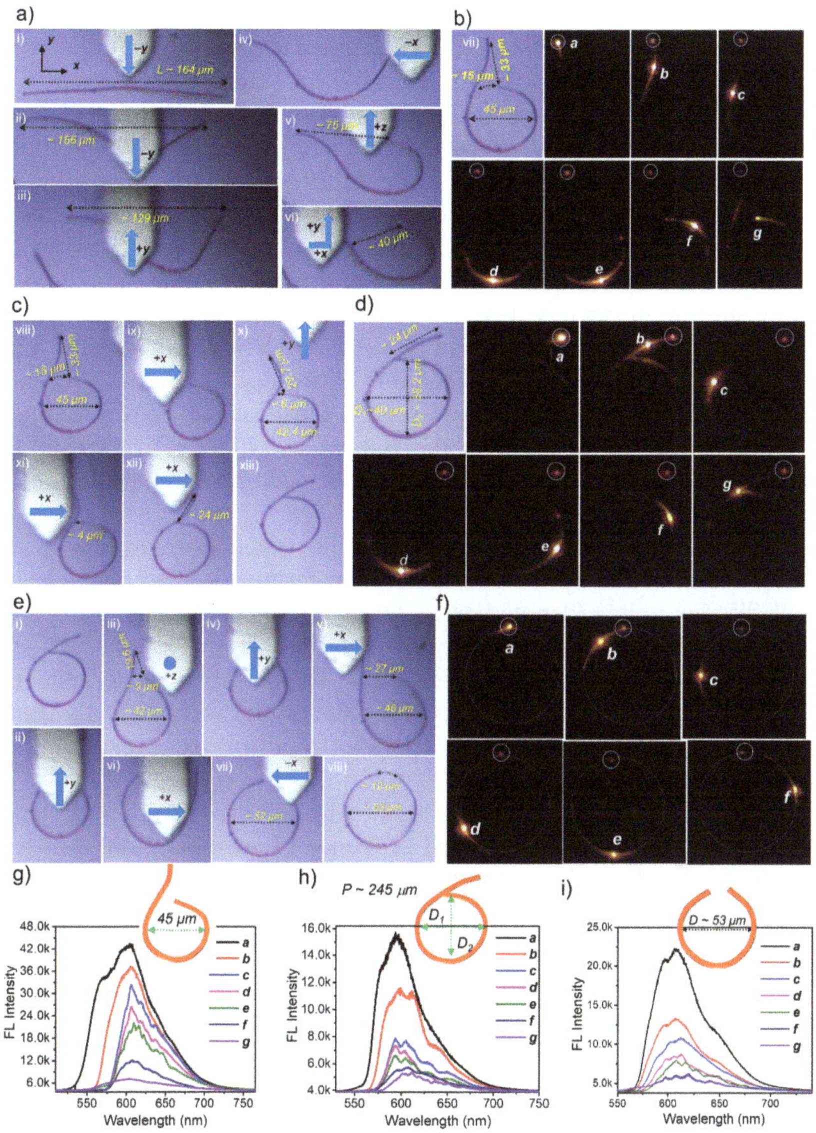

Figure 3.8. Confocal optical microscope images depicting the mechanical micromanipulation of a DPIN crystal with AFM, (a, b) from straight to near-circular geometry with a tail (i–vii), (c, d) from a near-circular DPIN crystal with a tail into an elliptical loop with a tail (viii–xiii), (e) from an elliptical loop shape (with a tail) to a truncated circular shape. (b) Confocal optical/FL images of a near-circular geometry with a tail crystal shown in (vii) excited at different positions (g) the corresponding FL spectra. (d) Confocal optical/FL images of a bent crystal shown in (xiii) excited at different positions (*a–g*), and (h) the corresponding FL spectra. The white circles show the FL spectra recorded sites. (f) FL images of a truncated circular-shaped crystal shown in (viii) excited at different positions (*a–g*). (i) The corresponding excitation position-dependent FL spectra. The white circles show the FL spectra recorded site. [11] John Wiley & Sons. Copyright (2021).

The crystal was methodically manipulated using an AFM cantilever, watching through the objective to achieve the desired curved morphology. Initially, a gentle force was exerted in the forward direction ($-y$), causing the crystal to bend, effectively reducing the tip-to-tip distance to approximately 156 μm (figure 3.8(a, ii)). Subsequently, applying a reverse cantilever force in the ($+y$) direction diminished the tip-to-tip distance to around 129 μm (figure 3.8(a, iii)). Further cantilever maneuver in the ($-x$) direction at the right tip of the crystal resulted in the formation of a nearly half-circular configuration, featuring a distinctive tail while maintaining a tip-to-tip distance of approximately 75 μm (figure 3.8(a, iv and v)). The resulting intricately bent crystal was then subjected to a series of meticulous cantilever manipulations in two distinct directions. Initially, a pushing force was employed on the left tail section of the crystal in the ($+x$) direction, followed by the application of force in the ($+y$) direction. These precise operations led to two significant reductions in the distance: firstly, between the right and left tips, reaching ≈33 μm, and secondly, between the right tip and the left body, with a distance of ≈15 μm (figure 3.8(b, vi and vii)).

The resulting extremely curved crystal, as illustrated in figure 3.8(b, vii), was subsequently subjected to optical waveguiding experiments. The crystal was excited at various discrete points (denoted as (*a*–*g*)) along its curved axis, using a 355 nm laser source. Later, the FL spectra were recorded at the left output terminal, indicated by the white circle in figure 3.8(b). Remarkably, the orange FL that emitted at different excitation positions along the crystal was efficiently transduced to both termini of the crystal. The FL spectra displayed excitation position-dependent F–P optical modes.

To engineer an elliptical crystal with a pathway conducive to the circular flow of light (figure 3.8(c, viii)), a finely tuned sequence of manipulations was precisely executed. The crystal's tail section, measuring approximately 33 μm, was gently displaced to the right side in the $-x$-direction, achieving a significant reduction in the gap from approximately 15 μm to a mere 6 μm (figure 3.8(c, ix and x)). An additional force applied in the $+x$-direction connected the left terminal with the right-side body of the crystal, yielding an elliptical loop configuration, with a trailing ≈24 μm long crystal segment (figures 3.8(c, xiii) and (d)). The elliptical loop exhibited long ($D1$) and short ($D2$) axes, spanning 40 μm and 38.2 μm, respectively, with a perimeter totaling approximately 245 μm. Upon optical excitation at various positions along the elliptical loop, the crystal exhibited an outcoupled light emission at the tail end (figure 3.8(d)). The corresponding FL spectra, depending upon the excitation position, were recorded at the terminal of the crystal, within the outlined white circle. These spectra also revealed optical modes (figure 3.8(h)) distinguishable from those in figure 3.8(g) proof of the effect of geometric transformation of the microcrystal on the optical modes.

Subsequently, the aim was to shape a truncated circular structure from the elliptical loop with a tail presented in figures 3.8(c, xiii) and (e, i). This intricate process unfolded in two exactly orchestrated steps. Initially, the cantilever was delicately lifted in the $-z$-direction and positioned inside the elliptical loop. Subsequently, to widen the loop, the crystal's tail was gently drawn outward in the $-y$-direction (figure 3.8(e, ii)), initially to a modest gap of 9 μm and later to an expanded distance of approximately 27 μm (figure 3.8(e, iii–v)). The right ($+x$) and left ($-x$) segments of the crystal were strategically aligned to form a truncated circular loop, featuring a tip-to-tip distance of 12 μm and a diameter of 53 μm (figure 3.8(e, vi–viii)). This truncated circular crystal aptly directed optical signals along its circular trajectory towards its termini, depending upon the excitation positions (figure 3.8(f)). The corresponding spectra, collected at the left terminal, showed F–P modes within the emitted FL signal, underscoring the crystal's adaptability to manipulate light in a circular path, thus holding promising implications for advanced photonic circuit applications.

The manipulation was further continued to create a circular ring-shaped geometry wherein the two termini of the waveguides touch each other side by side. The details of utilizing this ring for creating a hybrid circuit are discussed in this chapter (in section 3.2.6) and their photonic properties are presented in section 4.3.3 of the next chapter.

3.2.3.3 Ring-shaped crystal waveguide/cavity

9,10-Dibromoanthracene (DBA) molecules grow into needle-shaped crystals in hexane solution (figures 3.9(a) and (d)) [16, 17]. The crystals exhibit stability through weak or dispersive forces, including $\pi \cdots \pi$ and C–H$\cdots\pi$ interactions (figures 3.9(b) and (c)). In a study conducted by Hayashi *et al* the mechanical elasticity of DBA macrocrystals was investigated using a three-point bending technique, providing a comprehensive understanding of their structural basis for elasticity [17]. Notably, these long acicular crystals displayed multiple reversible bends without breaking (perpendicular to 010), showcasing remarkable mechanical flexibility. However, our observations of long DBA microcrystals grown on coverslips revealed pseudo-plastic behavior (figure 3.9(f)). The FESEM image of the microcrystals depicted the square-shaped cross-sectional geometry of the elongated crystals and their hollow feature (figure 3.9(d)).

To create a ring resonator from a nearly straight pseudo-plastic crystal, we mechanically manipulated a 175 μm long waveguide of $\alpha' = 0.0628$ dB μm^{-1} with an AFM cantilever tip (figure 3.9(f)) [18]. Initially, the left side, followed by the center of the crystal, was gently moved from its original position in the $+y$-direction to create a bent crystal (figure 3.9(f, i–iii)). Subsequently, a force was applied alternately to the left and right sides of the crystal in the $-y$-direction to create a U-shaped crystal (figure 3.9(f, iv–viii)). Finally, applying force in the $-x$-direction brought the left and right sides of the crystal together with a contact distance of about 13 μm (figure 3.9(f, ix–x)). The resulting ring is elliptical, with horizontal and vertical diameters of $D1 \approx 36$ μm and $D2 \approx 48$ μm, respectively (figures 3.9(f, xi) and (g)). The mechanical bending-induced strain of the curved ring along the horizontal and vertical sides was 11.5% and 12.7%, respectively. Under UV irradiation, the strained ring exhibited cyan FL outcoupling at the crystal termini as bright light, indicating that the waveguiding nature is retained by the strained crystal (figure 3.9(h)). The strained ring showed an α' value of 0.0693 dB μm^{-1} due to

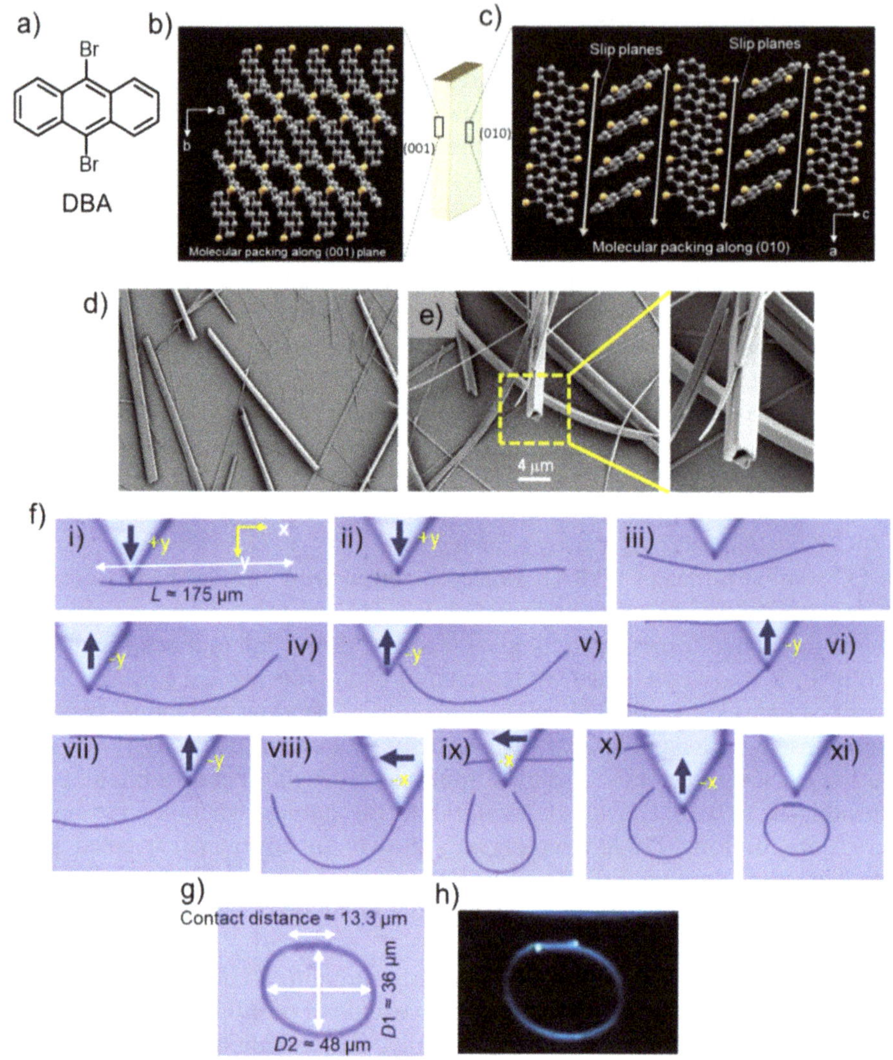

Figure 3.9. (a) Molecular structure of DBA. (b, c) Molecular packing arrangement viewed along *c* and *b* axes. The inset depicts the presence of slip planes. (d, e) FESEM image of acicular DBA single crystals. The crystal's tubular nature is shown in zoom view. (f, g) Micromanipulation of a straight crystal waveguide into a ring-shaped cavity. (h) FL image of the ring. [18] John Wiley & Sons. Copyright (2023).

bending-induced loss. As one can see, the bending-induced optical loss is not very significant indicating the high structural integrity of the strained crystal.

3.2.4 Slicing/separation of an optical crystal waveguide from a bundled aggregate

Usually, microcrystals are grown by the slow evaporation of 2–3 drops of crystallization mother liquor kept on a coverslip. It is normal to observe that some of the grown crystals physically aggregate together, creating bundles. We have observed this phenomenon in another derivative of *N*-benzylideneaniline, namely, NBA-3 crystals (figure 3.10(a)) [7]. These crystals are mechanically elastic crystals analogous to NBA-2 crystals (figures 3.6(a) and 3.10(b) and (c)). To separate individual crystals from the bundle my group developed *slicing* and *bending* operations using an AFM cantilever tip.

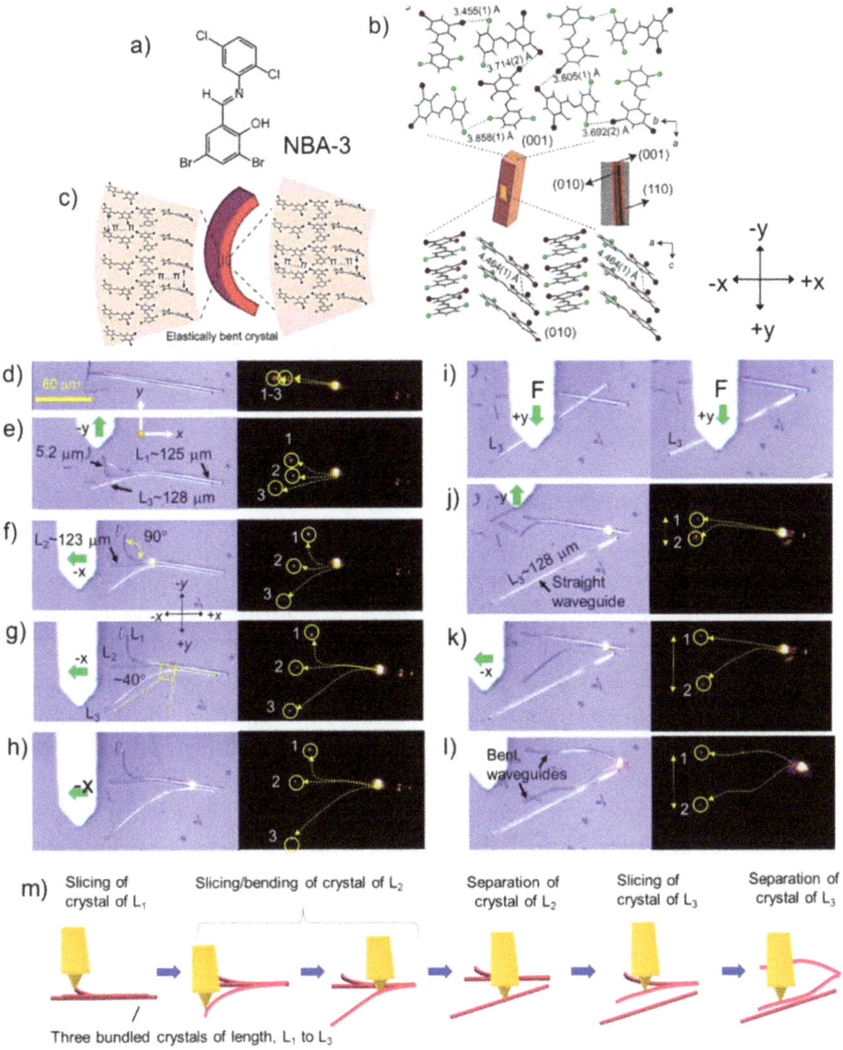

Figure 3.10. (a) Molecular structure of NBA-3. (b) Molecular packing in the crystal viewed perpendicular to their (010) and (001) faces. (c) Schematic showing the suggested mechanism of elastic bending of the crystals. (d) Confocal optical microscopy images (left) and respective FL images (right) of three crystals that were bundled initially and were later partially or wholly separated by slicing, cutting, and bending. (i–l) Complete separation of a crystal from the bundle, cutting of one of the component crystals and bending at multiple points to obtain a crystal with an undulated shape. The broken yellow arrows in all FL images indicate the direction of light propagation. (m) Illustration of slicing and bending operations. [7] John Wiley & Sons. Copyright (2020).

A demonstrative microbundle containing three crystals was taken for this study (figure 3.10(d)). The presence of three axially aggregated crystals of varying lengths (labeled L_1 to L_3) was confirmed by optically exciting the middle of the bundle, which displayed three outcoupled FL spots (see the FL image shown in figure 3.10(d), right). To segregate the individual crystals, the cantilever tip was placed between the bundle gap, and the force was applied by moving the tip back and forth perpendicular to the bundle axis. This slicing operation partly detached the crystals at the bundle's left side. This operation required a careful execution, for example, excessive tip movement cuts one of the separated crystals into a piece of length $\approx$5.2 μm (figure 3.10(e)). Further, as enough room is available at the left side of the bundle, the *bending* operation can be implemented to bend the left terminal of L_1 to $\approx$90° (figure 3.10(f)), and L_3 to $\approx$40° and subsequently to $\approx$50° in such a way that L_1 and L_3 termini face opposite to each other (figures 3.10(g) and (h)). This is evident from the corresponding FL images depicting FL outcoupling directions. The crystal L_3 was detached from the bundle and the substrate by applying cantilever force in the $+y$ coordinate (figures 3.10(i–k)). As a result, less adhesion to the substrate crystal L_3 straightened from its bent geometry. As the right ends of crystals L_1 and L_2 are attached, L_2 was made into a wavy geometry to validate its flexible nature (figure 3.10(l)).

3.2.5 Moving/aligning/integrating operations in monolithic crystal optical waveguides

The fabrication of monolithic OPICs necessitates the relocation of multiple micro-crystals from the substrate to the circuit fabrication site, in the same substrate. This process entails the identification and precision cutting of optical crystals of appropriate sizes near the circuit fabrication site (typically within 500 μm^2). Once identified, these crystals can be *moved* (without *lifting* them) to the fabrication site. Subsequently, diverse crystal geometries can be engineered at the fabrication site. Integration of these selected crystals with micro-precision is vital for transferring light from one crystal to another via evanescent coupling in OPICs.

An illustrative example is a directional coupler (DC) [10], constituting a 4-port optical device designed for light division and the distribution of varying intensities. The blue FL compound, namely, dithieno[3,2-a:2′,3′-c]phenazine (DTPz), produced an acicular flexible crystal (figure 3.11(a)). In the course of crystal bending, the *herringbone arrangement* of molecules allow the rotation of DTP_Z towards planarity, enabling compression at the concave region of the bent crystal. Conversely, the reverse rotation facilitates extension at the convex region of the crystal (figure 3.11(b)). The bent crystals exhibited a notable strain tolerance ranging from 2.5–2.8%. The elastic modulus (**Y**) and hardness values ($\mathbf{H_c}$) of DTP_Z were determined to be 5.83±0.62 GPa and 0.26±0.01 GPa, respectively. Unlike bulk DTP_Z crystals, self-assembled microcrystals displayed pseudo-plasticity at the SiO_2 surface during mechanical micromanipulation using an AFM tip.

The fabrication of a 2×2 DC involves a sequential process, commencing with the creation of two bent crystals through a *bending* operation. Subsequently, an *aligning* operation is employed to achieve physical contact or *integration* between the convex

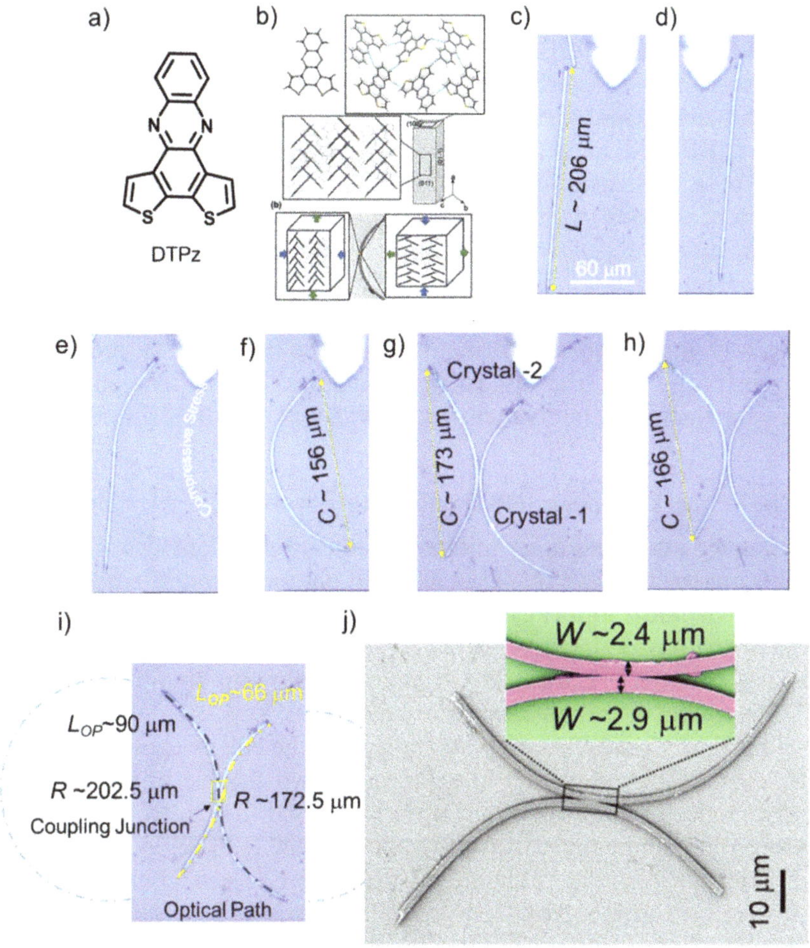

Figure 3.11. (a) Molecular structure of DTPz. (b) Molecular packing and movement in the straight and bent crystal states, (c–f) Confocal optical microscope images showing the micromanipulation of a straight crystal-1 into an arc shape with an AFM cantilever tip. (g) A second arc-shaped crystal-2 was made and moved adjacent to the crystal-1. (h, i) Radius (R) of curvature. Directional coupler showing optical path length difference between two coupled waveguides. The yellow rectangle shows the coupling junction. (j) FESEM image of a directional coupler. The inset shows the widths of the waveguides at the junction. [10] John Wiley & Sons. Copyright (2020).

segments of the crystals. Initially, a straight waveguiding crystal-1 ($L \approx 206$ μm) was chosen (figure 3.11(c)). Utilizing a cantilever tip force, both termini of the crystal were bent into an arc geometry, resulting in a terminal-to-terminal distance of 156 μm (figures 3.11(d)–(f)). Similarly, another bent waveguiding crystal-2, with a terminal-to-terminal distance of 173 μm, was fabricated using similar *bending* operations. The critical phase of integrating these two bent crystals was executed by *moving* crystal-2 near crystal-1 with the cantilever tip, facilitating physical contact or optical junction. A minor *alignment* was made to reduce the tip-to-tip distance of crystal-2 to approximately 166 μm (figure 3.11(h)). The radii of curvature (R) for crystal-1 and -2 in the directional coupler were approximately $R \approx 172.5$ μm and 202.5 μm, respectively (figure 3.11(i)). This directional coupler has four ports or termini with two different optical path lengths (L_{OP}) of 90 and 66 μm, respectively. Furthermore, FESEM images confirmed (i) physical contact between the convex portions of both crystals, and (ii) the structural stability of the circuit even under high vacuum conditions (figure 3.11(j)).

The application of the *moving/aligning/integrating* methods described above holds significant potential for the fabrication of DCs across multiple hierarchical levels, rendering them particularly apt for integration into neuromorphic computing technologies. Furthermore, these methods can be extended to produce intricate add-drop filters (ADFs). The optical functionalities that are inherent in these DCs and filters are elucidated in chapter 4, specifically in sections 4.2 and 4.3.

3.2.6 Transferring and integrating crystal optical waveguides for hybrid circuits

The *transferring* and *integrating* micromechanical operations described herein are very significant for the development of hybrid photonic circuits employing crystals with distinct chemical, optical, or mechanical properties. The crystal *transferring* operation encompasses the *lifting* of a crystal from one substrate and subsequently *releasing* it. The latter is achieved by bringing the tip into proximity to the surface, allowing the crystal to establish contact.

In 2020 [11], my research group demonstrated the effectiveness of the mechano-photonic technique through the integration of two flexible crystals with distinct chemical and electronic properties: DPIN (discussed in chapter 2, figure 2.8) and (*E*)-1-(4-bromo)iminomethyl-2-hydroxyl-naphthalene (BPIN), in the fabrication of a first hybrid and reconfigurable circuit (figure 3.12(a)). The crystal structure of BPIN showed hydrogen bonding between CH=N and OH units, C–H···π, and π···π stacking interactions (≈3.4 Å) (figure 3.12(b–e)). The molecules exhibited a criss-cross packing arrangement along the (001) plane (figure 3.12(f)). It is anticipated that the mechanical flexibility of BPIN crystals stems from the contraction and expansion movements of the crisscross-packed molecules along the inner and outer regions of the bent crystal, respectively.

To realize a hybrid OPIC, the integration of a straight BPIN waveguide with a truncated circular-shaped DPIN waveguide, as discussed in (section 3.2.3), was systematically planned [11]. Firstly, a BPIN crystal waveguide was meticulously *lifted* in the $-z$-direction from substrate-1 utilizing an AFM cantilever tip (figure 3.12(g, i)). Subsequently, it was carefully *released* adjacent to a pre-fabricated truncated circular-shaped DPIN crystal on substrate-2 (figure 3.12(g, ii)). Prior to this, a 164 μm long straight DPIN crystal waveguide was bent into a truncated circular geometry ($D \approx 53$ μm) through a series of micromanipulation steps. The deposited BPIN crystal was then *moved* toward the convex side of the DPIN crystal in two sequential moves ($-y$ and $-x$) and ultimately *integrated*, resulting in the realization of a hybrid OPIC-I (figure 3.12(g, iii–iv)).

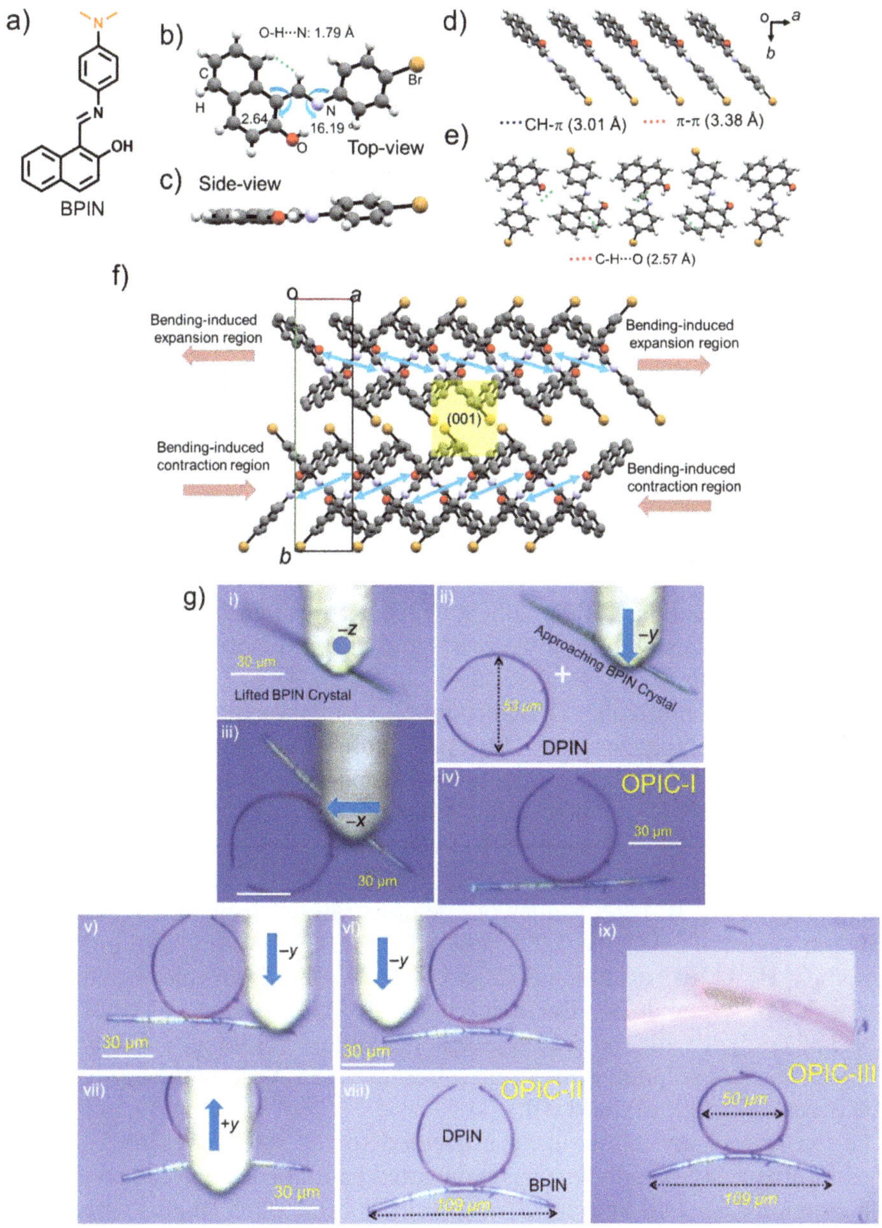

Figure 3.12. (a) Molecular structures of BPIN. (b, c) X-ray crystal structure of BPIN showing top and side views. (d) Formation of BPIN chain facilitated by C–H···π and π···π interactions. (e) C–H···O hydrogen bonding interactions between BPIN along the chain. (f) Molecular packing viewed along the crystallographic *c*-axis displaying the crisscross pattern of molecular chains. (g) Confocal microscope images depicting (i–iv) the mechanical manipulation steps involved in the integration of straight BPIN crystal with truncated circular DPIN crystal to create an OPIC-I. (v–viii) Mechanical bending of BPIN crystal to bow-shape to produce OPIC-II, and (ix) construction of a ring from truncated circular-shaped DPIN-OW crystal creating OPIC-III. [11] John Wiley & Sons. Copyright (2021).

Also, for the geometric reconfiguration of OPIC-I, the initially straight BPIN crystal waveguide underwent mechanical bending, achieved by moving both crystal termini in the $-y$-direction (figure 3.12(g, v,vi)). The subsequent step involved a final pull of the BPIN crystal toward the DPIN crystal, accomplished by moving the cantilever in the $+y$-direction, resulting in the formation of OPIC-II. The terminal-to-terminal distance of the bent BPIN crystal was approximately 109 μm (figure 3.12(g, viii)). Further modification of OPIC-II involved connecting the two termini of the truncated semi-circular DPIN crystal, forming a closed-circular loop with a diameter of approximately 50 μm, resulting in the creation of OPIC-III (figure 3.12(g, ix)). This experiment confirmed the reconfigurability of the OPICs. The specific details of the photonic experiments are thoroughly discussed in chapter 4, section 4.3.3.

3.2.7 Rolling of crystal optical waveguides

The arrangement of facets in acicular crystals, particularly those with square, rectangular, or octagon cross-sections, inherently prevents crystal rolling by conventional means. However, in the case of an acicular crystal with a square or rectangular cross-section, rolling can be achieved through twisting along its long axis. Notably, crystals with multiple twists are naturally occurring and are referred to as *twisted crystals*.

For example, molecule (*E*)-2-bromo-6-(((4-methoxyphenyl)imino)methyl)-4-nitrophenol (BMPIN) grows as a 1D crystal with an inherent twisted morphology (see figure 3.13(a, e) [13]. BMPIN crystals exhibit a broad absorption band spanning approximately 250–575 nm and emit orange FL with a peak centered around 590 nm (bandwidth approximately 510–750 nm). The x-ray crystal structure reveals an intramolecular N-H···O hydrogen bond with a distance of 1.722 Å (figure 3.13(b)). The BMPIN framework demonstrates near-planarity, characterized by torsion angles of 3.58° and 1.20° between the methoxyphenyl and imine groups. Through C–H···O hydrogen bonds (2.389–2.691 Å), C–H···Br interactions (approximately 2.298 Å), and Br···Br interactions (4.465 Å), each BMPIN molecule establishes contact with six nearest neighbors (figure 3.13(d–g)). These molecules form π···π stacked columns with a separation of 3.885 Å along the crystallographic *a*-axis (figures 3.13(c), (d), and (f)).

The twisting of crystals is likely attributed to the rotation of (001) and (010) planes within orientationally uneven aggregated nanofibers along the crystal growth [100] direction [22]. This unevenness results in variable pitch lengths within a single twisted crystal. The twisted geometry of the crystal enables mechanical rolling on the substrate by applying force at its center using an AFM cantilever tip. The confocal optical microscopy and FESEM images of an illustrative twisted crystal, with dimensions of length (L) ≈ 212 μm, width (W) ≈ 2.2 μm, and height (H) ≈ 1.8 μm, revealed a dark contrast in the twisted region compared to other areas (see figures 3.13(h) and (i)).

In the course of the crystal's mechanical locomotion, the dark optical contrast of the twist changes to brightness and vice versa. Intriguingly, the twisted-crystal

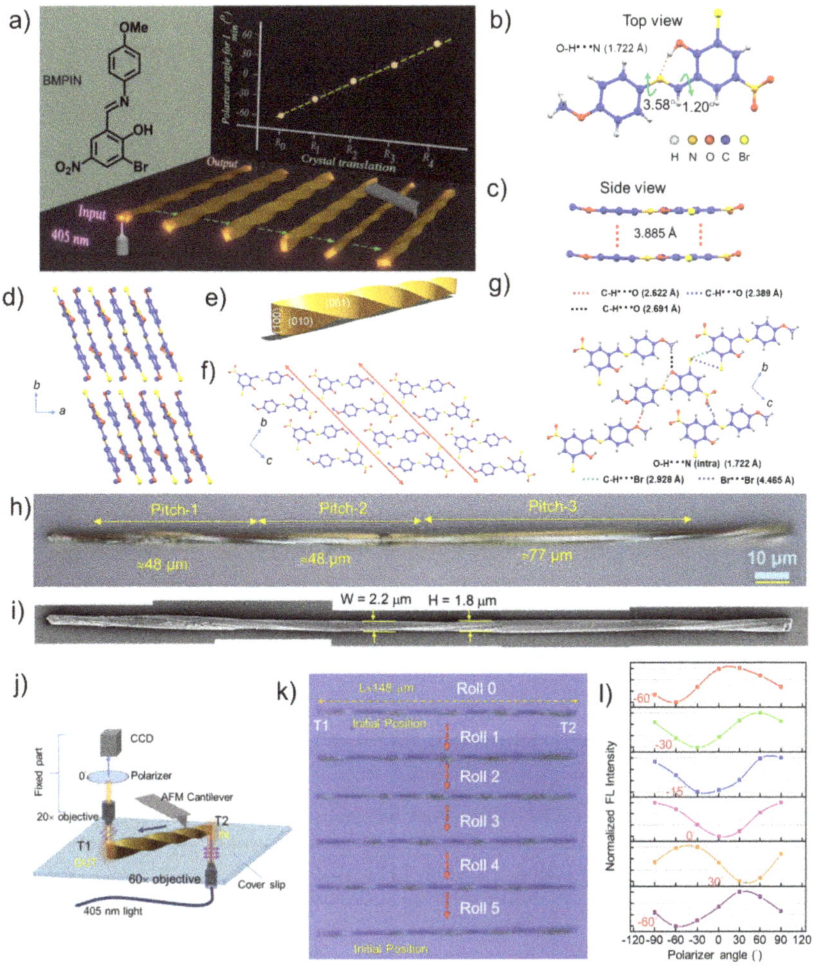

Figure 3.13. (a) Illustration of mechanically-powered rolling translational locomotion of a BMPN twisted-crystal optical waveguide cavity rotating linear polarization of output light. (b) Top view of a BMPIN x-ray crystal structure. (c) Side-view two π···π stacked BMPIN molecules. (d, f, g) π···π stacked chains viewed along the crystallographic *c*-, *a*-, and *b*-axes, respectively. The orange dotted line represents the intramolecular N···O–H hydrogen bond. The red, blue, and black dotted lines represent C–H···O interactions. The green and purple dotted lines represent C–H···Br and Br···Br interactions. (e) Graphical representation of a twisted BMPIN single crystal and its *hkl* planes. (h, i) Optical and FESEM images of a twisted crystal, respectively. (j) Schematic of experimental set-up demonstrating a micromechanically-powered twisted-crystal waveguide acting as a polarization rotor. (k) Optical images of 148 μm long twisted crystal before and after a series of rolling (R). (l) FL intensity recorded at terminal T2 for the input given at terminal T1 by changing polarizer angle (−90° to +90°) for the crystal before and after each roll. [13] John Wiley & Sons. Copyright (2022).

exhibits bending behavior when subjected to mechanical force. In both geometric configurations, these twisted crystals manifest optical waveguiding characteristics with a mere difference in the α' value of 0.06826 dB μm^{-1}. Additionally, the crystals function as F–P cavities, revealing optical modes in the FL spectra due to mirror-like light-reflecting facets. Leveraging the dual role of the twisted crystal as a waveguide cavity, the intended exploration involves the polarization rotation of outcoupled guided light as a function of crystal rotation.

For that, a stationary straight twisted crystal with a length of 148 μm was selected and designated as Roll 0 position (figure 3.13(k)). The crystal was incrementally moved by the AFM cantilever tip, recording the outcoupled light intensity at T2 when excited at T1 in a stepwise manner (figures 3.13(j) and (k)). Throughout each rotation of the twisted crystal (from Roll 1 to Roll 4), the analyzer angle was systematically adjusted from −90° to +90°, capturing FL spectra at 30° intervals (figure 3.13(l)). Notably, the intensity minimum of the FL spectrum progressively shifted at polarization angles of −60°, −30°, −15°, 0°, 15°, and 30° for Roll 0 to Roll 4, respectively. Upon further rotation to Roll 5, the intensity minimum reappeared at −60°, indicating the crystal had returned to its original Roll 0 position (figures 3.11(k) and (l)). These experimental findings substantiate the deliberate modulation of the linear polarization direction of outcoupled guided light through the mechanical rolling of a twisted-crystal waveguide.

Furthermore, during the rolling motion of the twisted-crystal cavity, the wavelength of the F–P modes exhibited significant variation. These noteworthy results proved that the polarization state of the optical modes also could be willingly rotated using a rolling twisted-crystal waveguide cavity.

3.2.8 Standing of twisted-crystal optical waveguides

Scientists working in the domains of materials science, solid-state chemistry, and crystal engineering are actively exploring new mechanically-triggered or controlled dynamic phenomena within soft crystals, going beyond previously documented observations. The discovery of unique crystal dynamics, including *standing*, *leaning*, *stacking*, and *interlocking*, marks a substantial leap forward in comprehending material behavior at the microscale.

Recently, my group observed a range of exceptional and unusual mechanically-triggered dynamic effects in three dimensions (3D) in a novel twisted crystal of 2,4-dibromo-6-(((2-bromo-5-fluorophenyl)imino)methyl)phenol (BFIMP) (figure 3.14(a)) [14]. This section explores how this twisted crystal can stand at the surface at different inclination angles (figure 3.14(b)). Within the crystal lattice, BFIMP molecules displayed intramolecular O–H···N hydrogen bonding (1.863 Å, top view) and π···π stacking interactions (3.683 Å, side view) (figure 3.14(c, i and ii)). This resulted in a nearly planar geometry with minimal torsion angles between its phenyl rings (0.64° and −2.15°). Each molecule was surrounded by eight others with two types of Br···Br (3.589 Å, 3.690 Å) and F···Br (3.175 Å) intermolecular interactions, and two molecules were in a π···π stacked manner (figure 3.14(c)). Crystallographic analysis of BFIMP revealed an antiparallel molecular orientation within the *b* and *c* planes. When observed along the crystallographic *c* and *b* axes, the molecules adopted a slip-stacked arrangement involving π···π interactions (figure 3.14(c, v, and vi)).

The twisted microcrystals can stand vertically on the substrate by leveraging differences in tip-to-substrate height profile and surface attachment (figure 3.14(b)).

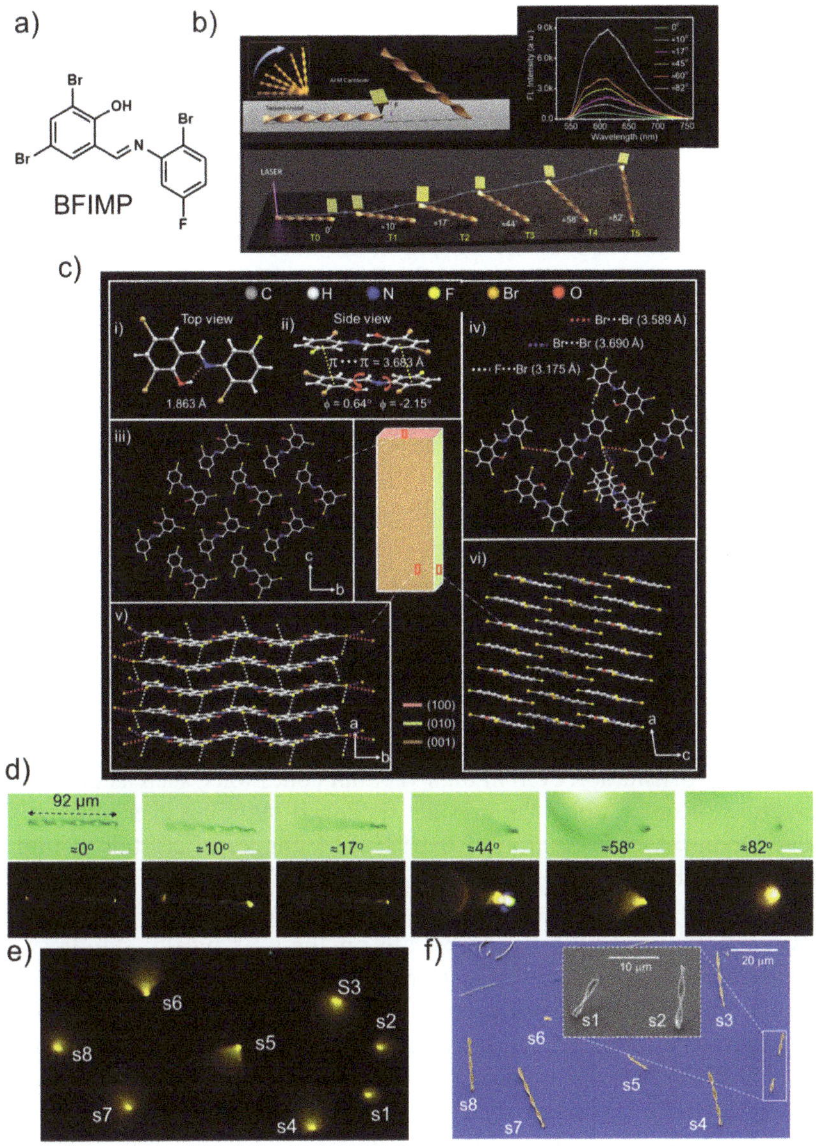

Figure 3.14. (a) Molecular structure of BFIMP. (b) Illustration of standing/leaning using the AFM cantilever tip. (c) X-ray crystal structure of BFIMP. (d) The sequential confocal and FL images show the slanted twisted microcrystal at various angles. The FL spectra are shown in the right-side inset of (b). (e) FL images of eight standing twisted crystals with various tilt angles. (f) The corresponding color-coded FESEM image. The inset shows the zoomed-in image of standing crystals s1 and s2. Reproduced from [14]. CC BY 4.0.

Specifically, the (010) and (001) facets of the linear twisted crystal align parallel to the substrate, while the (100) facet with a rectangular cross-section is positioned perpendicular to the substrate plane. A twisted waveguiding crystal, approximately 92 μm in length, was chosen for controlling the crystal's slanting angle and exploring its photonic characteristics. To attain this, the AFM tip pressed down on the right terminal ($-z$-direction), causing the left end of the crystal to lift at a 10° angle in the $+z$-direction. The crystal's leaning angle was adjusted by moving the AFM cantilever tip beneath the microcrystal and gradually shifting it in the $+z$-direction (figure 3.14).

The microcrystal, inclined at different angles of approximately $\approx 10°$, $\approx 17°$, $\approx 44°$, $\approx 58°$, $\approx 82°$, was exposed to 405 nm laser light by exciting the crystal at the bottom terminal. With an increase in the tilt angle, the recorded FL intensity at the top terminal also increased (as illustrated in figure 3.14(b), inset FL spectra). The respective FL images of the crystal depict the defocused top terminal of the leaning crystal with various tilt angles (shown in figure 3.14(d)). Through these innovative mechanical micromanipulation techniques, we achieved the ability to focus guided microlight at various angles within the 3D space. This result offers precise control over the orientation of guided light, opening up possibilities for tailored light manipulation and steering in diverse applications.

Furthermore, a long twisted microcrystal was selected and cut into smaller twisted crystals of varying sizes (s1 to s8; where 's' denotes standing). Subsequently, these microcrystals were methodically arranged in a vertical sequence at different angles, as per the abovementioned mechanical procedures (depicted in figure 3.14(e)). The FL image, captured under UV light exposure, reveals a vivid illumination of the standing crystals at different heights. The corresponding FESEM image, taken at a 40° tilt-view, distinctly illustrates the vertically positioned twisted microcrystals on the substrate (as shown in figure 3.14(f)). Intriguingly, these standing twisted crystals maintained their orientation even in high vacuum conditions, such as those within an SEM chamber. Remarkably, the crystals remained standing even after a span of 2 months. Upon closer inspection of the FESEM images of crystals s1 and s2, the contact of the rectangular cross-section (facet 100) of the crystal's bottom terminal with the substrate is evident as indicated in the inset of figure 3.14(f).

To evaluate the impact of the substrate on these atypically standing crystals, experiments were conducted by transferring twisted microcrystals (assisted by an AFM tip) from the original borosilicate glass substrate onto different substrates. These alternative substrates included indium tin oxide-coated polyethylene terephthalate (ITO-PET), silicon, gold-coated glass, and aluminum foil. Subsequently, using the previously described mechanical manipulation procedures, the microcrystals were vertically arranged on these diverse substrates, thus confirming the adaptability of this unique crystal behavior. These experiments shed light on the influence of different substrates on the standing behavior of the crystals. The ability of the microcrystals to maintain their vertical orientation on various substrates underscores the robustness and versatility of this distinctive crystal behavior.

3.2.9 3D Stacking of twisted-crystal optical waveguides

Organic crystal waveguides can be stacked in 3D using the *stacking* operation to create compact 3D organic crystal circuits. However, this challenging operation is assisted by other previously discussed mechanical manipulation operations viz, *moving*, *cutting*, *lifting*, *releasing*, *aligning*, *integrating*, etc.

My group demonstrated this operation to stack BFIMP crystals in 3D [14], which necessitated precise handling of the microcrystals to prevent damage during the lifting, dropping, and integration stages. This layer-by-layer stacking process involves incorporating BFIMP crystals with different aspect ratios onto a borosilicate glass substrate. In the first layer of the stacking process, meticulous attention was given to the selection and arrangement of three crystals, denoted as C1, C2, and C3. These crystals, each with approximate lengths of 94, 89, and 106 μm underwent a precise alignment procedure. The alignment was conducted with utmost care, ensuring that the crystals were arranged in a parallel orientation to each other along the *y*-direction. This precision was achieved through the delicate manipulation facilitated by an AFM cantilever tip (figure 3.15(a, i)). As we proceeded to add more layers, the precise placement of microcrystals on top of the existing layer demanded increasingly meticulous mechanical control. The complexity of aligning the crystals at specific angles intensified with each layer, presenting a significant challenge in the stacking process.

In the construction of the second layer, successful stacking involved two twisted microcrystals, namely C4 and C5. These microcrystals, each measuring approximately 77 μm and 75 μm in length, were precisely aligned and vertically stacked along the *x*-direction on the foundation provided by the first layer (figure 3.15(a, ii)). For the third layer, a pair of crystals, C6 and C7, each with lengths of approximately 72 and 67 μm, were precisely positioned. These crystals were placed perpendicular to the second layer while maintaining parallel alignment with the first layer along the horizontal (*y*-direction) axis (depicted in figure 3.15(a, iii)). Notably, the stability and integrity of the stacked crystals persisted even after undergoing a gold coating process, facilitating the examination of their microstructural details using FESEM (figure 3.15(b) and (c)).

3.2.10 Interlocking of twisted-crystal optical waveguides

The intricate interlocking of two different twisted crystals termini at the microscale requires precise handling and the careful selection of suitable crystals [14]. For this experiment, two crystals, denoted as crystal 1 and crystal 2, each with approximate lengths of 80 and 84 μm, were chosen (as illustrated in figure 3.15(d)). These crystals were strategically positioned head-to-head in an axial manner, aligning the dark contrast region of crystal 2 near the bright contrast region of crystal 1.

Subsequently, a meticulous manipulation process ensued. The orientation of the crystals was adjusted by gently maneuvering them first in the $-y$-direction and then in the $-x$-direction using an AFM cantilever tip. This maneuvering resulted in the entwining of the two crystals into a single unit. A close examination of the crystal under FESEM imaging evidently reveals the axial interlocking of the termini of crystals 1 and 2, as depicted in figure 3.15(g). Moreover, when exposed to UV light, the crystal manifests a vibrant yellow FL at its termini, providing clear evidence of its capability to function as a waveguide under interlocked conditions (figure 3.15(d) and (e)).

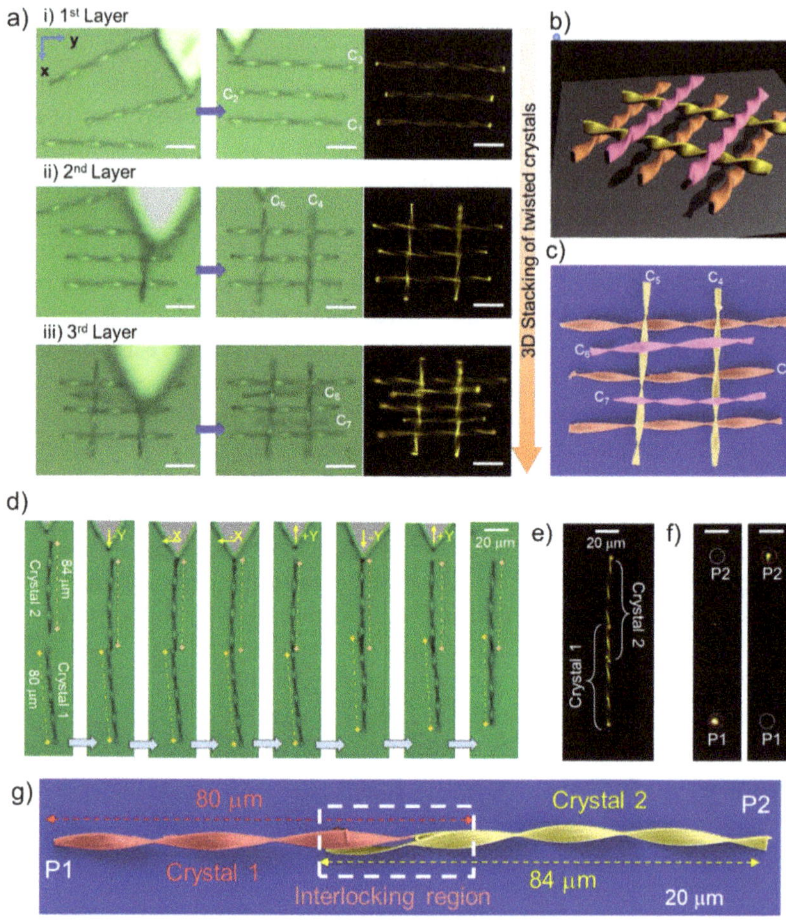

Figure 3.15. (a, i–iii) The sequential confocal images of twisted microcrystals show the stacking (crystals C1 to C7) operations on a glass substrate. The corresponding FL images after each layer addition. (b) Graphic and (c) color-coded FESEM image of 3-layered stacked microcrystals. (d) Confocal optical images displaying micromanipulation of two twisted microcrystals (crystals 1 and 2) to create an axially interlocked crystal. (e) FL image of interlocked crystal on exposure to UV light. (f) FL images of interlocked microcrystals excited at terminal P1 and P2 and recorded FL spectra at another end (dotted circle). (g) False color-coded FESEM image of axially connected twisted microcrystal. The white dashed box shows the interlocking region. Reproduced from [14]. CC BY 4.0.

3.3 Pathways to apply crystal micromanipulation towards OPICs

3.3.1 Crystal library and mechanophotonics

It is of paramount importance to create a comprehensive '*crystal library*' that systematically catalogs a diverse array of organic crystalline materials [1]. This library should encompass the essential properties of crystals, including but not limited to refractive index, optical absorption/emission maxima, optical absorbance/emission bandwidths (UV, Vis, NIR, Mid-IR), chirality, Young's modulus, hardness, and NLO attributes like TPL, SHG, and third harmonic generation (THG), terahertz generation (THz), etc.

Furthermore, the library should not only compile individual material characteristics but also establish connections between the optical properties of different crystals, considering factors such as reabsorption and energy transfer properties. This interlinked information will provide valuable insights into the collective behavior of organic crystals, enabling a deeper understanding of their performance in various hybrid device applications.

The inclusion of crystals with distinctive features, such as thermochromism, photochromism, electrostriction, and self-healing attributes, is essential. These special properties contribute significantly to the versatility and applicability of these materials. By integrating such unique characteristics into the library, we can create a comprehensive resource that not only serves as a reference for researchers but also promotes the discovery and development of innovative applications in fields ranging from photonics to advanced materials.

Efforts to curate and expand this crystal library will undoubtedly fuel advancements in scientific research and technological applications, fostering collaboration and accelerating progress in the dynamic field of organic crystalline materials.

Organic crystal based circuits can be classified as either monolithic or hybrid, depending on specific application requirements. Monolithic OPICs are created by utilizing crystals (not polymorphs) derived from identical molecules on a substrate. In the case of monolithic OPICs, the use of crystals from the same molecules offers advantages in terms of uniformity and consistent material properties throughout the circuit. The fabrication of organic crystal based circuits is a deliberate process that requires a systematic approach that begins with a clear understanding of the required optical functions. The circuit design is a critical step, laying the groundwork for the successful integration of organic crystals into a functional and efficient photonic circuit (figure 3.16). The design process involves careful consideration of factors such as crystal properties, configurations of photonic modules, and the desired interaction of light within the circuit and output. The success of a crystal-based photonic circuit relies on the precision of the circuit design, ensuring that the optical functions are not only met but optimized to achieve desired optical outcomes.

For the construction of a monolithic circuit, microcrystals need to be grown on a silica substrate ($n = 1.5$) to create various optical elements such as straight and bent waveguides, resonators, couplers, etc. The process begins by selecting suitable single microcrystals, with the option to modify their dimensions using crystal-cutting techniques. Although bundles of crystals are not frequently formed, crystal-slicing techniques can be employed for segregation if necessary. An essential step in OPIC fabrication involves choosing the appropriate construction site for different optical modules on the substrate. The selected waveguiding crystal is then lifted from one substrate and released or dropped on another substrate. Since the circuit requires several crystals, multiple crystals can be transported to the site by repeating the process. The released crystal(s) can be manipulated into curved waveguides or ring resonators through crystal bending operations. Finally, these straight, bent, and ring-shaped crystals can be precisely aligned and integrated to produce a monolithic OPIC.

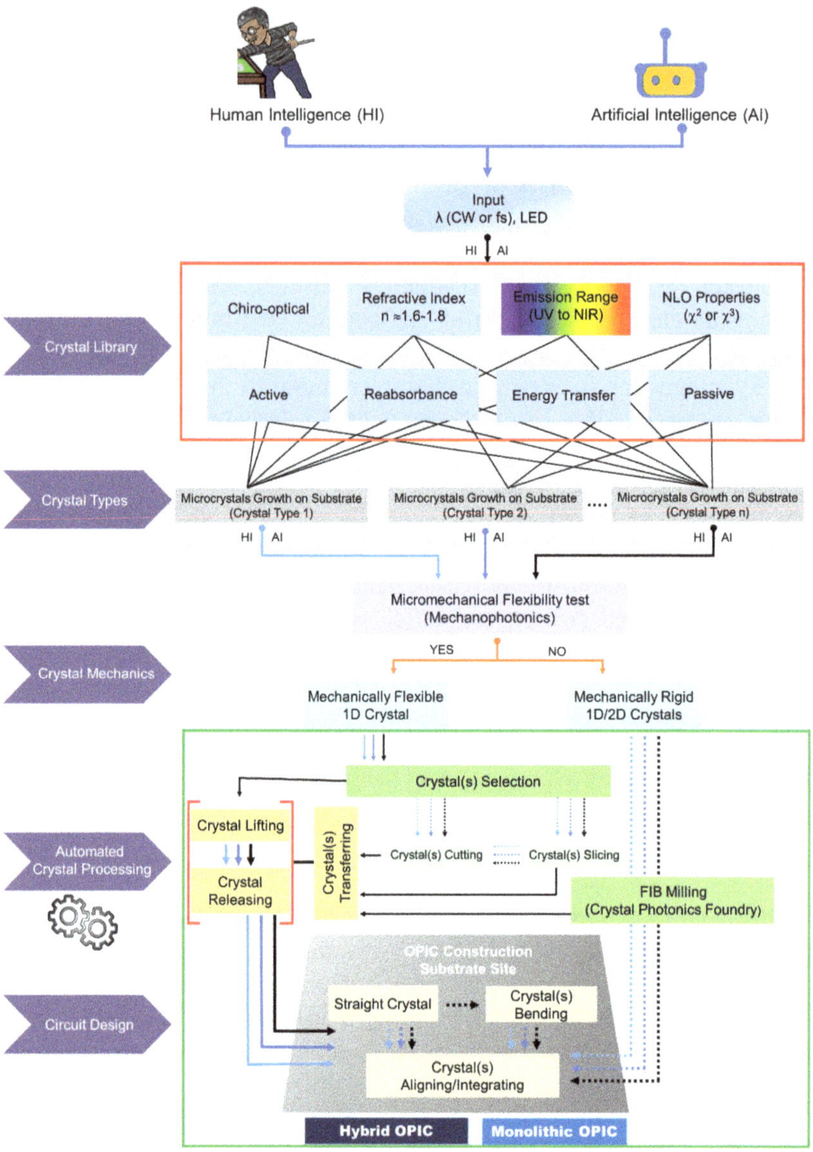

Figure 3.16. Schematic representation of pathways for applying crystal micromanipulation techniques for fabricating monolithic and hybrid crystal OPICs. HI stands for human intelligence and AI stands for artificial intelligence.

The crystal lifting/releasing techniques presented here provide a unique opportunity for fabricating hybrid OPICs. These hybrid circuits can be assembled by selecting microcrystals that differ chemically, electronically, and mechanically. The required protocols (cutting, bending, aligning, integrating) must be followed through the pathways outlined in figure 3.16.

3.3.2 Focused ion beam milling of organic crystal into photonic components

Ensuring highly reproducible modification of naturally grown organic optical crystals with precise geometry and dimension is crucial for the realization of industrially relevant all-organic micro optoelectronic and nanophotonic components, as well as the development of OPICs. My group demonstrated that focused ion beam (FIB) milling is a precise and versatile technique suitable for the fabrication of organic single-crystal photonic modules [19]. This technique involves using focused gallium ions, to mill or etch materials with nanoscale precision. The exceptional precision and spatial resolution of FIB renders it a valuable tool for creating photonic modules using organic crystals. These modules may include waveguides, bent waveguides, disc resonators, ring resonators, and even complete miniature circuits. Moreover, after fabrication, the integration of these photonic modules can be achieved through mechanophotonic techniques [20].

A detailed discussion on the FIB-assisted fabrication of various organic single-crystal photonic modules and their integration is provided in chapter 5 of this book.

3.3.3 Automation of mechanophotonics and artificial intelligence

The automation of mechanophotonics can be done by involving relevant technologies, where mechanical forces are used to control and manipulate FIB-milled crystals remotely (figure 3.16). This can include the use of robotics, sensors, and control systems to perform crystal manipulation actions. Automation in mechanophotonics can enhance the precision, efficiency, and reproducibility of experiments or applications, making it a promising area for advancements in both research and industrial applications.

Artificial intelligence (AI) systems can also significantly diminish the trial-and-error duration typically expended by scientists in laboratory settings. Depending upon the nature of the task, both human intelligence (HI) and AI can be employed.

AI can play a significant role in the design and optimization of organic crystal photonic circuits [21]. Here are several ways in which AI is applied to commercial photonic circuit design:

1. *Circuit topology optimization*: AI algorithms can be used to optimize the topology of photonic circuits. These algorithms explore a vast design space to find configurations that meet specific performance criteria, such as minimizing optical loss or maximizing bandwidth.
2. *Automated design*: AI can automate the design process by learning from successful designs and applying that knowledge to generate new, improved designs. This can significantly speed up the design cycle and enable the exploration of more complex and innovative circuit architectures.
3. *Material selection*: AI can aid in the selection of materials for photonic components from the *crystal library*. By analyzing material properties and their impact on circuit performance, AI algorithms can suggest optimal combinations of crystals for achieving desired outcomes.
4. *Performance prediction*: Machine learning models can be trained to predict the performance of photonic circuits based on their designs. This allows

designers to quickly assess the potential of a given configuration without the need for extensive and expensive simulation or prototyping.

5. *Error detection and correction*: AI can be employed to identify and correct errors in the circuit designed by HI. This is particularly useful in complex photonic circuits where manual error checking may be time-consuming and prone to oversights.
6. *Efficient parameter tuning*: Tuning the parameters of a photonic circuit for optimal performance can be a challenging task. AI algorithms, such as genetic algorithms or reinforcement learning, can be employed to efficiently search the parameter space and find configurations that meet specified objectives.
7. *Inverse design problems*: AI can tackle inverse design problems, where the goal is to find a design that exhibits a specific behavior. For instance, given a desired spectral response, AI algorithms can determine the photonic circuit configuration that achieves that response.
8. *Scaling and integration*: AI can assist in the scaling and integration of photonic circuits into larger systems. This involves optimizing how individual photonic components interact and ensuring compatibility with other electronic or photonic elements.

The integration of AI in photonic circuit design has the potential to revolutionize the field by enabling faster and more efficient design processes, facilitating the exploration of novel circuit architectures, and improving overall performance. As AI technologies continue to advance, their impact on the field of photonics is likely to grow.

References

[1] Chandrasekar R 2022 Mechanophotonics—a guide to integrating microcrystals toward monolithic and hybrid all-organic photonic circuits *Chem. Commun.* **58** 3415–28

[2] Ghosh S and Reddy C M 2012 Elastic and bendable caffeine cocrystals: implications for the design of flexible organic materials *Angew. Chem. Int. Ed.* **51** 10319

[3] Huang R, Wang C, Wang Y and Zhang H 2018 Elastic self-doping organic single crystals exhibiting flexible optical waveguide and amplified spontaneous emission *Adv. Mater.* **30** 1800814

[4] Chandrasekar R 2021 Mechanophotonics—mechanical micromanipulation of single-crystals towards organic photonic integrated circuits (OPICs) *Small* **17** 2100277

[5] Chandrasekar R 2023 Advanced all-organic microphotonic components and integrated circuits *Adv. Opt. Mater.* **11** 2301124

[6] Basak S and Chandrasekar R 2014 Passive optical wave guiding organic rectangular tubes: tube cutting, controlling light propagation distance and multiple optical out-puts *J. Mater. Chem.* **2** 1404–8

[7] Annadhasan M, Karothu D P, Chinnasamy R, Catalano L, Ahmed E, Ghosh S, Naumov P and Chandrasekar R 2020 Micromanipulation of mechanically compliant organic single-crystal optical microwaveguides *Angew. Chem. Int. Ed.* **59** 13821–30

[8] Venkataramudu U, Venkatakrishnarao D, Chandrasekhar N, Mohiddon M A and Chandrasekar R 2016 Single-particle to single-particle transformation of an active type organic microtubular homo-structure photonic resonator into a passive type hetero-structure resonator *Phys. Chem. Chem. Phys.* **18** 15528–33
[9] Venkatakrishnarao D and Chandrasekar R 2016 Engineering the self-assembly of DCM dye into whispering-gallery-mode micro-hemispheres and Fabry–Pèrot type micro-rods for Vis–NIR (600–875 nm) range micro-optical cavities *Adv. Opt. Mater.* **4** 112–9
[10] Annadhasan M, Agrawal A, Bhunia S, Pradeep V V, Zade S S, Reddy C M and Chandrasekar R 2020 Mechanophotonics: flexible single-crystal organic waveguides and circuits *Angew. Chem. Int. Ed.* **59** 13852–8
[11] Ravi J, Annadhasan M, Kumar A V and Chandrasekar R 2021 Mechanically reconfigurable organic photonic integrated circuits made from two electronically different flexible microcrystals *Adv. Funct. Mater.* **31** 2100642
[12] Ravi J, Kumar A V, Karothu D P, Annadhasan M, Naumov P and Chandrasekar R 2021 Geometrically-reconfigurable, two dimensional, all-organic photonic integrated circuits made from two mechanically and optically dissimilar crystals *Adv. Funct. Mater.* **31** 2105415
[13] Rohullah M, Pradeep V V, Ravi J, Kumar A V and Chandrasekar R 2022 Micromechanically-powered rolling locomotion of twisted-crystal optical-waveguide-cavity as a mobile light polarization rotor *Angew. Chem. Int. Ed.* **61** e202202114
[14] Rohullah M, Pradeep V V, Singh S and Chandrasekar R 2024 Mechanically controlled multifaceted dynamic transformations in twisted organic crystal waveguides *Nat Commun.* **15** 4040
[15] Annadhasan M, Kumar V, Venkatakrishnarao A, Mamonov D and Chandrasekar E A 2020 Mechanophotonics: precise-selection, assembly and disassembly of polymer optical microcavities via mechanical manipulation for spectral engineering *Nanoscale Adv.* **2** 5584–90
[16] Xiao J *et al* 2011 Preparation, characterization, physical properties, and photoconducting behaviour of anthracene derivative nanowires *Nanoscale* **3** 4720–3
[17] Hayashi S, Ishiwari F, Fukushima T, Mikage S, Imamura Y, Toshiro M and Katouda M 2020 Anisotropic Poisson effect and deformation-induced fluorescence change of elastic 9,10-Dibromoanthracene single crystals *Angew. Chem. Int. Ed.* **59** 16195–201
[18] Kumar A V, Mamonov E, Murzina T and Chandrasekar R 2023 Race-track type resonator integrated active add-drop filter from flexible organic crystals: experiments and numerical calculations *Adv. Opt. Mater.* **11** 2201507
[19] Pradeep V V and Chandrasekar R 2022 Micromanufacturing of geometrically- and dimensionally-precise molecular single-crystal photonic micro-resonators via focused ion beam milling arXiv:2203.14218v1 [physics.optics] *Adv. Opt. Mater.* **10** 2201150
[20] Pradeep V V, Kumar A V and Chandrasekar R A 2023 Tandem approach to fabricating a hybrid, organic-add-drop filter using single-crystal disk-resonators and pseudo-plastic crystal waveguides *Laser Photonics Rev.* **17** 2300552
[21] Piccinotti D, MacDonald K F, Gregory S A, Youngs I and Zheludev N I 2020 Artificial intelligence for photonics and photonic materials *Rep. Prog. Phys.* **84** 012401
[22] Shtukenberg A G *et al* 2020 Crystals of Benzamide, the first polymorphous molecular compound, are helicoidal *Angew. Chem. Int. Ed.* **59** 14593–601

Chapter 4

Organic single-crystal photonic components and integrated circuits

4.1 Curved (micro) organic waveguides

4.1.1 Bent waveguides

A bent waveguide is an important circuit element useful in directing the light path at different angles. An example of a bent waveguide fabrication technique at different angles is discussed in section 3.2.3. Mechanophotonic micromanipulation of a straight pseudo-plastic NBA-2 crystal waveguide ($L \approx 210$ μm) with a rectangular square cross-section provide a bent crystal geometry (figure 4.1) [1]. The same crystal, in both geometries, display a bright yellow FL when electronically excited with a 488 nm laser confirming their active-type optical waveguiding propensity (figures 4.1(a) and (b)). The excitation position-dependent optical waveguiding experiments performed along the crystal's long axis at different positions (210, 183, 137, 108, 49, 22 μm) revealed an increase of the FL when the distance between the excitation and spectral recording position decreases (figures 4.1(c) and (d)). The waveguides showed α' values of 0.00937 and 0.00943 dB μm^{-1} before and after crystal bending, respectively, indicating negligible bending-induced optical loss due to curved crystal geometry (figures 4.1(e) and (h)). The crystal waveguides also displayed F–P optical modes in the FL spectra which arise due to the crystal cavity's light-reflecting facets (figures 4.1(f) and (g)).

4.1.2 Bending strain-dependent optical loss

Investigations of a flexible and twisted BFIMP waveguide crystal (figure 4.1a; $L \approx 136$ μm; pitch = 23 μm) in both straight and curved geometries under 405 nm light revealed guided orange FL towards the crystal tips, confirming its optical waveguiding capabilities (figures 4.1(i) and (j)) [2]. The self-absorption of FL, influenced by the overlap of absorption and emission spectra, resulted in a gradual cutoff in the 510–540 nm region. The optical loss of the waveguide was estimated

doi:10.1088/978-0-7503-5336-6ch4

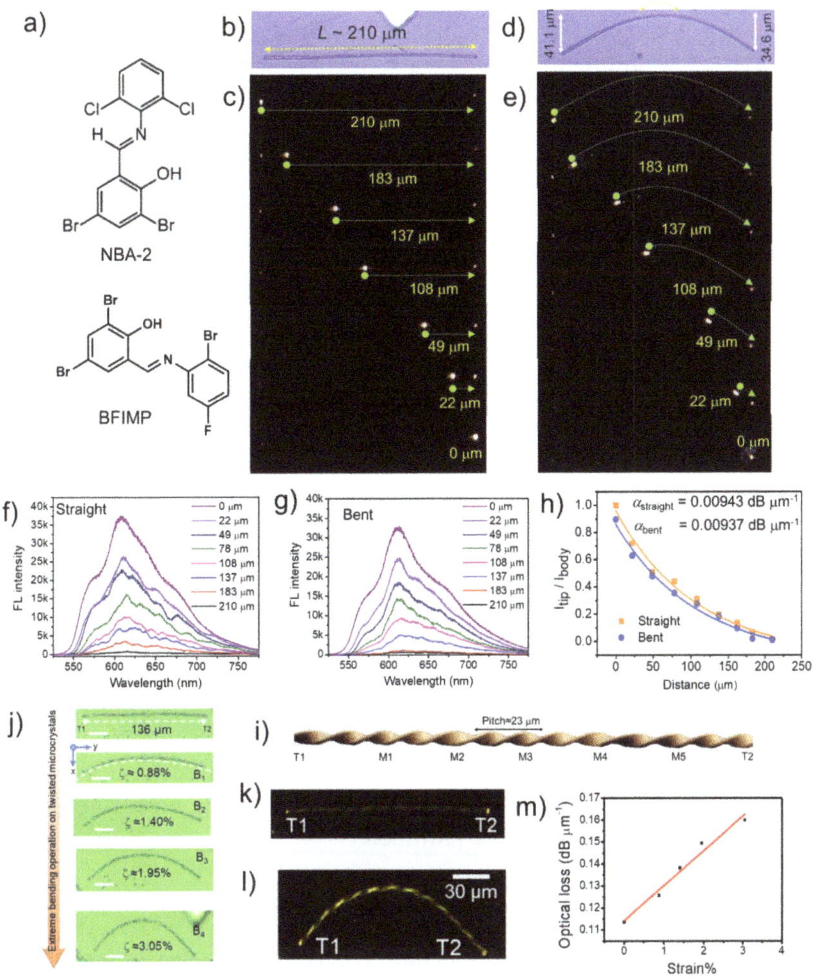

Figure 4.1. (a) Molecular structure of NBA-2 and BFIMP. (b, d) Confocal optical images of straight and bent crystals of NBA-2, respectively. (c, e) FL images of straight and bent crystals of NBA-2 transducing light when excited at different positions, respectively. The green dotted arrows indicate the direction of light propagation. (f, g) FL spectra of straight and bent crystals of NBA-2 collected at the end of the crystals excited at different positions. (h) Plots of the intensity ratio at the tip and the body of the crystal (I_{tip}/I_{body}) versus the distance of the light propagation path used to estimate the optical loss coefficient (α') for NBA-2 and BFIMP, respectively. (a–h) [1] John Wiley & Sons. Copyright (2020). (i) Graphical representation of the twisted crystal. (j) Optical images of a straight twisted microcrystal of length ≈ 136 μm with its subsequent bent geometries (B1–B4) of BFIMP and respective mechanical strain (ζ). (k, l) FL images of microcrystal of BFIMP in straight and bent geometries, respectively. (m) The plot of optical loss versus % strain caused due to bending of the crystal waveguide. (i–m) Reproduced from [2]. CC BY 4.0.

from the plot against $\frac{I_{tip}}{I_{body}}$ and the light propagation length D using the following equation,

$$\frac{I_{tip}}{I_{body}} = e^{-\alpha' D} \tag{4.1}$$

where α' is the optical loss coefficient, usually given in μm^{-1}, I_{body} is the FL intensity at each excitation position, and I_{tip} is the FL intensity at T2. From the fit value of α', optical loss in dB μm^{-1} (dB loss = 4.34 α') was estimated to be 0.11361 for the straight twisted microcrystal.

To evaluate bent crystal's mechanical flexibility at different strains and the impact on light-guiding efficiency, my group applied mechanical force to the straight twisted crystal, using an AFM cantilever tip to create a curved geometry (figures 4.1(j)–(l)). The retention of this curved shape after removing the force highlighted the pseudo-plastic properties of these microcrystals (figure 4.1(j)). To examine the influence of the radius of curvature on optical loss, we performed stepwise mechanical bending followed by estimation of strain and optical loss (figure 4.1(j)). After each bend (B1–B4), the respective strain was calculated using the following equation:

$$\zeta = \left(\frac{\text{thickness}}{\text{diameter}}\right) \times 100 \tag{4.2}$$

For the crystal with the ζ of 0.88%, 1.40%, 1.95%, and 3.05%, optical waveguiding experiments were performed. A linear correlation between mechanical strain and the percentage of optical loss is evident in the plotted data (figure 4.1(m)). We observed lower optical losses when exciting the crystal at its widest faces as opposed to the narrow faces within the same crystal. This observation is attributed to the scattering of FL away from the microscope objective, especially from the narrow faces. Experiments involving the AFM cantilever-assisted cutting of crystals with different aspect ratios and pitches revealed varying α' values, indicating sensitivity to changes in the crystal's geometrical parameters.

4.1.3 Effect of waveguide-cavity bending on optical modes

Investigation of the impact of crystal bending on optical spectra is crucial to control the signal modulation in the interferometers. To carry out the study, the BPyIN molecular crystals were chosen [3]. In the solid state, (*E*)-1-(((5-bromopyridin-2-yl) imino)methyl)naphthalen-2-ol (BPyIN) showed a broad optical absorption band ($\approx$300–550 nm). Its green FL appeared in the range of $\approx$500–650 nm with a lifetime of 0.89 ns (figure 4.2(a)). Further, BPyIN exhibited a high quantum yield of $\approx$16.7%. The x-ray crystal structure showed the interaction of pyridine's nitrogen with the vinylic hydrogen (C–N···H: 2.425 Å); thus, providing planarity and π-conjugation to the molecule (figure 4.2(b)). Further, weak interactions, such as hydrogen bonding (C–O···H: 2.576 Å; N–H···O: 1.914 Å; C–H···N: 2.595 Å), C–H···Br (3.006 Å) and π···π (4.352 Å) stacking stabilize the molecular packing in the solid state (figures 4.2(c) and (d)). Upon applying mechanical stress perpendicular to the (101) plane, the crisscross molecular chain (*c*-axis) possibly provides mechanical compliance letting back-and-forth movements of atoms from their equilibrium positions. The TEM and SAED studies conducted along the needle-like crystals [010] zone axis displayed diffraction spots, confirming the microcrystal's crystallinity (figures 4.2(i) and (j)).

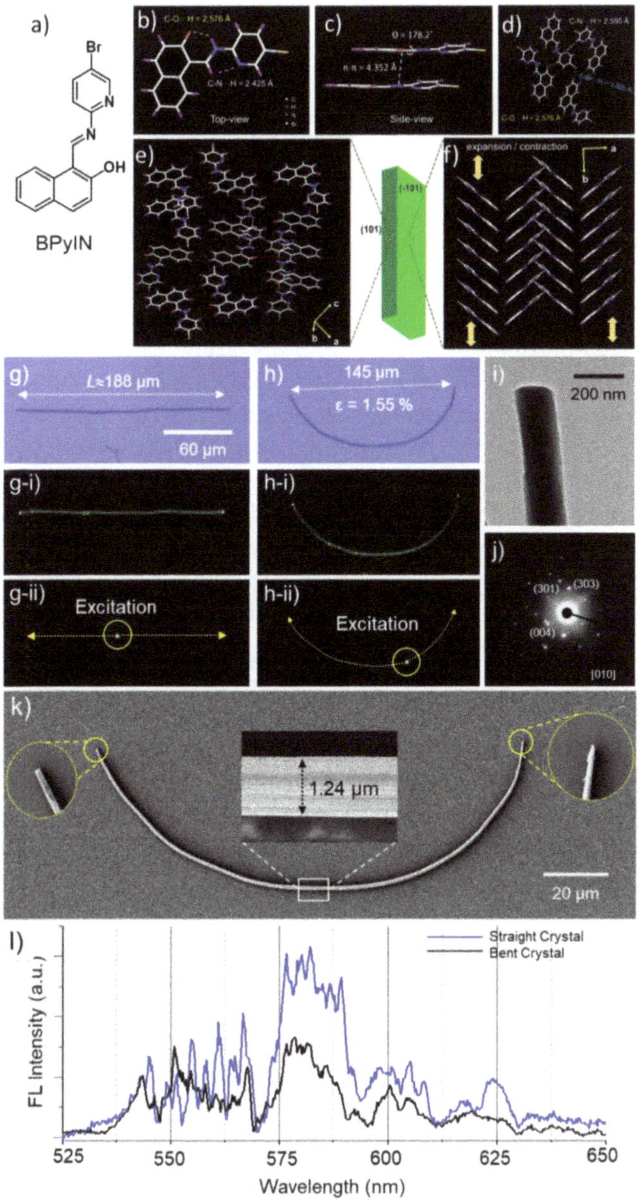

Figure 4.2. (a) Molecular structure of BPyIN. (b) Crystal structure of BPyIN (top view). (c) Dihedral angle between naphthyl and pyridyl moieties (side view). (d) Intermolecular interactions stabilizing the BPyIN crystal structure. Molecular packing along the (e) nonbending (101) and (f) bending (−101) planes, respectively (double-headed arrows indicate the plausible relative directional movement of atoms supporting mechanical flexibility). (g, h) Confocal optical, and (g-i, -ii, h-i, -ii) FL images of straight and bent microcrystals of BPyIN, respectively. (i, j) TEM image and electron diffraction patterns obtained along the [010] zonal axis of BPyIN crystals, respectively. (k) FESEM image of bent BPyIN microcrystal. (l) Optical modes in FL background-subtracted spectra obtained for excitation at the left tip and spectral collection at the right tip of straight and bent crystals. [3] John Wiley & Sons. Copyright (2022).

To investigate the optical modes before and after the bending of the same crystal, a straight microcrystal of $L = 188$ μm was selected and its FL image is shown in (figures 4.2(g)–(i)). Optical excitation of the crystal showed its waveguiding tendency with an α' of 0.1082 dB μm^{-1}. The FL spectrum exhibited optical modes due to the F–P cavity effect (figure 4.2(l)). The same straight crystal waveguide was deliberately bent into a curved shape with a ζ of 1.55% using an AFM cantilever tip by applying force on either terminus of the crystal. The bent crystal retained its curved geometry due to pseudo-plasticity (figure 4.2(h)) even under FESEM measurement conditions. The FESEM topography showed the smooth surface morphology and rectangular cross-sectional geometry of the crystal waveguide cavity (figure 4.2(k)). The crystal revealed its waveguiding nature even in the strained state with an α' of 0.1288 dB μm^{-1}. Strikingly, the modes and their intensities obtained for the bent waveguide cavity were different from the unstrained waveguide (figure 4.2(l)). This result revealed the perceptible effect of the cavity's bent geometry on the optical modes. The modulation of the optical mode confirms the change in the wave interferences in the bent cavity.

Additionally, the bending-induced molecular movements in the convex and concave regions of the rectangular cavity might alter the refractive index of the cavity slightly in these regions. The following equation provides the relation between mode wavelength shifts with shape and refractive index changes.

$$\frac{d\lambda}{\lambda} = \frac{da}{a} + \frac{dn}{n} \tag{4.3}$$

Consequently, the change in the crystal F–P cavity's size (a) and the refractive index (n), which leads to the resonant wavelength shifts, could also be taken into account. Yamamoto *et al* reported the mechanical bending-induced blue shift of F–P modes in a cyano-substituted oligo(phenylene-vinylene) flexible crystal resonator due to the Poisson effect [4].

4.1.4 Spiral waveguides

In photonic circuits, an effective way to densely pack long waveguides in a small area is by creating their 2D spiral versions called spiral waveguides (SWs) [5]. The exclusive geometry of SWs allows routing of the optical signal clockwise or anticlockwise providing direction-specificity of the guided optical signal in PICs. These SWs are important components in photonic devices to be used as delay lines [6, 7], spectrometer components [8, 9], sensors [10], and PIC components [11].

Two-dimensional (2D) spiral shapes can be mathematically described using polar coordinates as $r = r(\theta)$, where the radius r is the monotonous continuous function of angle, θ. Amongst known 2D spiral configurations, Archimedean, Cornu, Fermat, Fibonacci, hyperbolic, logarithmic, and polygons with three or more sides are well-known [12]. The realization of organic crystal based SWs is challenging as it requires an extremely flexible crystal displaying pseudo-plasticity and unchanged optical properties in highly strained geometries.

My research group reported an organic charge-transfer crystal, namely (Z)-3-(4-(9H-carbazol-9-yl)phenyl)-2-(3,5-bis(trifluoromethyl)phenyl)acrylonitrile (Cz-2CF_3)

shown in figure 4.3(a). Cz-2CF_3 crystal is extremely flexible, blue-emitting ($\lambda_{max} \approx$ 482 nm; FL bandwidth $\approx$ 410–650 nm; reabsorption range $\approx$ 410–490 nm; Φ_{FL} = 36.7%; FL lifetime of 2.6 ns), and withstands mechanical fabrication condition

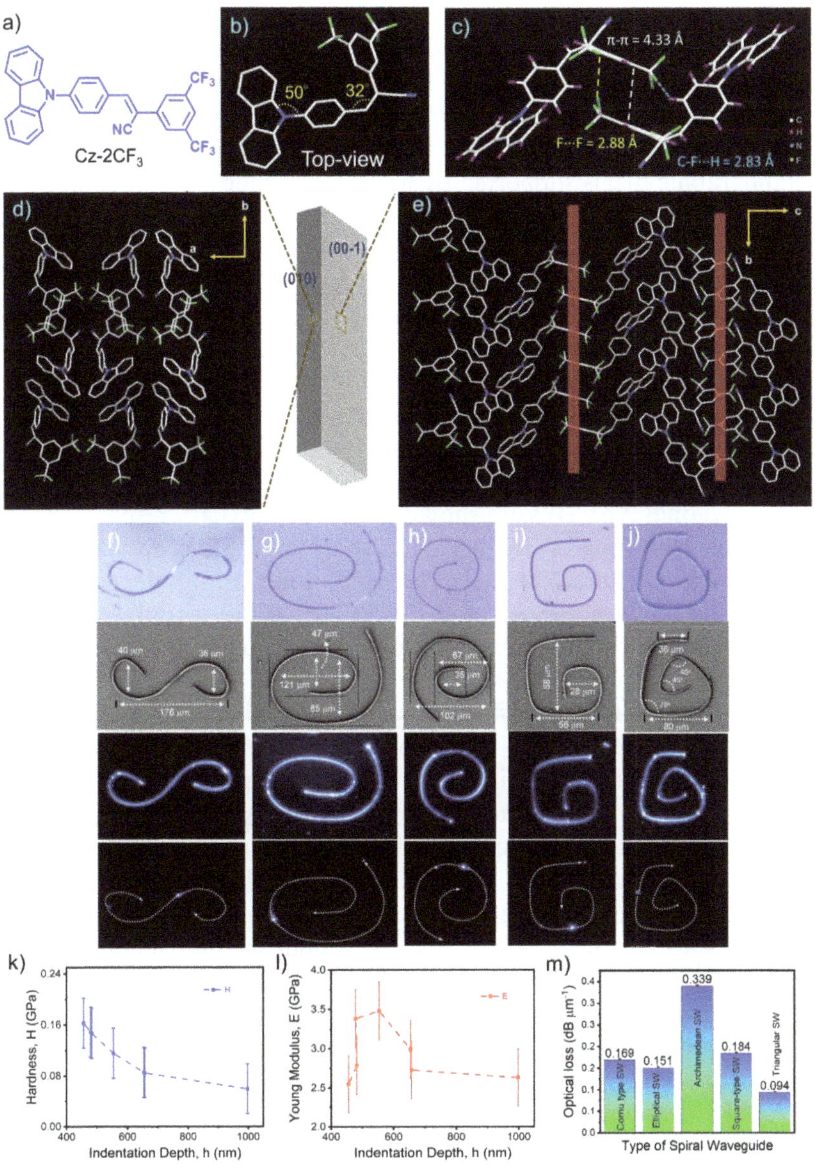

Figure 4.3. (a) Molecular structure of Cz-2CF_3. The x-ray crystal structure (b) top view, (c) various noncovalent interactions stabilizing the crystal. (d) Molecular stacking along the (010) plane. (e) Molecular arrangement and the presence of slip planes along (001) direction of Cz-2CF_3 crystal. (f–j) Confocal optical, field emission scanning electron microscopy, FL microscopy, FL waveguiding images of Cornu, elliptical, Archimedean, square-type, and triangular polygonal SWs. (k and i) Young's modulus and hardness and plots, respectively. (m) Optical loss of various SWs. [13] John Wiley & Sons. Copyright (2023).

forming various spiral configurations [13]. Single-crystal x-ray diffraction studies reveal that Cz-2CF_3 molecules crystallize in a monoclinic system. The structure features a π-spacer phenyl group that separates the carbazole (electron donor) and 3,5-bis(trifluoromethyl)phenyl (electron acceptor) units, each twisted by 50° and 32°, respectively, relative to the spacer phenyl moiety (figure 4.3(b)). Noncovalent interactions stabilizing the crystal packing include type-II halogen bonding (F···F ≈ 2.88 Å), weak hydrogen bonding (C–F···H ≈ 2.83 Å), and π···π stacking (distance ≈ 4.33 Å, figure 4.3(c)). The packing along (010) and (00–1) is depicted in figures 4.3(d) and (e), forming an extended layered molecular arrangement along the crystallographic *b*-axis supported by halogen bonding. The twist angle between donor and acceptor units, resulting from F···F interactions along [001], facilitates stress absorption under mechanical stress perpendicular to this plane.

The load-displacement curve revealed large residual depths upon unloading, indicating the crystal's elasticity. The **Y** and $\mathbf{H_c}$ of the crystal are 2.94±0.366 GPa and 0.115±0.039 GPa, respectively suggesting the super-soft and super-complaint nature of Cz-2CF_3 crystals (figures 4.3(k) and (l)). On the silica surface, Cz-2CF_3 microcrystals exhibited their pseudo-plasticity. As a result, different SW architectures were fabricated using the mechanophotonics approach. The confocal optical, FL, and FESEM images of Cornu, elliptical, Archimedean, square, and triangular polygonal type SWs are shown in figures 4.3(f)–(j), respectively. The FL images displayed glowing crystal termini, underscoring the optical waveguiding characteristics of the spiral crystals. The excitation position-dependent emission spectra of SWs revealed the α' of 0.169, 0.151, 0.339, 0.184, 0.094 dB μm^{-1} (figure 4.3(m)) for Cornu, elliptical, Archimedean, square, and triangular-shaped SW's, respectively.

Fermat's spiral is another important spiral type, also known as the parabolic spiral [12]. This spiral manifests a distinctive behavior whereby the radial separation between consecutive turns varies inversely with their angular displacement from the spiral center. In polar coordinates, Fermat's spiral is expressed as $r_d = s\sqrt{\theta}$, elucidating the relationship between the radial distance (r_d), a constant scaling factor (s), and the angular parameter (θ). This mathematical relation distinguishes Fermat's spiral from Archimedean and logarithmic spirals. Further, as two termini are pointed outside the spiral region, they can be used as light input and output points. This type of spiral waveguide is used in Si photonics as delay lines and sensors [10].

The fabrication of Fermat's spiral involves a mechanophotonics strategy and the utilization of a millimeter-long crystal [14]. My group used a highly flexible BpyIN crystal of length 1.618 mm and meticulously crafted it into a hairpin-like bent crystal through AFM-assisted mechanical micromanipulation (figure 4.4(a)). Optical experiments, confirmed the crystal's effective light guidance through the 180° bend, demonstrating minimal optical loss of 0.0100 dB μm^{-1} (figures 4.4(b) and (f)). Subsequent mechanical micromanipulation operations, depicted in figure 4.4(c), facilitated the transformation of the hairpin-like bent crystal into a Fermat's spiral crystal waveguide (figure 4.4(d)). The resulting Fermat-type spiral crystal exhibited remarkable light guidance with a minimal optical loss of 0.0111 dB μm^{-1} (figures 4.4(d) and (g)). The FESEM investigation revealed the thickness of the

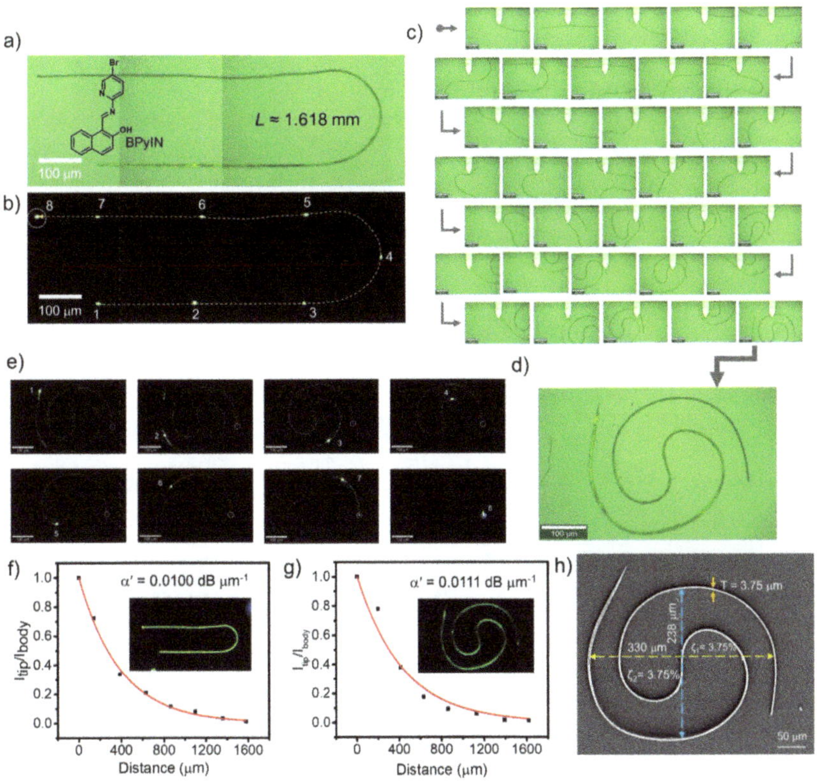

Figure 4.4. (a) Confocal optical (stitched) and (b) FL images (stitched) of a hairpin-like bent crystal waveguide. The inset shows the molecular structure of BPyIN. (c) Confocal optical images depicting the mechanical micromanipulation of a hairpin-like bent waveguide into a Fermat's spiral waveguide shown in (d). (e) FL images showing excitation position-dependent optical waveguiding experiments. (f, g) Plots of the intensity ratio at the tip and the body of the crystal (I_{tip}/I_{body}) versus the distance of the light propagation path used to estimate the optical loss coefficient (α') for the hairpin-like bent crystal waveguide and spiral waveguide. (h) FESEM image of the Fermat's spiral waveguide. Reproduced from [14]. CC BY 4.0.

crystal is about 3.75 μm and the horizontal and vertical lengths of the Fermat's spiral crystal are about 330 and 238 μm, respectively (figure 4.4(h)).

4.2 Directional couplers (DCs)

4.2.1 A monolithic organic 2×2 DC

A DC is a crucial four-port photonic device designed to split light into two separate beams. The construction of a simple monolithic 2×2 DC coupler involves connecting the convex regions (evanescent coupling regions) of two curved waveguides. Dong *et al* [15] exemplified this by fabricating a silicon-on-insulator DC tailored for broadband mid-IR photonics technologies. In their design, two curved waveguides, positioned in parallel with a 30 μm separation, were brought into proximity within a few hundred nanometers through S-shaped bends in the middle (figures 4.5(a) and (b)).

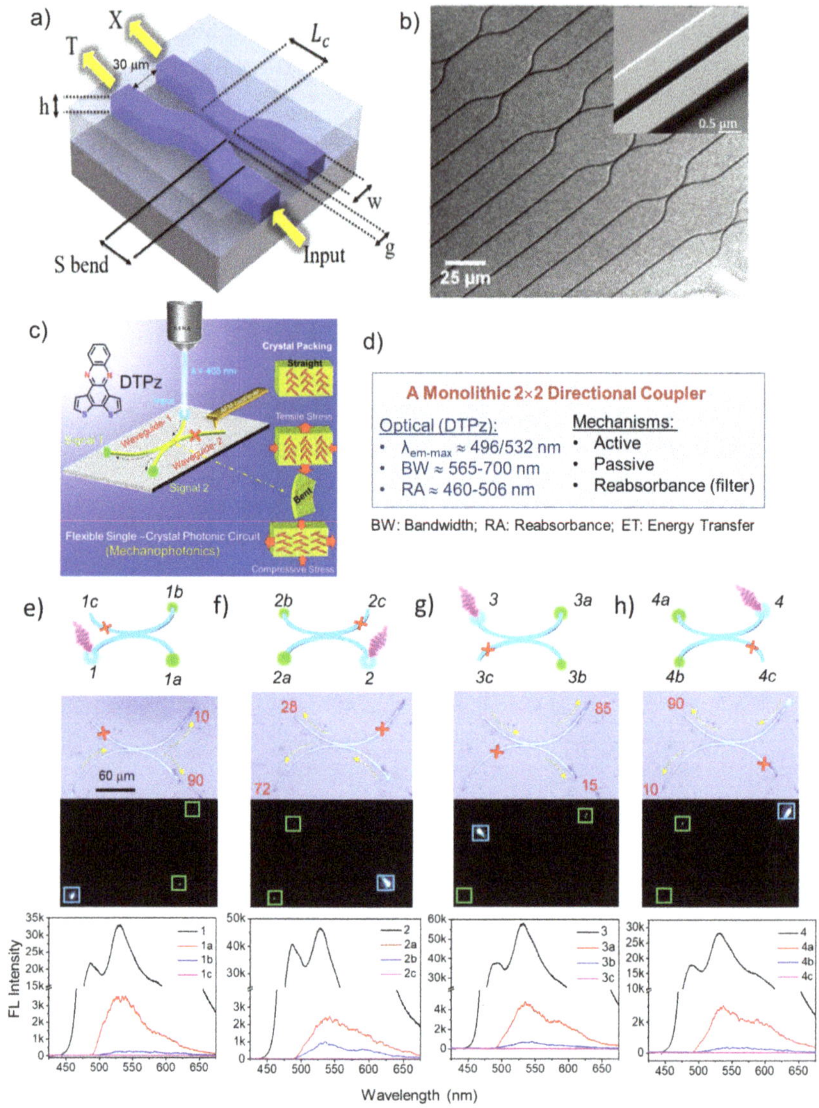

Figure 4.5. (a) Schematic of the designed directional coupler (DC) for power coupling coefficient extraction. T and X are the transmitted and coupled power, respectively. (b) SEM image of an array of Si-on-insulator DCs with varying coupling length *Lc*. The inset shows the zoom-in gap of a DC. (c) Cartoon depicting the monolithic organic-on-insulator 2×2 DC. (d) Optical properties and mechanisms of light transport in a DTPz molecule. (e–h) Schematic of splitting of propagating FL signal for an input, the corresponding (a, b) optical and FL images of DC, and FL spectra, collected at different termini are shown vertically, respectively. © [2017] IEEE. Reprinted, with permission, from [15]. (c, e–h) [16] John Wiley & Sons. Copyright (2020).

The transmitted light power (T) through the waveguide is evanescently coupled to the neighboring waveguide in the curved region. By adjusting the gap (g) between the two waveguides, they could variably control the output powers at T and X. The fabrication process of a 2×2 DC coupler based on pseudo-plastic DTPz crystals with distinct optical path lengths is detailed in section 3.2.5 (figures 4.5(c) and (d)). For an optical input at any one of the termini, the propagating signal undergoes splitting into two beams, exiting at the opposite termini with varying intensities. Importantly, no light is directed toward the terminal adjacent to the input terminal.

Equivalent to SOI DC, my group fabricated the first organic crystal DC for light-splitting applications at the University of Hyderabad (figure 4.5(c)). The mechanophotonic fabrication of DTPz crystal DC is given in section 3.2.4 [16]. Here, it is important to pay attention to the optical data of DTPz crystal to interpret the photonic functioning of fabricated DCs (figure 4.5(d)). The FL maxima of DTPz centered at 495/532 nm, revealing the wide spectral bandwidth (BW) of 565–700 nm. In the short wavelength region (460–506 nm) the FL overlaps with the optical absorbance of DTPz; thus, facilitating FL reabsorbance (RA). In this fabricated DC, the FL light is transported both *actively* and *passively*. Further, due to RA, bandwidth-reduced FL signals of varying intensity outcouple at the circuit's output termini. The extent of band reduction is dependent upon optical path length. Longer optical path length facilitates complete RA of the FL in the overlap region.

The optical excitation (input) of terminal-*1* of the DC with a 405 nm laser produces FL, which is *actively* guided towards terminal-*1a*. The propagating FL reabsorbed and a narrowband signal outcouples at this terminal-*1a*. Before reaching terminal-*1a*, the FL passively enters the adjacent waveguide via evanescent coupling and outcouples at terminal-*1b*. Due to circuit geometry, no signal arrives at terminal-*1c* (figure 4.5(e)). The ratio of the intensity of the split signals at terminal-*1a* and -*1b* is 90:10. The optical path length (L_{op}) from *1* to *1b* is 90 μm and from *1a* to *1c* is 66 μm. Similarly, excitation of any one of the other termini (2–4) produced split optical signals (passive and active) at the opposite two termini with varying split ratios (figures 4.5(d)–(f)). For example, excitation at terminal-*2* produces two split active and passive outputs at *2a* and *2b* with a splitting ratio of 72 and 28, respectively (figure 4.5(f)). Optical excitation of terminal-*3* or -*4* also produced consistent outcoupling of split signals at *3a*/*3b* or *4a*/*4b*, with split ratios of 85/15 and 90/10, respectively (figures 4.5(g) and (h)). The variation of the output intensities is related to the L_{op} of the propagating light in DC. The longer the path length the lower the output due to the increase of light propagation loss.

4.2.2 A monolithic organic 2×2 asymmetric DC coupler

The demonstration of an organic crystal based DC discussed in section 4.2.1 underscored the significance of the convex section of the bent waveguide. This convex region plays a crucial role in efficiently coupling distinct optical components, such as waveguides and resonators, by utilizing the evanescent coupling region. The construction of OPICs with intricate geometries necessitates the integration of waveguides featuring multiple bends at various angles.

My group employed a molecular crystal, namely, (*E*)-1-(4-(iodo)phenyl)iminomethyl-2-hydroxyl-naphthalene (IPIN) (figure 4.6(a), inset) [17]. This crystal demonstrates notable mechanical flexibility, emits a bright green FL ($\lambda_{max} \approx 530$ nm), featuring a broad signal with a bandwidth spanning approximately 475–700 nm, and exhibits reabsorption from 475 to 516 nm (figure 4.6(d)).

The mechanical flexibility of IPIN microcrystals is attributed to intermolecular hydrogen bonding and $\pi\cdots\pi$ stacking interactions (figure 4.6(b)). Through precise mechanical micromanipulation, we fabricated a triply-bent crystal waveguide ($L \approx 291.2$ μm) characterized by three convex bends ($\zeta \approx 1.98\%$). Subsequently, integration of this waveguide with a singly-bent crystal waveguide at their convex regions resulted in an asymmetric, monolithic 2×2 DC featuring four termini, denoted as T1–T4 (figures 4.6(c) and (e)).

Similar to the behavior observed in the DTPz DC, optical excitation at T1 with a 405 nm laser led to the active guidance of the generated broadband FL (signal 1A)

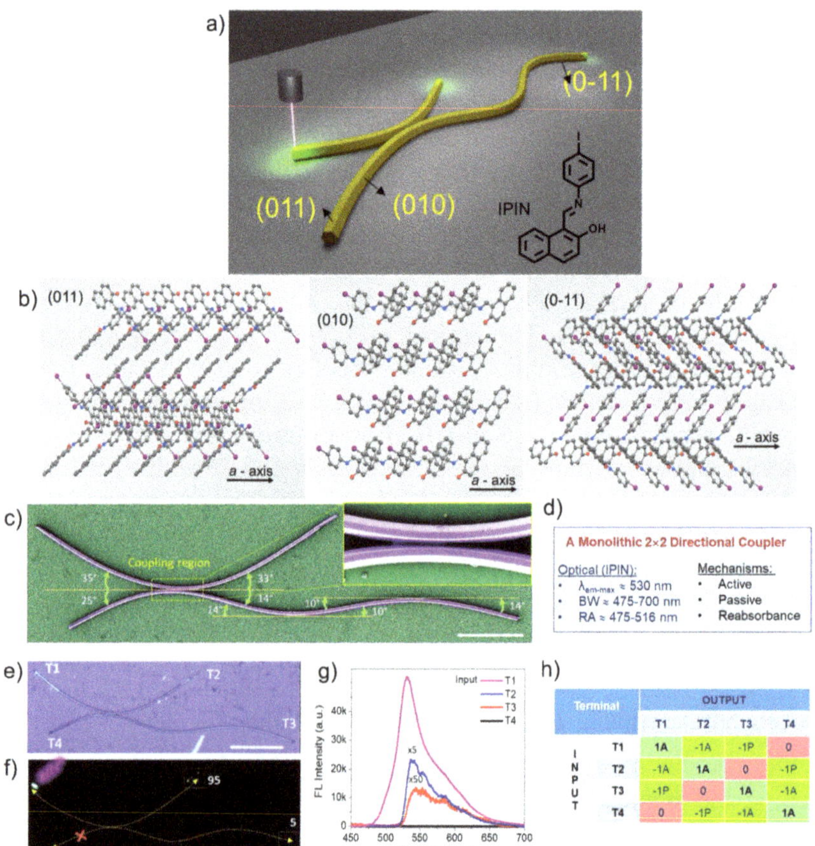

Figure 4.6. (a) Graphical representation of a monolithic 2×2 asymmetric DC made with singly- and triply-bent microcrystals. (b) Molecular packing in the crystal lattice. (c) Color-coded FESEM image of DC and the inset shows the coupling region's close-view. The scale bar is 60 μm. (d) Optical properties and mechanisms of circuit operation. (e) Optical image of the splitter. (f) FL image displaying excitation position-dependent, directional-specific input and output signals observed at different terminals. (g) FL spectra collected at various termini. (h) Summary of signal outputs at different terminals for various inputs. (b–h) [17] John Wiley & Sons. Copyright (2021).

through the singly-bent waveguide (figures 4.6(e) and (f)). This process emitted a long-pass filtered FL (-1A signal) at T2. Moreover, the propagating FL evanescently coupled at the convex region to a triply-bent crystal waveguide and passively (P) emitted at T3 as a -1P signal. As expected, no light (signal 0) was detected at T4 due to circuit geometry. The intensity ratio of signals for T2:T3 is 95:5 (figure 4.6(g)). Similarly, by varying the excitation terminal, direction-specific active/passive narrow-band optical signals can be transported to DC's opposite termini (figure 4.6(h)).

4.2.3 A hybrid organic 2×2 DC

For the construction of a 2×2 hybrid DC (HDC), crystal waveguides based on BPyIN and (*Z*)-3-(3′,5′-bis(trifluoromethyl)-[1,1′-biphenyl]-4-yl)-2-(4-methoxy-phenyl)acrylonitrile (CF_3OMe) molecules, possessing distinct optical properties, were chosen (figure 4.7(a)) [18]. In the solid state, CF_3OMe exhibits optical absorption up to approximately 475 nm, peaking at 425 nm. The FL spectral bandwidth of CF_3OMe ranges from 420 to 660 nm, with a maximum of 475 nm (figures 4.7(b) and (c)). Given the overlap of CF_3OMe FL with the 420–562 nm absorption region of BPyIN, substantial radiative ET from CF_3OMe to BPyIN becomes feasible. The FL of BPyIN extends to 560 nm with a λ_{max} at 525 nm. Importantly, no ET was observed from BPyIN to CF_3OMe due to the absence of spectral overlap between the absorption of CF_3OMe and the FL of BPyIN.

Upon applying stress, the single crystal of CF_3OMe revealed its pseudo-plasticity with a ζ of about 11% (figure 4.7(d)). The molecular packing of CF_3OMe in the crystal showed slip planes along [010] (figure 4.7(f)), which perhaps assist the gentle to-and-fro motion of molecules or reversible sliding to support stress disbursement. Application of mechanical force perpendicular to the (001) plane shows reversible elastic deformation of the crystal. Due to a lack of stress disbursal mechanisms, the crystal breaks when force is applied perpendicular to the (010) plane (figure 4.7(g)).

The pseudo-plasticity of CF_3OMe and BPyIN (see section 4.1.3) crystals were creatively utilized to fabricate 2×2 HDC via mechanophotonics by connecting the convex regions of the bent waveguides (figures 4.8(a) and (b)). To check the light-routing competency of HDC, a 405 nm light input was given at any one of the crystal termini or ports (abbreviated as P) and the FL signal response at the other three ports was recorded (figures 4.8(c)–(f)). For instance, when the input was given at P1, it produced an active (A) blue FL signal (λ_1(A); ≈420–670 nm) of CF_3OMe (figures 4.8(c) and (g)). This active signal propagates towards P2, after optical reabsorption, outcouples as a narrow bandwidth signal ($-\lambda_1$(A); ≈440–670 nm) at P2. Moreover, this propagating λ_1(A) signal evanescently excites the BPyIN wave-guide via ET generating an active green FL signal ($-\lambda_2$(A); ≈532–700 nm) and outcouples at P3. As expected, no light was recorded at P4.

Similarly, when P2 was input, signals $-\lambda_1$(A), $-\lambda_2$(A), and 0 were recorded at P1, P4, and P3, respectively (figures 4.8(d) and (i)). On the contrary, providing input at P3 generated a bright green FL (λ_2; ≈482–700 nm) signal (figures 4.8(e) and (h)). At P4, due to reabsorption, a narrowband active signal ($-\lambda_2$; ≈532–700 nm) out-coupled. As there is no ET possibility from BPyIN to CF_3OMe, the same signal (-λ_2)

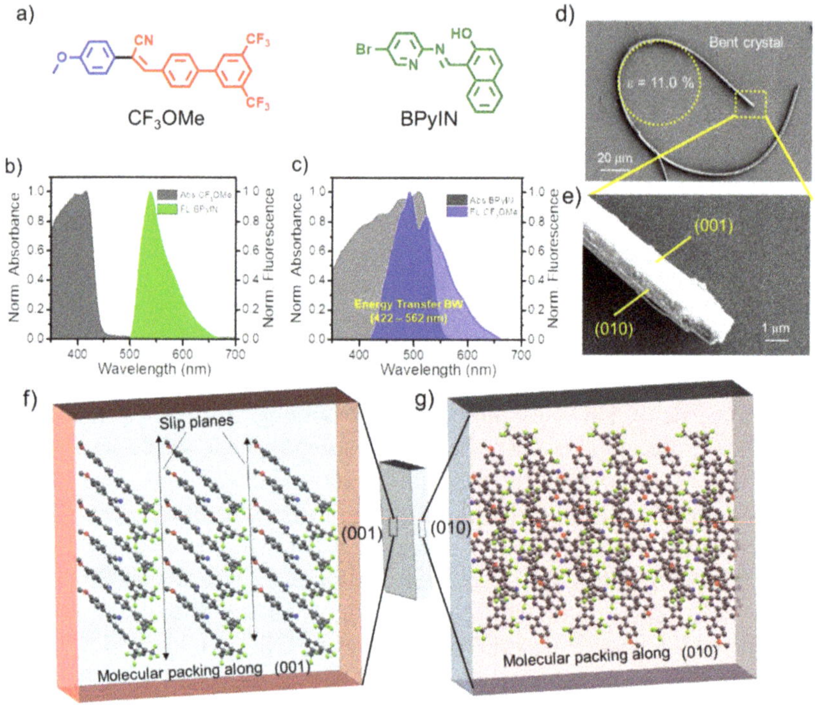

Figure 4.7. (a) Molecular structures of CF_3OMe and BPyIN. (b, c) Solid-state absorption and emission spectra of CF_3OMe and BPyIN. (d) FESEM image of a bent CF_3OMe crystal. (e) Close-up FESEM view of the crystal shown in (d), depicting the facets. (f, g) Molecular orientation along (001) and (010) planes in the crystal packing of CF_3OMe. Reproduced from [18] with permission from the Royal Society of Chemistry.

passively couples to the BPyIN waveguide and emits as $-\lambda_2$(P) signal at P1. Further, no signal was detected at P2. Likewise, providing input at P4 generated $-\lambda_2$(A), $-\lambda_2$(P), and 0 signals at P3, P2, and P1, respectively (figures 4.8(f) and (j)). In one case, the HDC functions as signal chromaticity modulator-and-splitter (λ_1 and λ_2; blue and green) and in another case as a simple beam splitter (λ_2; green).

4.2.4 A hybrid 3-layer DC for optical neural networks

Like the biological brain, artificial neural networks (ANNs) are human-made structures that process the input signal using several complex interconnected neurons (nodes). In ANNs, each connection performs as a synapse, like in the biological brain. PICs are, in a way, related to ANNs. A systematically arranged and multiple interconnected optical waveguide, generating, guiding, and splitting multiple optical signals can function as an ANN. The initial step to achieving an organic crystal based ANN is via systematically and hierarchically interconnected DCs.

Firstly, implementing the mechanophotonic approach on a BPyIN crystal allows the creation of a highly strained waveguide with five bends (figure 4.9(a)). Due to different radii of curvature, the ζ values of each bend varies from ≈3.8 to 5.3%. The physical integration of one of the convex bends of the BPyIN waveguide to

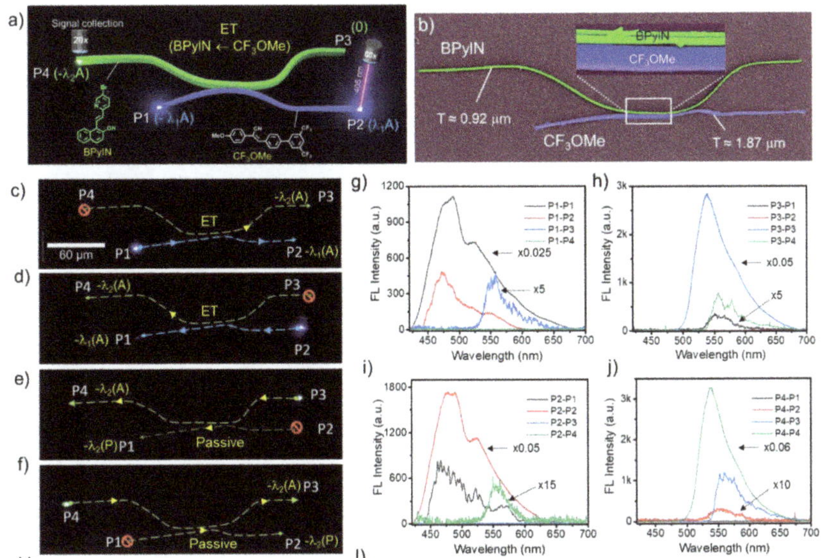

k)

Port		Output P1	P2	P3	P4
Input (405 nm)	P1	λ_1(A)	$-\lambda_1$(A)	$-\lambda_2$(A)	0
	P2	$-\lambda_1$(A)	λ_1(A)	0	$-\lambda_2$(A)
	P3	$-\lambda_2$(P)	0	λ_2(A)	$-\lambda_2$(A)
	P4	0	$-\lambda_2$(P)	$-\lambda_2$(A)	λ_2(A)

l)

Signal	λ_1(A)	$-\lambda_1$(A)	λ_2(A)	$-\lambda_2$(A)	$-\lambda_2$(P)	$-\lambda_2$(A)	0
Type	Active	Reabsorbed	Active	Reabsorbed	Passive	Energy Transfer	NIL
BW (nm)	≈ 420 – 670	≈ 440 – 670	≈ 482 – 700	≈ 532 – 700	≈ 532 - 700	≈ 532 - 700	-

Figure 4.8. (a) Cartoon representation of a 2×2 hybrid DC (HDC). Confocal optical image of a hybrid directional coupler (HDC). (b) Color-coded FESEM images of HDC. (c–f) FL images of HDC for 405 nm input light given at P1–P4, respectively. (g–j) Corresponding FL spectra obtained as optical output response at various ports of HDC for respective light input at P1–P4, respectively. (k) Optical performance of the HDC tabulated according to the signal observed at various ports against given input. (l) Different optical signal types and bandwidths operational in HDC. Reproduced from [18] with permission from the Royal Society of Chemistry.

another singly-bent BPyIN waveguide's convex region generated the first optical synapse (S-1; evanescent coupling point) that can transfer signals from one waveguide (neuron-1) to another (neuron-2). Lastly, the interconnection of another singly-bent BPP waveguide (S-2; second synapse that functions via ET) to another available bend of BPyIN waveguide provided a novel 3-layer HDC with six termini, T1 to T6. The optical data showed that BPP and BPyIN have well-separated absorbance bandwidths of 400–600 and 500–680 nm, respectively (figure 4.9(b)). The possibility of ET from BPyIN to BPP arises when the crystals are in evanescent contact, facilitated by the spectral overlap in the range of 478–570 nm. However, in the reverse direction, from BPP to BPyIN, there is no observed ET. BPP shows no reabsorbance, while BPyIN exhibits reabsorbance characteristics. The absolute quantum yields (Φ_{FL}) of BPP and BPyIN were found to be <1% and 16.2%, respectively. Here, the high Φ_{FL} of BPyIN facilitates the long-distance propagation of optical signals through synapses.

The FESEM and FL images of 3-layer HDC with six terminals are shown in figures 4.9(c) and (d). When T1 was excited with a 405 nm laser, the resulting green FL (λ_1(A)) emanated by the BPyIN crystal (figures 4.9(f), (l), and (n)) actively

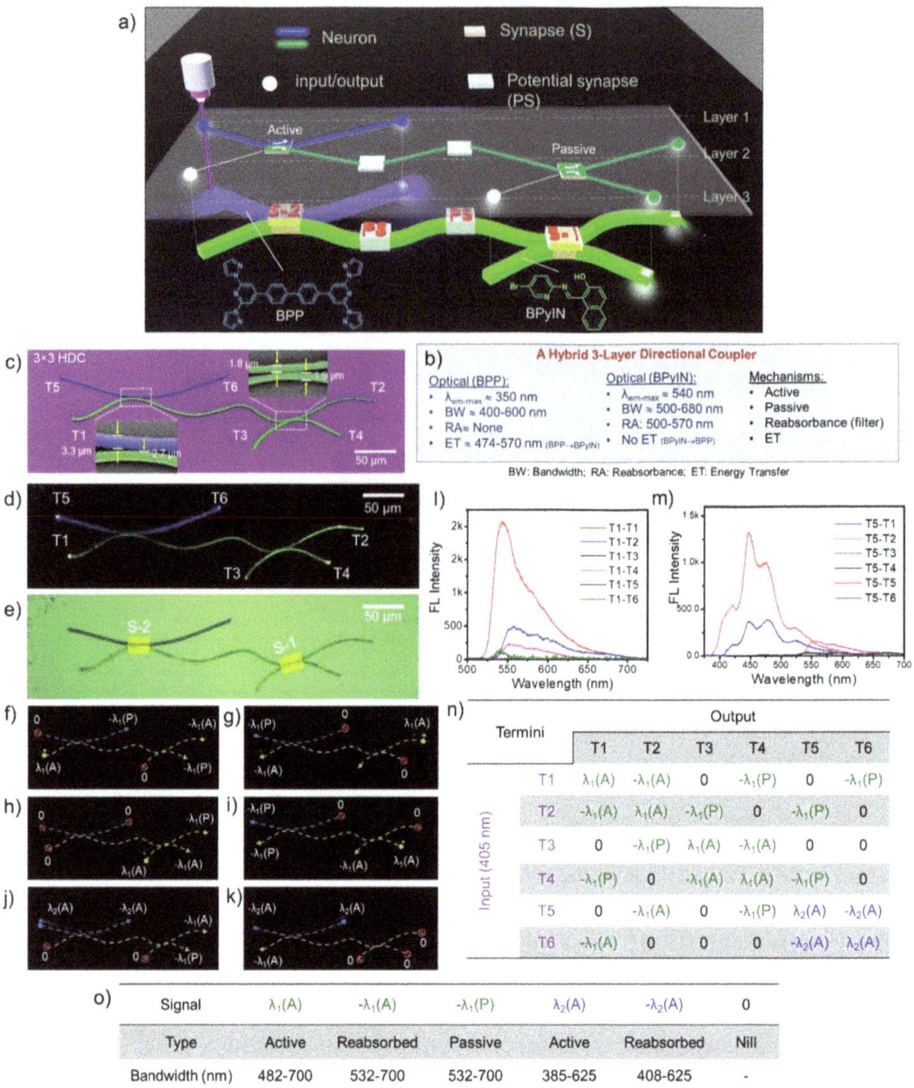

Figure 4.9. (a) Graphical representation of an organic crystal based 3-layer hybrid DC acting as an optical artificial neural network (ANN). Insets present the chemical structures of BPP and BPyIN molecular crystals used as optical waveguides. (b) Optical data BPP and BPyIN molecules. (c) FESEM image of a 3-layer hybrid DC. (d, e) FL and confocal optical microscopy images of a 3-layer DC. (f–k) FL images when laser input was provided at T1-T6 termini, respectively. (l, m) Representative FL spectra recorded at various termini for light input at T1 and T5, respectively. (n) Optical performance of hybrid 3-layer HDC. (o) Various optical signals and mechanisms involved in 3-layer HDC. (c-o) [19] John Wiley & Sons. Copyright (2024).

propagated to T2 as a reabsorbed signal ($-\lambda_1$(A)). As expected, a passive signal ($-\lambda_1$(P)) was observed at T4 via the optical contact point (synapse, S-1), while no signal was observed at T3 and T5. Fascinatingly, through S-2, a third passive signal $-\lambda_1$(P) outcoupled at T6 as there is no ET from BPyIN to BPP crystal (figures 4.9(f), (l), and (n)). This experiment demonstrated the signal splitting ability (one input signal is split into three) and efficient signal processing through the optical synapses (S-1 and S-2) of 3-layer DC.

When T5 of BPP was optically excited, the generated blue FL corresponding to signal $\lambda_2(A) \approx 385$–625 nm transduced to T6 as a reabsorbed signal $-\lambda_2(A) \approx 408$–625 nm (figures 4.9(j), (m), and (n)). However, the evanescently seeped blue FL at synopsis S-2 excites the adjacent BPyIN waveguide via ET. The resultant green FL outcouples as $-\lambda_1(A)$ green signal at T2. The same $-\lambda_1(A)$ signal once more undertakes evanescent coupling via S-1 and emits a passive $-\lambda_1(P)$ signal at T4. Further, no optical signal was detected at T1 and T3. During excitation of T6, the $-\lambda_2(A) \approx 385$–625 nm signal reaches T5. At the evanescent coupling junction, it couples to BPyIN and reaches T1 as $-\lambda_1(P) \approx 532$–700 nm signal. Interestingly, this time, there was no signal recorded at three termini (T2, T3, and T4).

The table presented in figure 4.9(n) illustrates the obtained results from the circuit when other termini (T2, T3, and T4) of the HDC were used as input. Each time, for the input, the constructed 3-layer HDC simultaneously generates multiple optical signals with variable split ratios and bandwidths traveling through three-level hierarchically placed waveguides. Thus, illustrating that the novel 3-layer organic HDC functions as an optical ANN to process different optical signals. The operational mechanisms and complex optical trajectories of the functioning circuit bear a resemblance to neural networks.

4.3 Add-drop filters (ADFs)

4.3.1 A monolithic organic ADF

In an add-on drop filter (ADF), a ring resonator (RR) is coupled to two optical waveguides to control the light propagation direction(s) by utilizing the resonator's clockwise and anticlockwise light-routing capability.

For the mechanical fabrication of a monolithic organic ADF, three flexible crystals are required. For that, three DPIN crystal waveguides (discussed in chapter 2, section 2.1.5 [20, 21]), namely, OW1–OW3, were used (figure 4.10(a)). The DPIN crystals' FL maximum appears at 607 nm. The bandwidth of FL is 540–760 nm with a reabsorption around the 525–580 nm region.

From the excitation position-dependent experiments along the waveguide's 234 μm ($T \approx 2.4$ μm) long axis, the output signals were collected at the right-side terminal of OW1 (figure 4.10(a) and (c)). The waveguide showed an α' of 0.0445 dB μm^{-1}. As the RR is a key element in ADF, OW1 was micromechanically bent into a ring shape of $D \approx 58.6$ μm by connecting the crystal termini (figure 4.10(d)). The calculated strain of the ring ($T \approx 2.4$ μm) is about 4.09%. The RR exhibited a slightly higher α' value of 0.0488 dB μm^{-1} owing to bending-induced optical loss (figure 4.10(e)).

Subsequently, two straight optical waveguides, denoted as OW2 (length, $L \approx$ 162 μm) and OW3 ($L \approx 285$ μm), were meticulously positioned in parallel, making contact with opposing convex sides of the ring resonator (refer to figure 4.10(g)). This configuration yielded an ADF system characterized by four termini or ports labeled P1–P4. Upon optically exciting one of the crystal termini of OW3 (specifically,

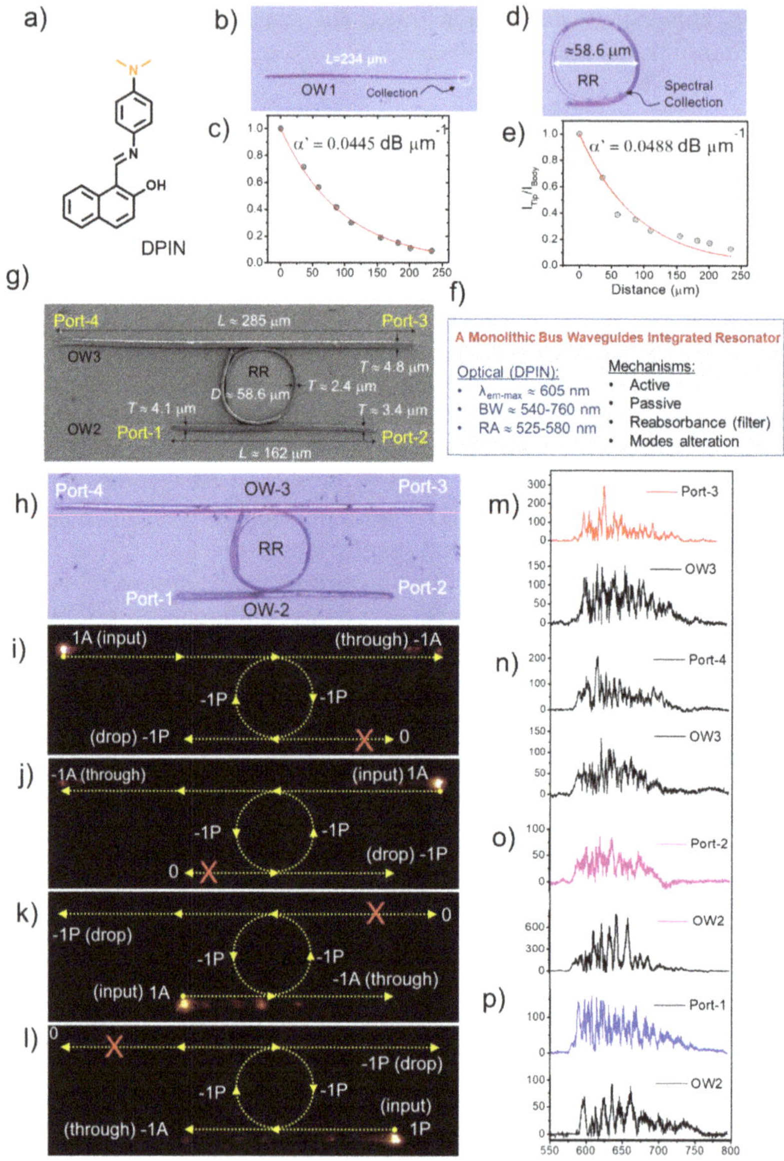

Figure 4.10. (a) Molecular structure of DPIN. (b, d) Optical images of the DPIN waveguide (OW1) in its straight and ring resonator geometries, respectively. (c, e) Their corresponding plots of the intensity ratio at the tip and the body of the crystal (I_{tip}/I_{body}) versus the distance of the light propagation path used to estimate the optical loss coefficient (α'). (f) Optical properties and mechanisms of light transport in DPIN crystal. (g) FESEM image of a monolithic micro add-on/drop filter fabricated via mechanical micromanipulation of DPIN microcrystals. (h) Confocal optical image of flexible crystal based dual waveguides coupled ring resonator. (i–l) FL images, (m–p) the corresponding spectra of dual waveguides coupled ring resonator excited at different ports displaying mechanism-selective and directional-specific optical outputs. [20] John Wiley & Sons. Copyright (2021).

P-4), the generated FL (labeled as signal 1A) actively propagated along the crystal's long axis (figures 4.10(h) and (i)). This signal subsequently outcoupled as a reabsorbed signal at P-3 (labeled signal 1A). As previously elucidated, the propagating FL evanescently enters the RR passively. The light then circulates the ring in a clockwise manner, coupling to OW2 evanescently and eventually outcoupling at P-1 as a -1P signal (as illustrated in figure 4.10(i)). Port-2 recorded no light signal (designated as Signal: 0). This observation serves as conclusive evidence affirming the directional specificity inherent in the light-guiding behavior of the circuit.

Notably, a captivating observation was made on the optical modes present in the FL spectrum recorded at P-3 when comparing it with the OW3 spectrum (figure 4.10(m)). The intensity and wavelength of these modes exhibited discernible alterations when compared to the modes at the same terminal for the isolated OW3 waveguide.

Likewise, by changing the optical excitation port (input), the propagating FL can be routed either clockwise or anticlockwise, accompanied by altered modes. Upon introducing light input at port-3 (as illustrated in figure 4.10(j)), the resulting generated FL signal, denoted as 1A, underwent active transduction. Following partial reabsorption, this signal then outcoupled at port-4 as a signal -1A. Notably, the optical modes observed in signal -1A exhibited significant distinctions from those observed at port-4 of the isolated OW2 before its integration into the circuit (refer to figure 4.10(n)). Simultaneously, the bandwidth-altered light signal (-1P) entered the RR, passively circulating in an anticlockwise direction. Subsequently, it selectively outcoupled at port-2 as a -1P signal, with no light signal (designated as signal: 0) detected at port-1. Furthermore, the excitation of port-1 resulted in signals -1A, -1P, and 0 at port-2, -4, and -3, respectively (as depicted in figures 4.10(j) and (o)). Similarly, for light input at port-2, signals -1A, -1P, and 0 were recorded at ports 1, 3, and 4, respectively (refer to figures 4.10(l) and (p)).

Remarkably, the optical modes exhibited by the signals at port-1 and port-2 of the circuit were notably altered compared to the isolated OW2 waveguide (figures 4.10(o) and (p)).

These alterations in optical modes, recorded at ports 1–4 of the circuit in comparison to individual waveguides, serve as conclusive evidence confirming the occurrence of optical interferences within the circuit.

4.3.2 A hybrid and reconfigurable organic ADF

To fabricate a simple hybrid ADF, a minimum of two chemically distinct flexible crystals is required. Two promising candidates for testing are DPIN and (*E*)-1-(4-(chloro)phenyl) iminomethyl-2-hydroxyl-naphthalene (CPIN) flexible crystals, exhibiting structural disparities solely in chloro and dimethylamine substitution (as illustrated in figure 4.11(a) and (b)) [22]. Despite their subtle structural differences, these modifications significantly influence the electron donor–acceptor properties, thereby manifesting in distinct optical absorption and emission spectra. In the solid state, DPIN emits orange FL (bandwidth ≈540–760 nm), while CPIN emits green FL (bandwidth ≈475–700 nm). Notably, the overlap of CPIN FL with DPIN absorption facilitates ET from CPIN to DPIN (figure 4.11(j)). Conversely,

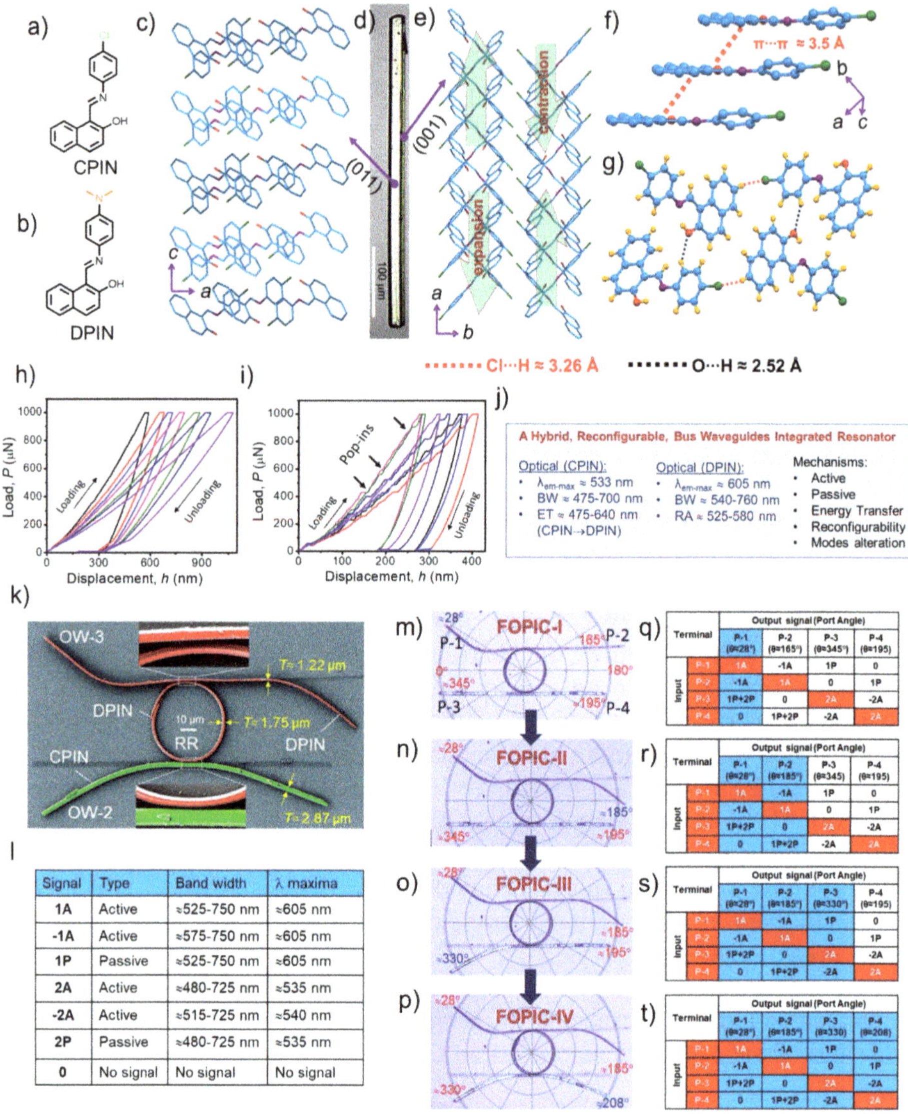

Figure 4.11. (a, b) Molecular structures of DPIN and CPIN. (c) Molecular packing of CPIN on the (011) face. (d) Optical image of a CPIN single crystal. (e) Molecular packing on the bendable (001) face showing crisscross molecular chains. The inset arrows show the molecule's possible expansion and contraction movement from the equilibrium position. (f, g) π···π stacking, C–H···O and C–Cl···H intermolecular interactions in the crystal structure. (h, i) Load versus displacement plots obtained from nanoindentation on the bending faces of DPIN and CPIN, respectively. (j) Optical data of DPIN and CPIN. (k) False-color FESEM image of the circuit. The inset shows a close-up view of the evanescent contact region between CPIN and DPIN crystals with RR. (l) Table explaining the signal codes, type, spectral bandwidth range, and FL λ_{max}. (m–p) Optical images of stepwise port-angle reconfigured FOPICs for the need-based altering of light output direction. (q–t) Optical performance tables demonstrating the light input-dependent output signals at P-1, P-2, P-3, and P-4 over various angles. [22] John Wiley & Sons. Copyright (2022).

due to the absence of overlap between CPIN absorption and DPIN emission, no ET from DPIN to CPIN is anticipated. Furthermore, the Φ_{FL} of CPIN (6.31%) was found to be higher than that of DPIN (0.567%) crystals. These distinctive optical properties of CPIN and DPIN reinforce their suitability as complementary components in the construction of a hybrid ADF system.

The mechanical compliance exhibited by these crystals is attributed to their distinctive molecular packing arrangements. The structural foundation for the mechanical flexibility of DPIN crystals has been elucidated earlier (refer to figure 2.8). Upon single-crystal x-ray investigation, CPIN revealed a closely similar structure to DPIN, with both crystals exhibiting an O–H···N hydrogen bond (at a distance of 1.77 Å). Notably, the CPIN molecule displayed a more planar conformation compared to DPIN. In the crystal lattice of CPIN (figure 4.11(f)), various weak interactions were identified, including C–Cl···H (approximately 3.26 Å), π···π stacking (around 3.5 Å), intermolecular hydrogen bonding (C–O···H; approximately 2.52 Å), and C–H···π interactions (approximately 2.83 Å) (refer to figure 4.11(f) and (g)). The π···π and C–H···π interactions gave rise to crisscross-packed one-dimensional chains, forming layers along the (001) plane (figure 4.11(e)). When a mechanical force was applied perpendicular to this (001) plane, CPIN crystals exhibited flexibility (figure 4.11(h) and (i)).

Despite comparable molecular packing, DPIN and CPIN demonstrated distinct load versus displacement plots during nanoindentation experiments. The **Y** and $\mathbf{H_c}$ values for DPIN were measured to be 1.02±0.4955 GPa and 0.1168±0.0307 GPa, respectively, indicating the crystal's super-soft and super-compliant nature. Conversely, CPIN exhibited **Y**=8.22±1.62 GPa and $\mathbf{H_c}$=0.3532±0.0959 GPa, signifying a moderate elasticity of the crystals. The unit cell volumes for DPIN and CPIN crystals were determined to be 1454.09 Å^3 and 1288.49 Å^3, respectively. This indicates that DPIN's relatively compact molecular packing, compared to CPIN, is reflected in their observed differences in mechanical properties. Furthermore, the observed 'pop-ins' in the curve of CPIN showed that the resistive response of the crystal is in-homogeneous [23] (figure 4.11(h) and (i)). Both crystals demonstrated their pseudo-plasticity on the coverslip.

Initially, employing bending and moving operations, a circular RR with a diameter of approximately 126 μm (ζ = 8%) was crafted from a 205 μm long low-loss (α' = 0.0837 dB μm^{-1}) DPIN crystal waveguide. Due to bending-induced optical loss, the RR exhibited an α' of 0.1098 dB μm^{-1}. Subsequently, on the lower convex side of this RR, a straight CPIN waveguide (OW2) measuring 199 μm in length and featuring α' = 0.0790 dB μm^{-1} was integrated. This integration involved transferring the CPIN waveguide from its original substrate to the substrate of the RR, employing *lifting*, *dropping*, *moving*, and *integrating* micromanipulation operations, resulting in a two-port (P-3 and P-4) device. Upon excitation of the P-3 terminal of OW2, a bright green FL was generated, transducing towards P-4. The green FL of OW2 excited the RR via the ET mechanism, producing DPIN FL with a λ_{max} of approximately 610 nm. The circular configuration of the RR facilitated the routing of light in a counterclockwise path. Alternatively, when P-4 was excited, the same ET process occurred, but this time the optical signal took a clockwise path, culminating in outcoupling at P-3.

Following the establishment of the two-port device function, a straight DPIN crystal (OW3) was integrated at the upper convex side of the RR, aligning OW2 and OW3 in a parallel geometry connected through the RR. The final hybrid ADF system was thus equipped with four ports, namely P-1 to P-4 (figure 4.11(m)) [22]. This configuration facilitated inspecting the performance of flexible OPIC I (FOPIC-I) by optically exciting any one of the specific ports and monitoring the spectral output of the remaining ports.

To achieve port direction reconfigurability in the *x*, *y*-plane, the tip of P-1 was mechanically moved away from the RR at an angle of $\approx 28°$ from its original position, resulting in the creation of FOPIC-I (figure 4.11(m) and (q)). Upon exciting P-1, it generated a red FL signal, denoted as 1A (where A signifies an active signal) with an emission bandwidth of approximately 525–750 nm and a λ_{max} of about 605 nm. This FL actively propagated through the bent OW3 and outcoupled at P-2 as a -1A signal (where the negative sign indicates reabsorbance) with a bandwidth of 575–750 nm. Further, the signal 1A evanescently coupled to the RR, transducing passively as a 1P signal (where P denotes a passive signal) in a clockwise direction. Subsequently, it again evanescently coupled to OW2, delivering a 1P signal at P-3. The absence of output light (Signal 0) at P-4 is attributed to the clockwise light path within the RR. Similarly, P-2, when excited, can deliver a -1A signal at P-1, a 1P signal at P-4, and a 0 signal at P-3.

Remarkably, the excitation of P-3 results in the production of a green FL (signal 2A) with a bandwidth of approximately 480–725 nm and λ_{max} of about 535 nm. This FL outcoupled as a -2A signal (bandwidth $\approx$ 515–725 nm) at P-4 due to reabsorption. Along its trajectory, signal 2A evanescently coupled to the RR, generating signal 1A via an ET mechanism. Consequently, due to partial ET, signal 2A also passively propagated (as a 2P signal) counterclockwise with signal 1A, ultimately outcoupling at P-1 as a mixed 1P+2P signal. As expected, at P-2 no light (signal 0) was detected. Excitation of P-4 results in the outcoupling of a signal -2A at P-3 and a signal 1P+2P at P-2, with no output signal (signal 0) at P-1. These outcomes affirm the capability of FOPIC-I to generate six different optical signals (1A, -1A, 1P, 2A, -2A, and 2P) and selectively and predictably deliver them to three ports.

In the subsequent step, deliberate manipulation of six different optical signal output directions in two dimensions was demonstrated by mechanically repositioning each port-angle orientation in FOPIC-I. Specifically, P-2 was selectively reconfigured to an angle of approximately 185° to fabricate FOPIC-II (as shown in figure 4.11(n) and (r)). In comparison to FOPIC-I, the signal output directions in FOPIC-II differ at one port. Analogously, the reconfiguration of P-3 in FOPIC-II by approximately 330° led to the creation of FOPIC-III (figure 4.11(o) and (s)). Further, altering P-4 by approximately 208° in FOPIC-III resulted in the formation of FOPIC-IV (figure 4.11(p) and (t)). A false-color FESEM image of FOPIC-IV exhibits the ADF's accurate dimension; in particular, the contact between RR and OWs (figure 4.11(k)). This deliberate manipulation showcases the reconfigurability of the ADF in precisely controlling the optical signal output directions through mechanical adjustments in a two-dimensional space.

4.3.3 A hybrid and reconfigurable organic circuit

Before, presenting a discussion on the hybrid and reconfigurable organic circuit, it is important to look at the structurally isomorphous silicon device. Xu *et al* reported a device composed of a silicon microring resonator ($R = 5$ μm) coupled to a straight waveguide of width and height 450 and 250 nm, respectively (figure 4.12(a)) [24]. In the device, the ring resonator and the straight waveguide are separated by a gap of 200 nm. The input (quasi-TE mode) was given at one of the ends of the waveguide with a fiber, and the fiber-to-fiber insertion loss was 10.4 dB. Due to the coupling of propagating light to the ring resonator, and subsequent interference, the transmission

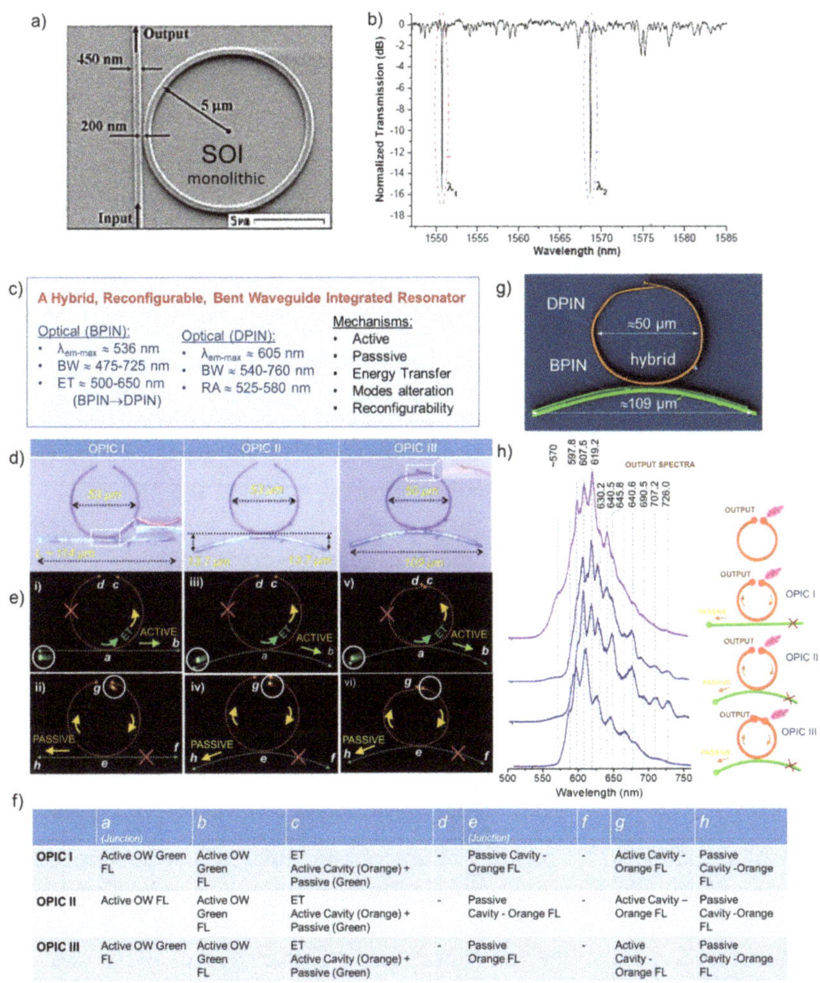

Figure 4.12. (a) SEM image of the silicon microring resonator before the SiO_2 cladding deposition. (b) The transmission spectrum of the resonator. (a, b) Reproduced from [24]. CC BY 4.0.(c) Optical data of BPIN and DPIN. (d) Confocal microscope (top) and (e) FL images (i–vi) of OPICs. (f) The input-dependent mechanism-selective and directional-specific optical outputs from OPICs. (g) FESEM image of hybrid OPIC III. (h) The change of the optical modes at the output (g) for different OPICs compared to an isolated semicircular waveguide. (c–f) [25] John Wiley & Sons. Copyright (2021).

spectrum showed two resonances ($\lambda_1 = 1550.7$ nm and $\lambda_2 = 1568.7$ nm). The output transmission of the waveguide reduces by $\approx$16 dB at these resonances (figure 4.12(b)). The quality factors of λ_1 and λ_2 are $Q_1 = 11\ 076$ and $Q_2 = 9804$, respectively. The estimated τ_P (refer to section 2.2.1.1 for more details on photon lifetime) of the ring resonator at λ_1 is 9.1 ps, which fundamentally limits the device operation speed.

To fabricate reconfigurable hybrid OPICs, two electronically different flexible organic single crystals DPIN and its structural analog BPIN were used. The fabrication of OPICs (I–III) via mechanical integration of DPIN and BPIN crystals is given in section 3.2.6 of chapter 3 [25]. The difference in the amino (strong electron releasing) and bromo groups of DPIN and BPIN is reflected in their optical properties (figure 4.12(c)). DPIN and BPIN exhibited orange ($\lambda_{max} = 600$ nm) and green ($\lambda_{max} = 536$ nm) FL, respectively, in the solid state, additionally allowing ET from the latter molecule to the former. These optical attributes of DPIN and BPIN crystals allow the operation of OPICs using ET, passive and active optical waveguiding characteristics.

OPIC I is a four-port device (figure 4.12(d)) [25], when excited with a 355 nm laser at its left terminal of the BPIN waveguide the generated green FL propagated till junction *a* (figures 4.12(e)–(i)), and split into two signals with the assistance of evanescent coupling. The active green optical signal ($\lambda_{max} \approx 540$ nm) transduced straight to the right terminal *b* of BPIN crystal as output. The evanescently coupled signal enters the junction of the truncated circular shaped DPIN in an anticlockwise manner and excites it via partial ET. The produced orange FL ($\lambda_{max} \approx 600$ nm) spectrum of DPIN outcoupled at terminal *c* revealed an intense active signal and a low-intensity passive green signal.

Due to the excitation terminal *c* of the DPIN crystal (figure 4.12(e-ii)) the orange signal traveled in a clockwise path and divided the signal at the junction *e*. One part of the signal, taking a clockwise path, actively goes to terminal *g*, and the FL spectrum exhibits FP-type modes indicating that the crystal acts as a waveguiding cavity (figure 4.12(h)). The split part of the orange signal ($\approx$600 nm) couples to BPIN–OW passively and outcouples at terminal *h*, without reaching port *f*. Here, clear optical modes were observed in the passively outcoupled signal at terminal *h*. These results confirmed the function of OPIC I to carry out complex optical functions via passive-, active-, and ET mechanisms.

The geometry of OPIC II is intentionally made to couple more light into the truncated ring by bending the BPIN waveguide (figure 4.12(d); see OPIC II). The optical experimental results are analogous to OPIC I.

The geometry of OPIC III is rather characteristic due to the ring-shaped geometry of the DPIN crystal allowing light circulation, and the function is quite like OPIC I and II. But, excitation of the point close to *c* of the DPIN ring delivered an active FL along with altered modes (see for example: disappearance of 640.5 nm mode) at terminal *g* of the ring (OPIC III of figure 4.12(e-vi, h)). These active FL also passively enter BPIN through a junction to reach terminal *h* with optical modes (OPIC III of figure 4.12(e-vi)).

Overall, the analysis of the FL spectra recorded at terminal *g* of OPICs I, II, III and truncated circular shape BPIN crystal (figure 4.12(h)) revealed variations in the optical mode positions and intensity variations. The alteration is due to a change in the interference pattern in geometrically varied OPICs.

4.3.4 A hybrid organic circuit with two ring resonators

Fabrication of a hybrid crystal circuit using two chemically and mechanically (plastic and elastic) different crystals is a challenging task (figure 4.13(a)). For the

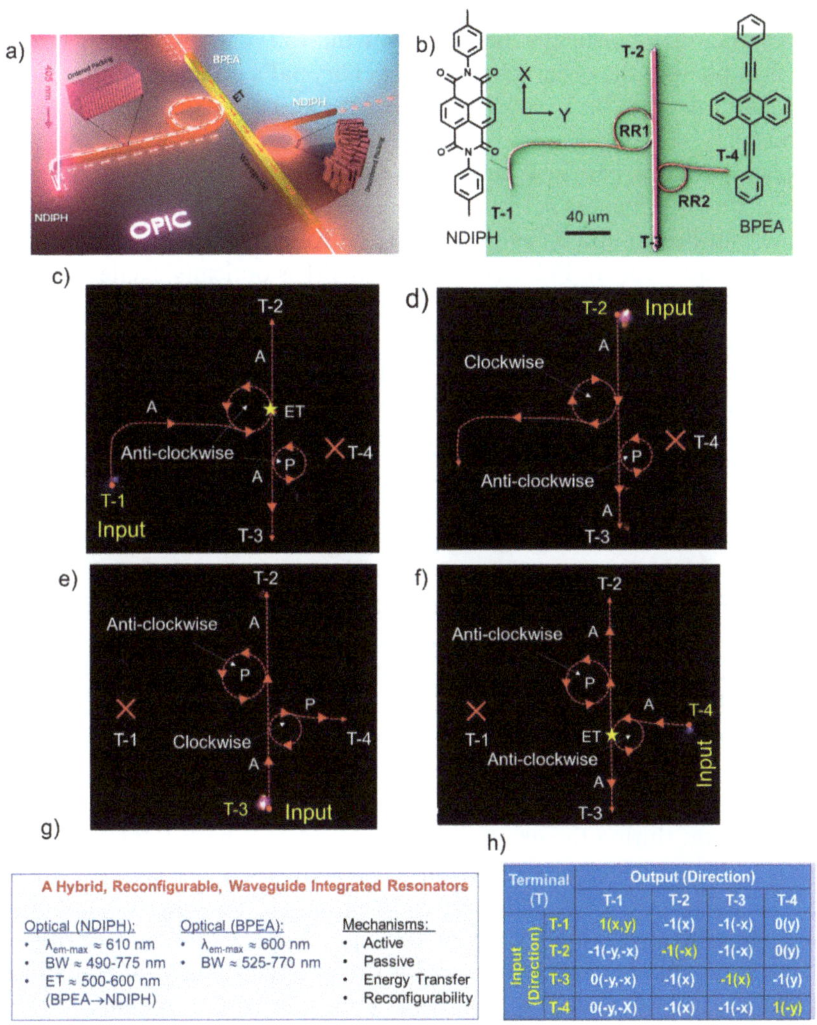

Figure 4.13. (a) Conceptual representation of an OPIC fabricated from NDIPH and BPEA crystals. (b) FESEM image of a reconfigured organic hybrid photonic integrated circuit with one input and three possible output termini. (c–f) FL images for inputs T-1, T-2, T-3, and T-4, respectively. The inset labels show the signal output's type (active/passive) and direction when input is given at terminals T1, T2, T3, and T4. (g) Optical properties and mechanisms of circuit operation. (h) Summary of signal outputs at three different terminals for a single input. [26] John Wiley & Sons. Copyright (2021).

construction, a plastic crystal of *N*,*N'*-bis(*p*-tolyl)-1,4,5,8-naphthalene diimide (NDIPH) and 9,10-bis(phenylethynyl)anthracene (BPEA) crystal were used (figure 4.13(b) insets). A circuit fabricated with two optically different crystals allows the manipulation of light via both active and passive mechanisms (figure 4.13(g)). The circuit facilitates optical ET from NDIPH to BPEA crystal via an evanescent field coupling. As discussed in chapter 3, the construction of a hybrid circuit necessitates selective picking, and transferring crystals from one substrate to another and carefully releasing them at the circuit-building site. NDIPH crystals are plastically bendable on the (001) whereas they break when the force is applied to the other facets. NDIPH crystal's elastic modulus on its (001) facet was determined to be $\mathbf{Y} = 0.95 \pm 0.02$ GPa, and its hardness was $\mathbf{H_c} = 0.30 \pm 0.02$ GPa.

The schematic and FESEM images of the fabricated hybrid circuit with four termini are shown in figures 4.13(a) and (b), respectively [26]. For any given input light, the circuit design allows signal output at two termini and no output at one of the termini (figure 4.13(a)). For example, when T-1 is optically excited with 405 nm light (figure 4.13(c)), the generated red FL (signal 1) reaches RR1 and circulates anticlockwise. This circulating light evanescently excites the BPEA crystal via ET. As a result, the produced FL of BPEA waveguide (signal -1) actively transduces along its *x*-axis to T-2 as -1(x) signal. At T-3, a weak -1($-x$) signal outcouples (figures 4.13(c) and (h). When the light travels towards T-3, part of the signal -1 also couples passively to RR2. However, due to the anticlockwise circulation of light, no light signal (0(y)) reaches T-4, redirecting the light towards T-3.

In the same way, providing light input at T-2 (figures 4.13(d) and (h)) delivers passive -1($-y,-x$) and active -1($-x$) output signals to T-1 and T-3, respectively, without directing any signal (0(y)) towards T-4 (figures 4.11(d) and (h)). When T-3 receives an input signal (figures 4.13(e) and (h)), due to the clockwise circulation of light within RR2, it pushes a passive light towards T-2 (signal -1(x)) and also outcouples at T-4 as a feeble passive -1(y) signal along with optical modes. No signal reaches T-1, due to anticlockwise light circulation at RR1. Also, excitation at T-4 (figures 4.13(f) and (h)) delivers output signals -1(x) and -1($-x$) to at T-2 and T-3, respectively, without allowing any light (0($-y,-x$)) to T-1.

As one can see, during the excitation of T-1 and T-4, both RR1 and RR2 route the signals in the anticlockwise direction. However, when T-2 and T-3 are excited, both RR1/RR2 routes of signals are clockwise/anticlockwise and anticlockwise/clockwise, respectively. This switch in the light-routing direction during specific excitations highlights the versatile behavior of the circuit. Such intricacies in signal routing demonstrate the adaptability of the optical structure, making it a dynamic and responsive component in the circuit.

Here, utilizing the plasticity of NDIPH crystal, any geometrical reconfiguration in 2D can be performed by mechanically bending the (001) facet. This circuit reconfigurability facilitates modifying the waveguide's optical output trajectories (x, $-x$, y, and $-y$) contingent upon the necessity without recreating the whole circuit.

4.3.5 A reconfigurable hybrid circuit made from spiral and straight waveguides

In an ADF, two waveguides and one ring resonator are used. Similarly, spiral waveguides discussed in (section 4.1.4) can be used to fabricate an ADF analog when it is coupled with straight waveguides [13]. One main difference is, that here, the ring resonator in an ADF (see sections 4.3.1 and 4.3.2) is replaced with a polygonal OSW.

As the device has to be hybrid to involve ET between the crystal components, four molecular flexible crystals, namely, Cz-2CF_3, 9,10-dibromo anthracene (DBA), and 4,7-bis(4-octylthiophen-2-yl)benzo[c] [1, 2, 5] thiadiazole (BTD), and BPyIN, were chosen (figure 4.14 (a)). The optical data show that Cz-2CF_3, DBA, BPyIN, and BTD FL maxima sequentially shifted to ≈498, 527, 538, and 595 nm (figure 4.14(b)). Interestingly, the Φ_{FL} and FL lifetime of blue-emitting Cz-2CF3 are 36.7% and 2.6 ns, respectively. Here, due to the careful selection of molecular crystals, Cz-2CF_3 crystals can transfer optical energy to DBA, BPyIN, and BTD. Similarly, ET from DBA to BTD is also possible.

At first, to fabricate a spiral circuit (SC) using three chemically different crystals, Cz-2CF_3's polygonal SW (PSW), red-emissive BTD ($L \approx 200$ μm; $\alpha' \approx 0.1064$ dB μm^{-1}), and cyan-emitting DBA ($L \approx 312$ μm; $\alpha' \approx 0.0558$ dB μm^{-1}) waveguides grown on different substrates were *transferred* to PSW containing substrate and *integrated* with PSW employing micromechanical operations (figure 4.14(c,d)) [13]. The FL image of the resultant SC displayed three operational colors (blue, cyan, and red) and six bright crystal termini (figure 4.14(e)). Interestingly, excitation of P-2 with a 405 nm light produced blue, cyan, and red FL signals at P-1, P-4, and P-6, respectively, via active and ET mechanisms covering from deep blue to the near IR part of the electromagnetic spectrum. Smartly, the circuit can limit the number (and wavelength) of signals that can be generated from one (red), two (cyan and red), and three (blue, cyan, and red) when the input is given at P-5, P-3, and P-2, respectively (figure 4.14(f)).

Circuit reconfigurability is a requirement for accomplishing programmable circuits. To demonstrate this, the DBA crystal waveguide was disconnected from the circuit and a BPyIN microcrystal waveguide ($L \approx 390$ μm; $\alpha' = 0.0669$ dB μm^{-1}) was integrated. Further, the left terminal of the BPyIN microcrystal was positioned in such a way as to make a tip-to-tip contact (junction J1) with the BTD left terminal. This simplistic approach helps in rechanneling the light to the right termini of BTD or BPyIN and also reduces the number of ports from six to three in the reconfigured spiral circuit (figures 4.14(g) and (h)). The excitation of P-2 with a laser beam produced blue FL (Signal:λ_1(A)) in the PSW, which transduced in a clockwise (CW) direction. On its way, this λ_1(A) signal excited the BPyIN waveguide via ET from Cz-2CF_3 to BPyIN to produce signal λ_4(ET), which outcouples at junction J1, generating another signal λ_3(ET), due to radiative ET from BPyIn to BTD. At P-6, a guided and slightly reabsorbed red signal $-\lambda_3$(ET) was detected (figure 4.14(j)).

An assessment of the experimental results indicates that the reconfigured spiral circuit effectively reduced the number of output signals and delivered input-specific outputs.

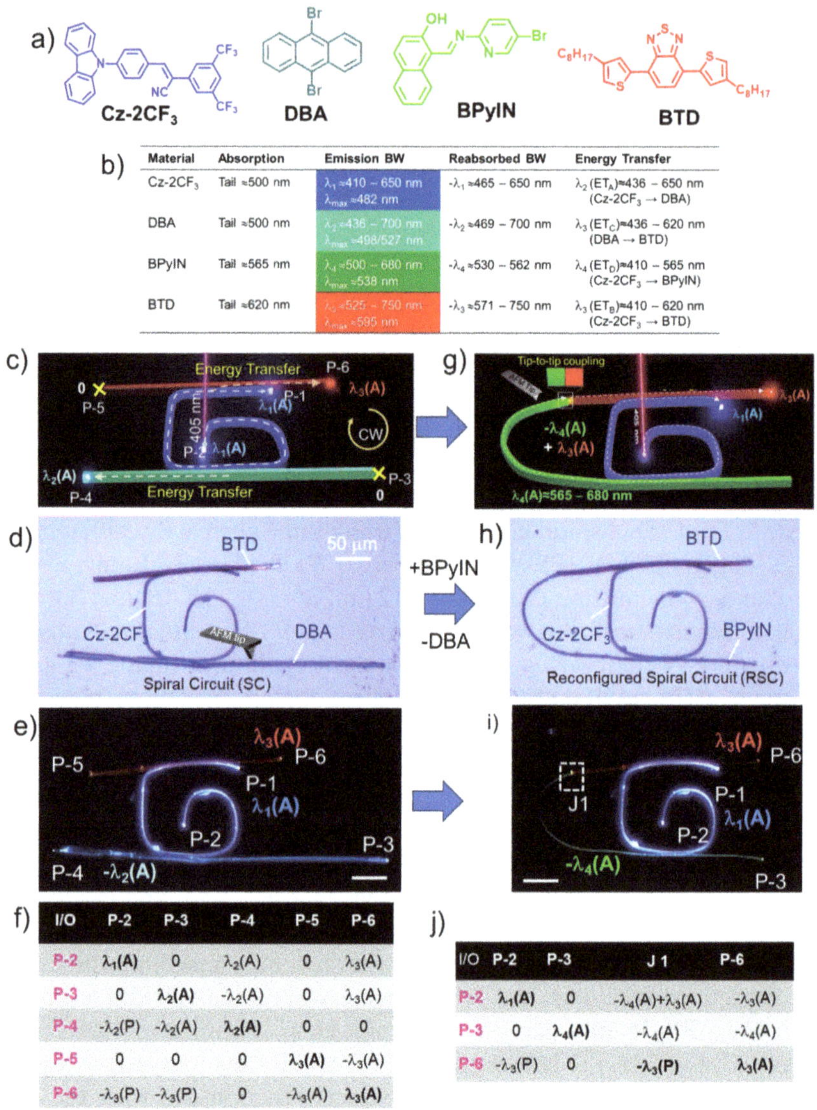

Figure 4.14. (a) Molecular structures of different materials used in this study. (b) Table demonstrating absorption tail, emission bandwidth (BW), reabsorbance BW, and intermolecular energy transfer. (c, g) Graphical illustration of optical functioning of the spiral circuit (SC) and reconfigured spiral circuit (RSC) for 405 nm laser light input. The corresponding (d, h) confocal optical and (e, i) FL images showing RSC. (f, j) Photonic response of SC and RSC for 405 nm laser input at various ports. The bold letters indicate excitation and collection at the same port. Labels A and P stand for active and passive, respectively. [13] John Wiley & Sons. Copyright (2023).

4.4 Organic crystal optical interferometers

Optical interferometers (IMs) allow interference (constructive and destructive) of two light signals of the same wavelength traveling in different path lengths producing wavelength and intensity modulated optical signals depending upon phase matching between the two signals.

Interferometers come in various types, namely, F–P and Michelson being well-known examples. Another widely utilized type is the Mach–Zehnder interferometer (MZI). MZIs find extensive applications as channel drop filters, modulators, and switching devices in OPICs.

Takazawa *et al* fabricated an asymmetric MZI interferometer (AMZI) using TC dye nanofibers (see chapter 2, section 2.1.4) as shown in figure 4.15(a) [27]. To manipulate the nanofibers at the microscale, they employed a glass tip attached to a micromanipulator, observed under a microscope (figure 4.15(b)). The use of a thin-oil layer ($n = 1.48$) prevented the nanofibers from adhering to the glass substrate, enhancing the manipulation process.

The purpose of the AMZI is to split the input light into two beams, which are subsequently recombined after traveling along paths of varying lengths. This path length difference, in this case, is 26.4 μm. Constructing this miniaturized AMZI using optical waveguides involved bending the waveguides (Fibers 1 and 2) at a small radius to achieve the desired path length difference (figures 4.15(c), (d), (f) and (g)). The interference resulting from the phase difference between the two beams propagating in fiber 1 and 2 generates fringes in the output FL signal in the AMZI (figure 4.15(h)) compared to uncoupled fibers (figure 4.15(e)). These fringes are F–P modes that occur due to reflection of the light at both end faces of the nanofiber. It is

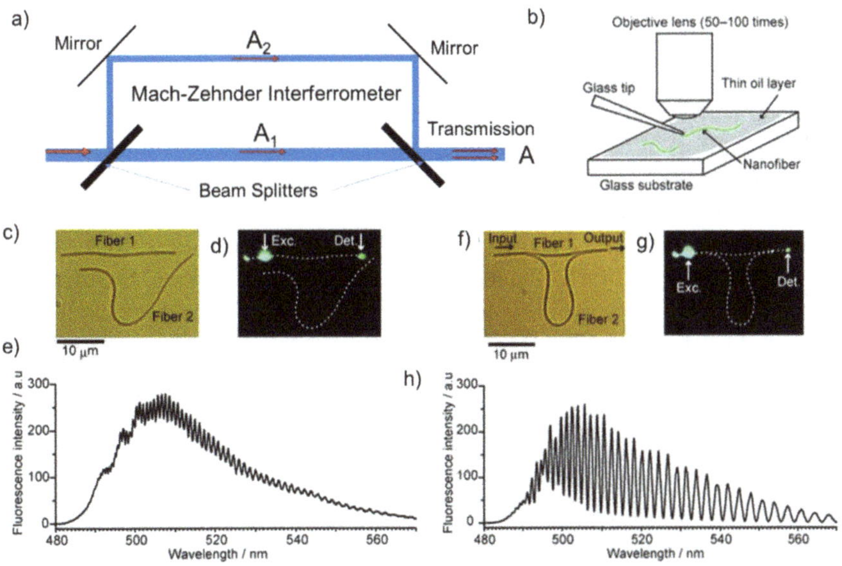

Figure 4.15. (a) Schematic of TC dye-based Mach–Zehnder interferometer. (a) Reproduced from [41] RSC. Copyright (2022). (b) Schematic illustration of micromanipulation of nanofibers. (c, f) Optical microscopy images of nanofibers (Fibers 1 and 2), which were used for the AMZI fabrication. (d, g) FL microscopy image of Fiber 1 recorded by exciting the position labeled 'Exc.' with a laser spot. Dashed lines representing fibers 1 and 2 are guides for the eyes. (e, h) Spatially resolved FL spectrum measured at the position labeled 'Det.' in (d, g). (b-h) [27] John Wiley & Sons. Copyright (2013).

noteworthy that the detectability of the output signal is influenced by the numerical aperture of the objective used.

A fiber loop mirror (FLM) is an example of a Sagnac interferometer [28], in which two counter-propagating optical signals inside the loop, with identical optical paths, optically interfere to generate a new signal. FLM is a slightly modified version of 2×2 DC and it can be created by connecting the two ports on one side, producing a loop geometry. FLMs are used as reflective optical elements in photonic devices, sensing [29, 30], and wavelength division multiplexing (WDM) [28, 31, 32]. Conventionally, materials with high-birefringence [33], maintaining polarization [34], or photonic crystals—are used for FLM fabrication [35].

My group constructed an organic crystal interferometer using a green-emitting BPyIN flexible crystal waveguide cavity of length, $L \approx 123$ μm and $T \approx 343$ nm (figure 4.16(a)) [3]. The optical properties of BPyIN are given in figure 4.14(b). Through multiple *moving* operations using an AFM cantilever tip a pre-FLM was shaped, wherein the gap between the two neck portions is 2 μm (figure 4.16(b)). Later, the gap was removed by physically connecting the two necks, thereby creating a FLM (figure 4.16(c)). For the fabrication of an IM using FLM, a straight microcrystal waveguide cavity ($L \approx 147$ μm) was micromechanically bent ($\zeta = 4.7\%$) on both sides of the termini and positioned in such a way that the convex part of the waveguide is parallel to FLM termini (figure 4.16(d)). Later, this doubly-bent waveguide cavity was physically integrated with FLM carefully to achieve an IM geometry (figure 4.16(e)). The FESEM image of the IM showed that the diameter of the FLM loop is ≈21 μm ($\zeta = 10.3\%$), and the neck region thickness is ≈512 nm (figure 4.16(n)).

The bending-induced effects on the photonic properties of the crystal were investigated for the straight waveguide cavity and pre-FLM (figure 4.16(a) and (b)). The optical modes extracted from the FL spectra are labeled as (a) 1–2 (denoting excitation at terminal-1 and spectral collection at terminal-2) and (b) 1–2 (figure 4.16(g) and (h)). Interestingly, both optical spectra were mostly comparable except for slightly intense modes (550.8, 573.1, 576.6, and 587.2 nm) and a noticeable new peak at 555.2 nm for pre-FLM compared to the straight waveguide cavity. Another observation is the lessening of FWHM of a broad (multiple modes) highly intense peak at ≈564 nm (indicated by a horizontal black double-headed arrow) from 11.7 nm (figure 4.16(g)) to 6.8 nm in pre-FLM (figure 4.16(h)).

Similarly, the spectra containing modes at output termini (2 or 3) for the same excitation at 1 or 4 were compared for the straight waveguide cavity, pre-FLM, FLM, doubly-bent waveguide cavity, and IM (figures 4.16(g)–(l)). In the FLM, its geometry allows transducing light to split at the evanescent coupling region into two beams propagating in opposite circular paths (clockwise and anticlockwise) causing interference for subsequent modulation of the output signal (figure 4.16(c) and (m)). As a result, distinct optical modes were obtained compared to pre-FLM (figure 4.16(h) and (i)). Further, the intensity increases in the 587.2 nm mode and decreases in the 573.1, 571.8, and 593.7 nm modes in FLM. A slight change in the FWHM of peak (multiple modes) at ≈564 nm was observed.

In the IM, a relative alteration of the mode intensities was observed. Pronounced effects could be seen when terminal 4 of IM was excited. A notable change is the

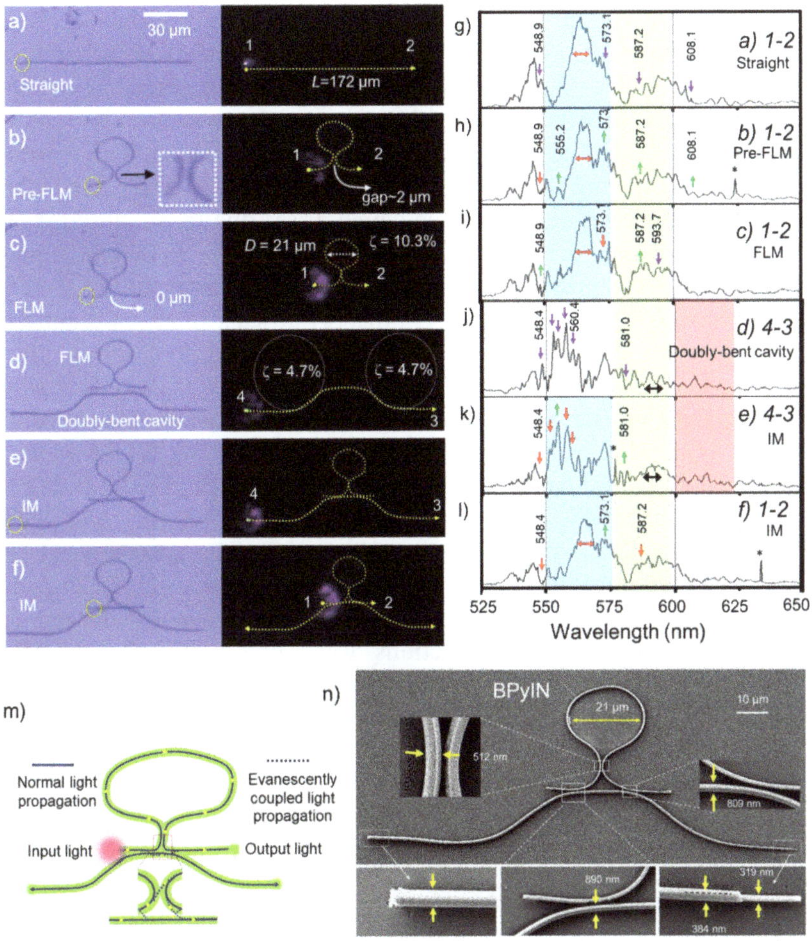

Figure 4.16. (a–f) Stepwise fabrication of an organic interferometer using BPyIN single crystals via mechanical micromanipulation. The yellow circles show the excitation area. (g–l) Optical modes extracted from the FL spectrum obtained for a–f, when the left tip was excited and emission was collected at the right tip (*indicates noise). (m) Schematic representation of a light propagating path inside the interferometer when excited at tip 1. (n) FESEM of fabricated organic IM (insets show a close-up view of the coupling junctions and the tips of the bottom waveguide). [3] John Wiley & Sons. Copyright (2022).

suppression or disappearance of 548.4 and 560.4 nm modes compared to the doubly-bent waveguide cavity (figure 4.16(k) and (l)). The mode intensities in the region of 550–575 nm have shown marked variations in the IM. The obtained results demonstrate the applicability of flexible crystal waveguide cavities for fabricating compact optical IM which can be integrated into OPICs.

4.5 Wavelength division multiplexers (WDMs)

For visible light communication (VLC)-related photonic technologies, the WDM is a vital element as different light signals (so-called *multiplexed* signals) have to be combined and transmitted in a single waveguide. The multiplexers based on inorganic optical fibers changed the way data can be transferred and received.

Liao *et al* used a core–shell crystal growth strategy for transporting signals of multiple colors [36]. Nonetheless, the fabrication of a waveguide that can intrinsically guide light to cover the entire visible region is challenging. The reason is that molecular crystals emitting white light are very scarce. He *et al* reported white light emission from a single organic molecule which arises due to the absence of room-temperature dual phosphorescence [37]. Generally, white light emission can be realized by mixing three primary colors or two complementary colors. To cover the full visible spectrum, a waveguide must transport at least three different emissions (λ_1, λ_2, and λ_3) which are self-produced or passively coupled to it, though the former is near to impossible with the FL materials. A plausible strategy is a waveguide emitting FL band (λ_1) that should guide two different passive signals of different wavelengths (λ_2 and λ_3), enabling a multiplexing ($\lambda_1+\lambda_2+\lambda_3$) WDM waveguide.

To realize single-crystal based WDM, my group used BPP, DBA, and (*Z*)-2-(3,5-bis(trifluoromethyl)phenyl)-3-(7-methoxybenzo[c][1, 2, 5]thiadiazol-4-yl)acrylonitrile ($BTD2CF_3$) flexible crystals (figure 4.17(a)) [38]. The x-ray crystal structure of $BTD2CF_3$ showed intramolecular hydrogen bonding (C–N···H: 2.425 Å) between the nitrogen group of the benzothiadiazole and adjacent vinylic hydrogen. Further, C–H···π, C–F···H and F···F interactions were present in the crystal lattice (figures 4.17(c) and (d)). Application of mechanical force perpendicular to the (110) plane of $BTD2CF_3$ crystal revealed its elasticity. It is expected that along the crystallographic *b*-axis, the channels of intermolecular hydrogen bonds provide stress relief. When the same experiments were performed in other directions, the crystal did not show elasticity. In the DBA crystal the slip planes principally control the mechanical flexibility [39]. Like DBA and BTD crystals, $BTD2CF_3$ also revealed its pseudo-plasticity.

For WDM construction, a BPP (FL bandwidth ≈ 420–700 nm; λ_1—blue) crystal waveguide was used as a main active/passive waveguide and laterally coupled to 177 μm long DBA (FL bandwidth 445–705 nm; λ_2—cyan; $\alpha' = 0.0752$ dB μm^{-1}) and 130 μm long $BTD2CF_3$ (FL bandwidth 480–755 nm; λ_3—yellow) flexible crystals as adjacent waveguides. These adjacent waveguides channel the corresponding emissions into BPP via evanescent coupling-mediated ET (figures 4.17(e)–(h)).

The micromechanical integration of DBA and $BTD2CF_3$ crystals with BPP microcrystals was achieved by lifting DBA and $BTD2CF_3$ crystals from their corresponding growth substrate and dropping them at the required construction site of another substrate adjacent to the selected BPP crystal. Finally moving, bending, and integrating micromechanical operations provided the required WDM (figures 4.17(e)–(h)). The FESEM image of the WDM revealed the smooth surface morphology of the crystals, and both left termini of DBA and $BTD2CF_3$ crystals are in direct physical contact with BPP crystal for the effective ET to take place (figure 4.17(k)). The FESEM image also revealed that the thicknesses of BPP, DBA, and $BTD2CF_3$ microcrystals are ≈2.9, 1.8, and 0.6 μm, respectively, though the thicknesses slightly varied along the crystal's long axis. The distance between the

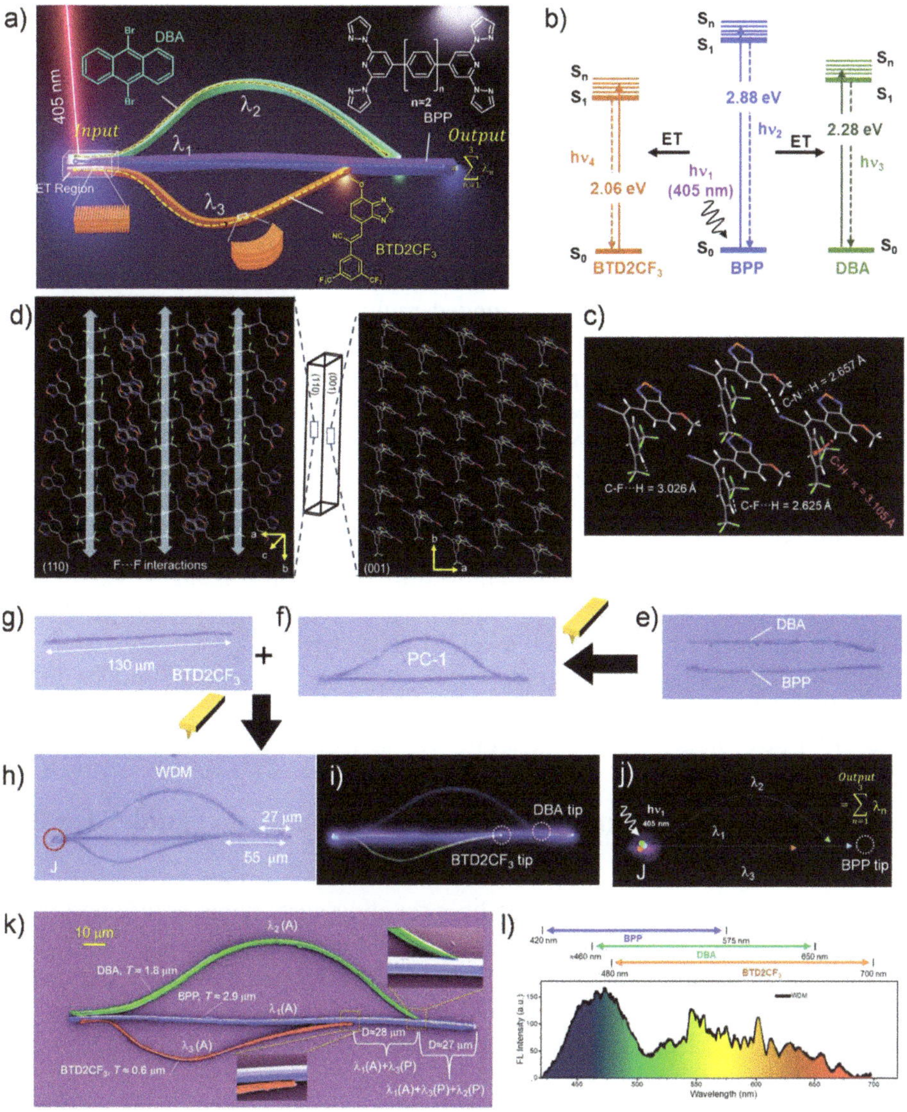

Figure 4.17. (a) Graphics depicting a wavelength divisional multiplexer (WDM) from flexible organic crystals. Insets show the molecular structures of BPP, DBA, and BTD2CF$_3$. (b) Jablonski diagram depicting the mechanism of ET between different crystals. (c) Crystal packing of BTD2CF$_3$ supported by different noncovalent interactions. (d) Molecular packing along (110) and (001) planes in BTD2CF$_3$ crystal, respectively. The double-headed arrows in d represent the intermolecular F···F interactions along [110]. Confocal optical images of DBA and BPP crystals (e) before and (f) after integration, respectively, to construct the photonic component (PC-1) shown in (f). (g) Micromechanical integration of BTD2CF$_3$ crystal and PC-1 to fabricate a WDM. (h, i) Confocal optical and FL microscope images of a WDM, respectively. (j) FL image of a WDM when the left end was excited with 405 nm laser light. (k) Color-coded FESEM image of an organic WDM. The insets show the close-up view of the junctions. (l) Optical spectrum (black line) detected at the BPP right tip possessing a multiplexed signal for collective excitation of the left side of a WDM. [38] John Wiley & Sons. Copyright (2022).

right termini of DBA and BPP crystals is 27 μm, while the distance between $BTD2CF_3$ and BPP is 55 μm.

In the solid state, due to the overlap of BPP FL within the optical absorbance of DBA and $BTD2CF_3$, efficient ET-triggered FL (λ_2 and λ_3) from the latter two molecular crystal waveguides is possible (figures 4.17(b) and (f)). The Jablonski diagram exemplifies the ET process in BPP, DBA, and $BTD2CF_3$ crystals (figure 4.17(b)). Due to the intermolecular interaction-assisted rigid molecular framework, the $BTD2CF_3$ exhibited a high absolute Φ_{FL} of 0.49.

During the simultaneous excitation of the three waveguide left termini with a 405 nm laser, the WDM effectively guided multiplexed ($\lambda_1+\lambda_2+\lambda_3$) FL signals as demonstrated by the FL spectra recorded at the right BPP waveguide terminal (figure 4.17(l)). The multiplexing takes place through the passive coupling of signals λ_2 and λ_3 from DBA and $BTD2CF_3$, respectively, into the BPP waveguide. The BPP waveguide transduced a composite of λ_1(A) and λ_3(P) signals spanning the 420–650 nm range over a length of approximately 28 μm. Subsequently, it facilitates the simultaneous transmission of three multiplexed signals (λ_1(A)+λ_2(P)+λ_3(P)) with a combined bandwidth of 420–700 nm, covering a length of 27 μm. This spectral outcome illustrates a gradual expansion in bandwidth, progressing from 420–575 nm to 420–700 nm, achieved through the multiplexing capabilities of the BPP crystal waveguide, supported by the crystal's WDM geometry.

Notably, the multiplexed output signal preserves the optical modes initially observed in the waveguide, even after their integration. These outcoupled optical modes offer advantages for multi-channel visible light data communication.

4.6 LED-powered organic single-crystal optical waveguides for VLC technologies

VLC technology harnesses natural light, providing sustainable solutions for both indoor and outdoor environments. Its adaptability to ambient light sources like sunlight and LEDs enhances its appeal. Therefore, it is crucial to develop innovative devices compatible with prevalent ambient light. Creative device-powering strategies, focusing on LEDs, are crucial for enhancing efficiency and reliability while promoting sustainability in photonics.

As explored earlier, advancements in flexible organic crystals have enabled the creation of photonics components and circuits driven by commercial lasers. Despite their benefits, lasers pose challenges such as high cost, power consumption, and material damage at high power. Transitioning to commercial LEDs as light sources provides a pragmatic remedy to the constraints associated with laser utilization and a cost-efficient alternative. Innovative design approaches, integrating appropriate organic materials and focused LED sources, are essential for realizing efficient photonic systems and circuits.

In this context, creating a proof-of-concept device capable of powering organic crystal waveguides, both actively and passively based on reabsorption and energy transfer principles, represents a crucial milestone in leveraging low-power LED sources for photonic systems. My research group has engineered focused

LED-powered organic waveguides and circuits, with full visible spectral operability. This feat was accomplished using three flexible crystals of blue-emissive CF_3OMe, orange-emissive BPEA, and yellow-emissive 2,2′-((1*E*,1′*E*)-hydrazine-1,2-diylidene-bis(methaneylylidene))diphenol (SAA), shown in figure 4.18(a). The **Y** and $\mathbf{H_c}$ of the SAA crystal were measured to be 0.22 and 6.49 GPa, respectively.

The solid-state absorption and emission data (figure 4.18(b)) revealed that the absorption of SAA and emission of CF_3OMe overlap within the 420–500 nm range, suggesting the potential for ET from CF_3OMe to SAA crystal. Similarly, ET is feasible from SAA to BPEA due to the absorption of BPEA overlapping with the emission of SAA within the 495–600 nm range. Additionally, ET from CF_3OMe to BPEA is possible within the 420–600 nm region. There is no ET from SAA to CF_3OMe.

To construct the LED-powered photonic circuit, we utilized a glass slide featuring a thin aluminum foil inscribed with a 40 μm diameter hole on its rear side to enable light illumination. A commercial microscale UV LED was tactically positioned on the foil, precisely aligning with the hole to allow focused light transmission (figure 4.19(a)). The focused light from the aperture powers (37 μW) the organic crystal optical waveguides and 2×2 hybrid directional coupler, strategically located on the reverse side of the glass slide to capture the LED illumination.

A CF_3OMe single crystal (length, $L \approx 2.69$ mm, and thickness, $T \approx 143$ μm) was positioned on the coverslip, with its left tip precisely aligned with the focused LED light (figures 4.19(e) and (f)). Upon activating the LED, the CF_3OMe crystal emitted blue FL ($\lambda_1 \approx 420$–665 nm), confirming its active-type waveguiding capability as the FL was transmitted to the opposite end (figure 4.19(f)). To determine the α' of the crystal waveguide, different zones along its long axis were exposed to LED light using a needle, with the resulting FL recorded at the right end of the waveguide (figures 4.19(f) and (i)). The α' was calculated to be 0.001 21 dB μm^{-1} (figure 4.19(k)). Similarly, an SAA crystal ($L \approx 1.11$ mm) was optically excited with focused LED light at its left tip. The generated yellow FL ($\lambda_2 \approx 495$–720 nm)

a)

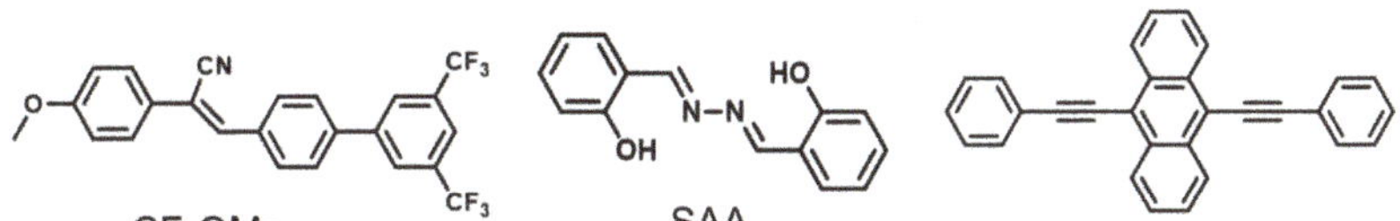

b)

Material	Absorption (nm)	Emission BW (nm)	PLQY (%)	Reabsorbed BW (nm)	Energy Transfer
CF_3OMe	Tail ≈ 470	$\lambda_1 \approx 420-665$ $\lambda_{max} \approx 492/524$	36	$-\lambda_1 \approx 435-665$	$\lambda_2(ET_A) \approx 420-500$ nm ($CF_3OMe \rightarrow$ SAA)
SAA	Tail ≈ 500	$\lambda_2 \approx 495-720$ $\lambda_{max} \approx 568$	23	$-\lambda_2 \approx 515-720$	$\lambda_3(ET_C) \approx 495-600$ nm (SAA → BPEA)
BPEA	Tail ≈ 600	$\lambda_3 \approx 550-770$ $\lambda_{max} \approx 595$	24	$-\lambda_3 \approx 580-770$	$\lambda_3(ET_B) \approx 420-600$ nm ($CF_3OMe \rightarrow$ BPEA)

Figure 4.18. (a) Molecular structure of CF_3OMe, SAA and BPEA. (b) Details of solid-state absorption maxima, emission bandwidth, photoluminescence quantum yield, reabsorption, and energy transfer. Reproduced from [40].

was transmitted to the opposite tip of the crystal (515–720 nm) with a slight reduction in bandwidth due to self-reabsorption of the optical signal (figures 4.19(g) and (h)). The α' for the SAA crystal was determined to be 0.0117 dB μm^{-1} (figures 4.19(j) and (k)).

Building photonic circuits hinges on the strategic utilization of active and passive waveguiding in the circuit design. Thus, to create a hybrid active/passive waveguiding optical system using CF_3OMe and SAA crystal waveguides, precise positioning of the two crystals near the LED light source was essential. Initially, both crystals were placed on the same glass slide and then meticulously mechanically repositioned using a sharp stainless-steel needle, resulting in the creation of the hybrid waveguide. Both crystals were arranged in a tip-to-tip configuration to facilitate efficient optical coupling between the waveguides (figure 4.20(a)). FL imaging of the hybrid system reveals the illuminated body and tips of both

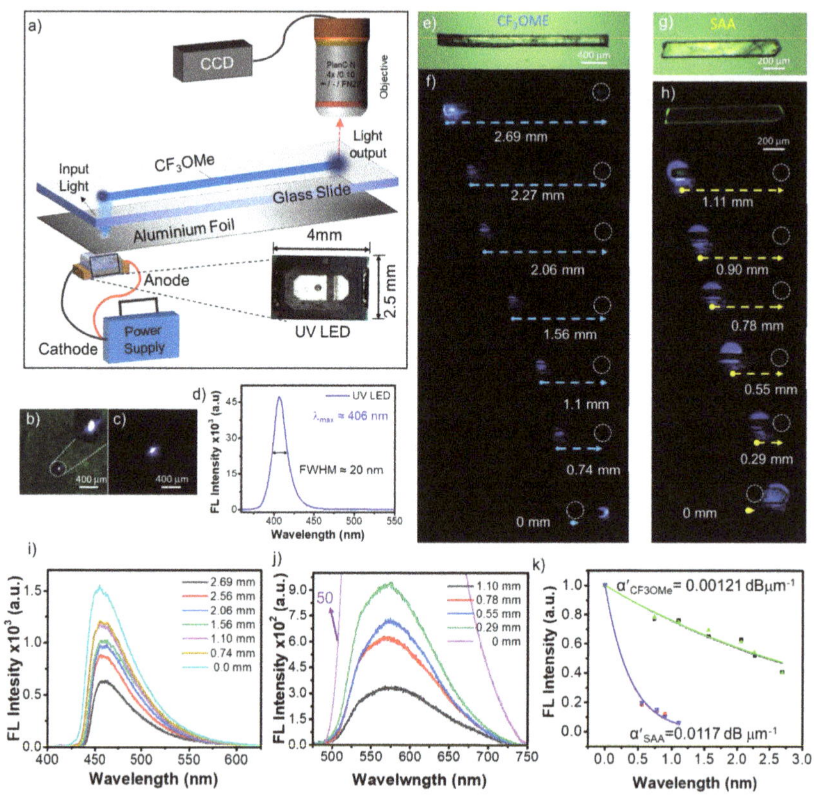

Figure 4.19. (a) Graphical representation of the device used for the demonstration of a focused UV LED-powered organic crystal waveguide. (The bottom right inset shows the photograph of a commercial UV LED). (b, c) Confocal optical and FL images of UV LED light emanating out of a 40 μm diameter hole in aluminum foil, respectively. (d) The optical emission spectrum of UV LED shows $\lambda_{max} \approx 406$ nm and FWHM ≈ 20 nm. (e, f) Confocal optical and FL images of CF_3OMe, respectively. (g, h) Confocal optical and FL images of SAA, respectively. (i, j) FL spectra of CF_3OMe and SAA, respectively. (k) I_{tip}/I_{body} vs distance plot for CF_3OMe and SAA. Reproduced from [40]. CC BY 4.0.

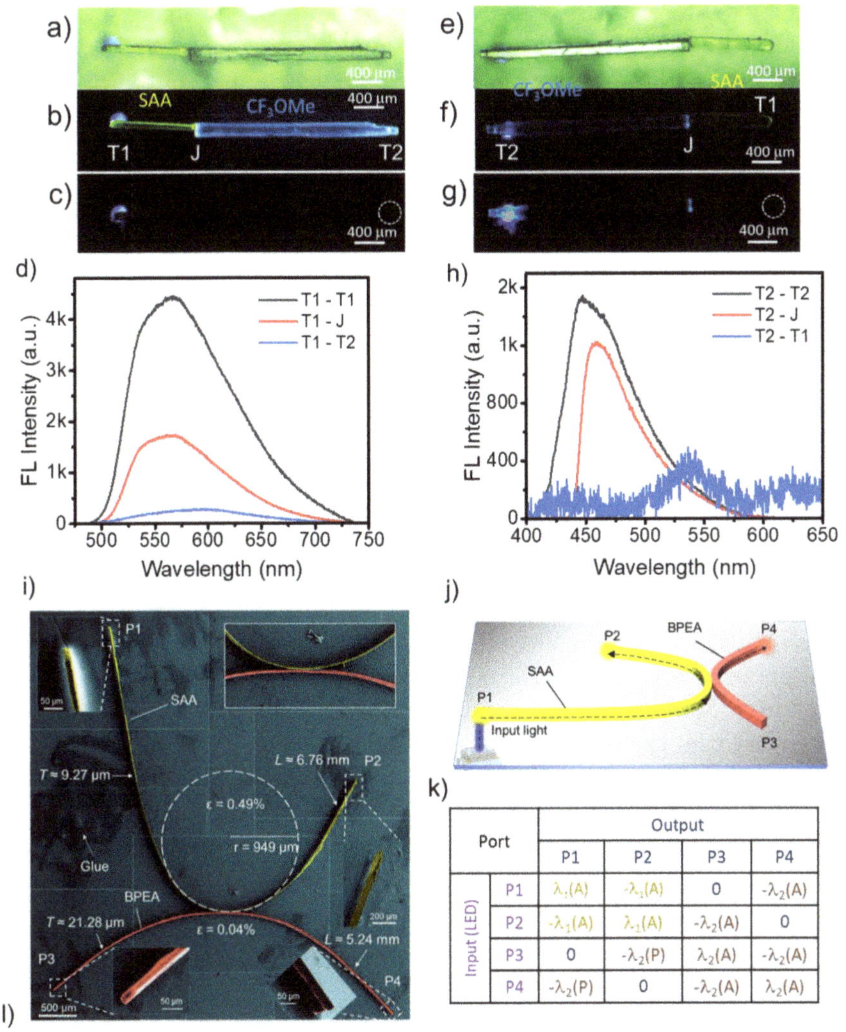

Port		Output			
		P1	P2	P3	P4
Input (LED)	P1	λ_1(A)	-λ_1(A)	0	-λ_2(A)
	P2	-λ_1(A)	λ_1(A)	-λ_2(A)	0
	P3	0	-λ_2(P)	λ_2(A)	-λ_2(A)
	P4	-λ_2(P)	0	-λ_2(A)	λ_2(A)

Signal	λ_2(A)	-λ_1(A)	λ_3(A)	-λ_2(A)	-λ_2(P)	-λ_2(A)	0
Type	Active	Reabsorbed	Active	Reabsorbed	Passive	Energy Transfer	NIL
BW (nm)	≈ 495 – 720	≈ 515 – 710	≈ 550 – 770	≈ 580 – 770	≈ 583 - 770	≈ 555 - 770	-

Figure 4.20. (a) Confocal optical, (b) FL (under UV torch excitation), and (c) FL microscope (when LED light input is provided at crystal's left end) images of an SAA-CF_3OMe photonic system. (e) Confocal optical, and (f, g) FL images of a CF_3OMe-SAA photonic system when LED light input is provided at the CF_3OMe tip. (d–h) FL spectra recorded at T1 and T2 when LED light is provided at (d) T1 and (h) T2. (i) Stitched color-coded FESEM images of the directional coupler. The inset shows the zoomed FESEM image of the junction of the waveguides. (j) Graphical representation of a directional coupler of SAA and BPEA flexible crystals excited with LED at P1. (k) Input-dependent directional-specific optical outputs obtained in the fabricated hybrid directional coupler. (l) The tabulated signal type and their specific wavelength ranges. Reproduced from [40]. CC BY 4.0.

SAA and CF_3OMe crystals when excited with a UV torch (figure 4.20(b)). Upon illumination with UV LED light at the left end of the SAA crystal, denoted as T1 (figure 4.20(c)), a vibrant yellow FL was emitted. This signal continued through the SAA crystal to junction J (figure 4.20(b)) and evanescently coupled into the CF_3OMe crystal and propagated passively to the opposite end of the CF_3OMe crystal at T2 (figure 4.20(d)).

Subsequently, the SAA and CF_3OMe waveguides were adjusted using a sharp needle to ensure optimal optical signal coupling via energy transfer from the former to the latter. By introducing a UV LED at the left end of the CF_3OMe crystal, a vibrant blue emission occurred at the excitation point and at its right terminal near the junction (figures 4.20(f)–(h)). This emitted blue FL ($\lambda_1 \approx 420$–670 nm) at the junction effectively excited the SAA crystal through ET from CF_3OMe to SAA, resulting in the generation of yellow FL ($\lambda_2 \approx 490$–720 nm) at the left end of the SAA crystal at J. Subsequently, the yellow FL propagated to the other end of the SAA crystal (T1), demonstrating active signal propagation at the right end of the SAA crystal. This experiment demonstrated the functionality of the LED-powered hybrid optical system, operating as active/passive waveguides with low-power LED light.

In constructing photonic circuits, directing light through various pathways is essential. Therefore, it is customary to demonstrate light propagation in curved waveguides, which can then be utilized in hybrid directional couplers (HDC) for practical applications (figure 4.20(j)).

Therefore, a SAA crystal was bent using a sharp metallic needle into a curved geometry. This bent geometry was retained by fixing it with glue on a glass coverslip and placed on the device set-up (figure 4.20(i)). The bent SAA crystal also guided light under LED illumination with an $\alpha' \approx 0.003$ dB μm^{-1}. This bent SAA was carefully integrated with a BPEA waveguide ($L \approx 5.24$ mm; $\alpha' = 0.004$ dB μm^{-1}) to create a HDC (figures 4.20(i) and (j)). The FESEM images reveal the thickness of 9.27 and 21.28 μm for the SAA and BPEA waveguides, respectively, and a clear physical contact between the two (figure 4.20(i), inset). Various ports in HDC are represented as P1–P4.

The HDC's optical performance was tested with LED light at P1 and signals recorded at other ports. At P1, the produced yellow signal (λ_1(A)) was transduced towards P2 as a reabsorbed $-\lambda_1$(A) signal. At P4, an orange signal (λ_2(A)) from ET from SAA to BPEA was observed. Limited signal propagation backward prevented light at P3. Optical input at P2 yielded yellow (λ_1(A)) at P1 and orange (λ_2(A)) at P3, no signal at P4 due to HDC's directionality. P3 or P4 input resulted in only an orange signal at P2 or P1 due to the absence of ET from BPEA to SAA (see figures 4.20(k) and (l)).

This advancement has the potential to revolutionize sustainable VLC technologies utilizing organic crystals.

References

[1] Annadhasan M, Karothu D P, Chinnasamy R, Catalano L, Ahmed E, Ghosh S, Naumov P and Chandrasekar R 2020 Micromanipulation of mechanically compliant organic single-crystal optical microwaveguides *Angew. Chem. Int. Ed.* **59** 13821–30

[2] Rohullah M, Vinay Pradeep V, Singh S and Chandrasekar R 2023 Mechanically-controlled multifaceted dynamic transformations in twisted organic crystal waveguides *Nat Commun.* **15** 4040
[3] Kumar A V and Chandrasekar R 2023 A mechanophotonic approach towards an organic, flexible crystal optical interferometer *Adv. Opt. Mater.* **11** 2201009
[4] Zhao S, Yamagishi H, Oki O, Ihara Ichiji Y N, Kubo A, Hayashi S and Yamamoto Y 2021 Mechanically flexible and optically tunable organic crystal resonator *Adv. Opt. Mater.* **10** 2101808
[5] Bogaerts W, Dumon P, Thourhout D V, Taillaert D, Jaenen P, Wouters J, Beckx S, Wiaux V and Baets R G 2006 Compact wavelength-selective functions in silicon-on-insulator photonic wires *IEEE J. Sel. Top. Quantum Electron.* **12** 1394–401
[6] Hayes R and Yap D 1993 GaAs spiral optical waveguides for delay-line applications *J. Light. Tech.* **11** 523–8
[7] Lee H, Chen T, Li J, Painter O and Vahala K J 2012 Ultra-low-loss optical delay line on a silicon chip *Nat. Commun.* **3** 867
[8] Redding B, Liew S F, Bromberg Y, Sarma R and Cao H 2016 Evanescently coupled multimode spiral spectrometer *Optica* **3** 956–62
[9] Simard A D, Painchaud Y and LaRochelle S 2013 Integrated Bragg gratings in spiral waveguides *Opt. Express* **21** 8953–63
[10] Tang S, Liu S, Yu X, Song Q, Gong Q and Xiao Y 2018 On-chip spiral waveguides for ultrasensitive and rapid detection of nanoscale objects *Adv. Mater.* **30** 1800262
[11] Chauhan N *et al* 2021 Visible light photonic integrated Brillouin laser *Nat. Commun.* **12** 4685
[12] Pickover C A 1988 Mathematics and beauty: a sampling of spirals and 'strange' spirals in science, nature, and art *Leonardo* **21** 173–81
[13] Kumar A V, Godumala M, Ravi J and Chandrasekar R 2023 A pick-and-place technology enabled mechanically programmable spiral optical waveguides integrated hybrid circuits from four extremely pseudo-plastic organic crystals for visible spectral band tunability *Adv. Opt. Mater.* **12** 2302807
[14] Chosenyah M, Kumar A V and Chandrasekar R 2024 Mechanophotonics: Pseudo-plastic organic crystal as a fermat spiral optical waveguide arXiv:2402.18878
[15] Dong B, Guo X, Ho C P, Li B, Wang H, Lee C, Luo X and Lo G-Q 2017 Silicon-on-insulator waveguide devices for broadband mid-infrared photonics *IEEE Photonics J.* **9** 4501410
[16] Annadhasan M, Agrawal A, Bhunia S, Pradeep V V, Zade S S, Reddy C M and Chandrasekar R 2020 Mechanophotonics: flexible single-crystal organic waveguides and circuits *Angew. Chem. Int. Ed.* **59** 13852–8
[17] Annadhasan M, Vinay Pradeep V, Vinod Kumar A, Ravi J and Chandrasekar R 2022 Integrating triply- and singly-bent highly flexible crystal optical waveguides for organic photonic circuit with a long-pass-filter effect *Small Structures* **3** 2100163
[18] Kumar A V and Chandrasekar R 2023 Mechanophotonics: fabrication of a 2×2 hybrid directional coupler from flexible organic crystals *J. Mater. Chem.* C **11** 7995–8001
[19] Rohullah M, Kumar A V and Chandrasekar R 2024 A hybrid organic-crystal-based curved waveguide network for producing, splitting, and transducing multi-color outputs *Laser Photonics Rev. early access* https://doi.org/10.1002/lpor.202400020
[20] Ravi J and Chandrasekar R 2021 Micromechanical fabrication of resonator-waveguides integrated 4-port photonic circuit from flexible organic single-crystals *Adv. Opt. Mater.* **9** 2100550

[21] Liu H, Lu Z, Zhang Z, Wang Y and Zhang H 2018 Highly elastic organic crystals for flexible optical waveguides *Angew. Chem. Int. Ed.* **57** 8448–52

[22] Ravi J, Vinod Kumar A, Annadhasan M and Chandrasekar R 2022 Realization of mechanically maneuverable circuit ports in organic hybrid photonic chip for 360° steering of bandwidth engineered signals *Adv. Opt. Mater.* **10** 2102545

[23] Varughese S, Kiran M, Ramamurty U and Desiraju G R 2013 Nanoindentation in crystal engineering: quantifying mechanical properties of molecular crystals *Angew. Chem. Int. Ed.* **125** 2765

[24] Xu Q and Lipson M 2017 All-optical logic based on silicon microresonator *Opt. Express* **15** 924

[25] Ravi J, Annadhasan M, Kumar A V and Chandrasekar R 2021 Mechanically reconfigurable organic photonic integrated circuits made from two electronically different flexible microcrystals *Adv. Funct. Mater.* **31** 2100642

[26] Ravi J, Kumar A V, Karothu D P, Annadhasan M, Naumov P and Chandrasekar R 2021 Geometrically-reconfigurable, two dimensional, all-organic photonic integrated circuits made from two mechanically and optically dissimilar crystals *Adv. Funct. Mater.* **31** 2105415

[27] Takazawa K, Inoue J-I, Mitsuishi K and Takamasu T 2013 Ultracompact asymmetric Mach–Zehnder interferometers with high visibility constructed from exciton polariton waveguides of organic dye nanofibers *Adv. Funct. Mater.* **23** 839

[28] Fang X and Claus R O 1995 Polarization-independent all-fiber wavelength-division multiplexer based on a Sagnac interferometer *Opt. Lett.* **20** 2146

[29] Frazao O, Baptista J M and Santos J L 2007 Recent advances in high-birefringence fiber loop mirror sensors *Sensors* **7** 2970

[30] Li S, Chiang K S and Gambling W A 2001 Gain flattening of an erbium-doped fiber amplifier using a high-birefringence fiber loop mirror *IEEE Photon Technol. Lett.* **13** 942

[31] Yuan L, Jin W, Zhou L, Hoo Y L and Demokan M S 2003 Enhancement of multiplexing capability of low-coherence interferometric fiber sensor array by use of a loop topology *J. Lightwave Technol.* **21** 1313

[32] Wang J, Zheng K, Peng J, Liu L, Li J and Jian S 2009 Theory and experiment of a fiber loop mirror filter of two-stage polarization-maintaining fibers and polarization controllers for multiwavelength fiber ring laser *Opt. Express* **17** 10573

[33] Dong X P, Li S, Chiang K S, Ng M N and Chu B C B 2000 Multiwavelength erbium-doped fibre laser based on a high-birefringence fibre loop mirror *Electron. Lett.* **36** 1609

[34] Hänsel W *et al* 2017 All polarization-maintaining fiber laser architecture for robust femtosecond pulse generation *Appl. Phys.* **B 123** 41

[35] Agrawal G P 2008 *Applications of Nonlinear Fiber Optics* 2nd edn (Boston, MA: Elsevier)

[36] Zhuo M-P, Wu J-J, Wang X-D, Tao Y-C, Yuan Y and Liao L-S 2019 Hierarchical self-assembly of organic heterostructure nanowires *Nat. Commun.* **10** 3839

[37] He Z, Zhao W, Lam J W Y, Peng Q, Ma H, Liang G, Shuai Z and Tang B Z 2017 White light emission from a single organic molecule with dual phosphorescence at room temperature *Nat. Commun.* **8** 416

[38] Kumar V A, Godumala M, Ravi J and Chandrasekar R 2022 A broadband, multiplexed-visible-light-transport in composite flexible-organic-crystal waveguide *Angew. Chem. Int. Ed.* **134** e202212382

[39] Hayashi S, Ishiwari F, Fukushima T, Mikage S, Imamura Y, Toshiro M and Katouda M 2020 Anisotropic Poisson effect and deformation-induced fluorescence change of elastic 9,10-dibromoanthracene single crystals *Angew. Chem. Int. Ed.* **59** 16195–201

[40] Khapre A, Kumar A V and Chandrasekar R 2024 Powering monolithic and hybrid organic optical waveguides via integrated focused micro-LEDs for sustainable photonic circuits arXiv:2402.18889 [physics.optics]

[41] Chandrasekar R 2022 Mechanophotonics–a guide to integrating microcrystals toward monolithic and hybrid all-organic photonic circuits *Chem. Commun.* **58** 3415–28

Chapter 5

Crystal photonics foundry

5.1 Crystal photonics foundry for organic photonic integrated circuits

The device characteristics of PICs are largely determined by the shape and size of the micro components, and it is necessary to precisely control them to achieve the desired performance. Until now, the specific control of microscale component shapes and sizes essential for PICs has been achieved to industrial standard in materials like silicon, silicon derivatives, and group III–V semiconductors using techniques such as electron-beam lithography, CMOS, and related etching processes.

However, as technology advances rapidly, PICs continue to demand novel materials platforms. The interest in harnessing molecular organic crystals as optical components for the development of entirely organic PICs (OPICs), made using economically viable precision manufacturing methods, is steadily increasing.

On the other hand, the fabrication of highly reproducible, dimensionally, and geometrically well-defined photonic modules from organic crystals is challenging. In 2022, my research group at the University of Hyderabad successfully milled organic single crystals into high-precision cavities or resonators using the focused ion beam (FIB) technique [1, 2] (figure 5.1). This technique also produced well-defined photonic modules such as circular- and square-shaped disk resonators, ring resonators, and ADFs with a nanometer precision. This next-generation technique is now termed the *crystal photonics foundry* [3], as it enables the design and manufacturing of high-performance organic single-crystal photonic modules and their integrated circuits for various applications.

5.1.1 FIB milling technology

FIB is an effective milling technique used in many fields, including materials science, photonics, semiconductors, and nanotechnology [4–7]. The process involves

doi:10.1088/978-0-7503-5336-6ch5

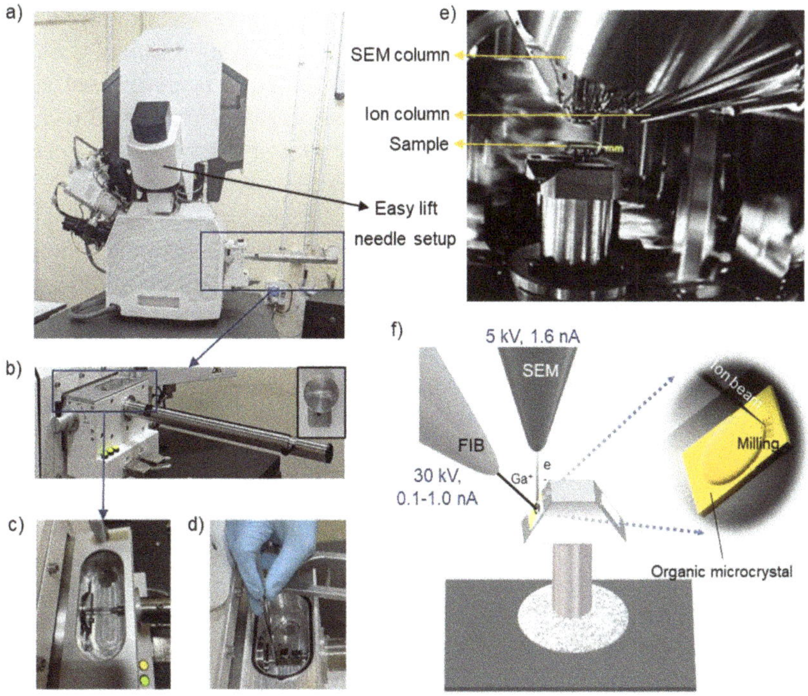

Figure 5.1. (a) Photograph of a typical FIB-SEM instrument. (b) Sample loading chamber with attached transfer rod. (Inset: sample holder stub.) Top view of a sample loading chamber, (c) without sample, (d) with sample. (e) Interior camera view of a FIB-SEM column with a loaded sample. (f) Graphical representation of a FIB set-up. (f) Reprinted with permission from [3]. Copyright (2023) Royal Society of Chemistry

removing or sculpting materials at the micro- or nanoscale by using a focused beam of ions. The FIB technology enables the controlled manipulation of very energetic ions at the nanoscale. These ions are accelerated to high energies and concentrated into a fine probe. When these intense and focused ions collide with the sample, they cause atoms to be ejected or sputtered from the surface through the transfer of momentum. This method allows for the precise removal of materials from the sample. Additionally, the highly localized nature of this ion beam enables the creation of precise patterns, the removal of unwanted material, and the deposition of substances with nanoscale accuracy.

Contemporary FIB systems are integrated with scanning electron microscopy (SEM), forming a dual-beam platform [8, 9] (figure 5.1(a-d)). These systems are designed with an electron beam column aligned vertically and an ion beam column positioned at a specified angle, often ranging from 52° to 55° with respect to the electron beam column (figure 5.1(e) and (f)). These systems also are equipped with a secondary ion, backscattered electron, and x-ray detectors and micromanipulation tools that enable precise micro- and nanoscale imaging, analysis, and manipulation of milled sample materials.

The capacity to attain nanometer-scale resolution is critical in various scientific and industrial applications, including sample preparation for advanced electron microscopy, modification and repair of integrated circuits in the semiconductor industry, and accurate nanofabrication procedures [10–12]. Thus, FIB technology is a cornerstone in materials science and nanotechnology, providing unprecedented control and precision down to the nanometer level.

The ion beam column consists of the ion source and optics column. Within the optics column, condenser lenses, objective lenses, apertures, deflectors, and scanning electrodes are precisely arranged. In FIB systems, various liquid metal ion sources (LMIS) with the capability to generate a beam spot as small as 10 nm are employed [13, 14]. Among these options, gallium (Ga) stands out as a particularly favored metal due to its metallic nature, low melting point (29.8 °C), low-temperature operability, high vacuum compatibility, and optimum momentum transfer capability.

Although gaseous ion beams, particularly helium, offer higher spatial resolution compared to Ga ion sources, their reduced mass imposes limitations, leading to longer milling durations. Recognizing the potential for improvement, recent studies explored the use of liquid metal alloy ion sources (LMAIS) featuring lower melting points, such as PdAs, PdAsB, AuSi, AuSiB, NiB, and NiAs. Key features of an LMAIS are long lifetime, high brightness, and stable ion current [15].

Further developments have resulted in the inclusion of plasma sources into FIB systems, which outperform the milling rates achieved by conventional Ga-FIBs. These advancements in FIB technology allow for more precision and efficiency in material manipulation and analysis at the nanoscale.

5.1.1.1 Factors influencing FIB milling [12, 16, 17]

The FIB technique is influenced by numerous factors, each playing a crucial role in achieving the precision and efficiency of the milling process.

The material properties under scrutiny bear significant weight in this context.

- The hardness, a fundamental material's characteristic, dictates the material's susceptibility to ion milling; softer materials tend to be more amenable to the milling process.
- The conductivity plays a pivotal role by facilitating efficient secondary electron imaging, low ion beam damage, and uniform and accurate material removal rate.
- The composition of the material is another critical determinant, as distinct elemental compositions exhibit unique responses to ion bombardment, thereby impacting milling rates and surface characteristics.

Further, the choice of ion species and their energy levels has considerable influence over FIB milling outcomes.

- The selection of ion species determines sputtering efficiency, the rate at which material is removed, and the precision and effectiveness of the milling process.
- The energy imparted to the ions is a critical parameter; higher ion energies lead to increased sputtering rates, thereby influencing the milling depth.

In essence, the convergence of material properties and ion-specific parameters demonstrates the intricacies of FIB milling, making it a meticulously controlled process with extensive influence on the milled material.

Let's examine the individual parameters that can influence the quality of sample milling:

Ion current: The ion beam's intensity plays a pivotal role in influencing the material removal rate during fabrication processes. This parameter becomes a key factor in controlling the aspect ratio of the fabricated material. The magnitude of the individual beam currents is directly correlated with the size of the beam spot, and the beam current can range up to 100 nA. This parameter serves as a critical element in illustrating how the spot size influences the milling operation.

For instance, at a 30 kV acceleration voltage, a 3 nm ion probe diameter can be attained by using a 0.001 nA beam current. Generally, a higher primary beam current accelerates the material removal process. The utilization of low-current ion beams facilitates the creation of high-quality features, albeit with the trade-off of significantly prolonged milling times. The sputtering frequency can be precisely controlled by adjusting the ion beam current or utilizing smaller spot sizes. When optimizing FIB processes, this delicate balance between milling duration, fabrication rate, and cost necessitates careful consideration.

Acceleration voltage: The acceleration voltage in FIB systems is a critical factor influencing ion species' momentum. Typically set in the range of 30–35 kV, an elevated acceleration voltage accelerates ion escape from the column, imparting higher energy to the specimen without affecting the probe current. This enhanced energy is advantageous for efficient processing. However, the pursuit of specific structural features demands a thoughtful consideration of the trade-off involved. While higher acceleration voltage brings benefits, such as a probe size of down to 1 nm, proper milling precision comes only at the expense of longer milling durations. Striking the right balance between precision requirements and processing efficiency is essential when optimizing FIB protocols.

Sputtering yield: The quantification of material removal processes is elucidated through the concept of sputtering yield, denoting the 'number of atoms ejected from the specimen surface per incident ion' under specific ion beam parameters. Typically falling within the range of 1–50 atoms per ion, this yield is intricately influenced by various factors, encompassing the mass of the specimen atoms and ions, their energies, the impact angle on the specimen, the specimen's temperature, and the ion flux. As ion energy increases, the sputter yield experiences an initial rise. However, beyond a critical energy threshold, where ions penetrate deeply into the substrate, the yield begins to decline. This reduction is attributed to ions becoming entrapped within the substrate, due to processes such as *implantation* or *doping*. Understanding these nuanced dynamics is crucial for optimizing material removal outcomes.

Optimizing beam scanning strategies is crucial for achieving precise and efficient FIB milling in various applications. The spacing between scanned lines, termed scan density, directly impacts surface roughness and milling precision. Gas-assisted etching, employing specific gases like XeF_2, enhances material removal rates and surface quality, with gas pressure playing a key role in modulating the etching process. Fine-tuning instrument parameters, including aperture size and precise focusing of ion optics, is essential for establishing a well-defined ion beam, a prerequisite for achieving accurate milling results.

Sample preparation: Applying a conductive coating (usually gold), significantly improves milling efficiency, especially for insulating materials.

The optimization of these factors enables the achievement of precise nanometer-scale geometries, providing control over both the local depth of the milled pattern and its lateral position.

FIB milling is a valuable tool in materials research, showcasing the capability to carve intricate structures and patterns such as lines, squares, rectangles, circles, etc. Whether employing inbuilt or imported bitmaps through the FIB user interface, this method ensures the consistent attainment of accurate and intricate results at both the micrometric and nanometric scales.

5.1.1.2 Nano/microscale effects on the sample during FIB milling

Ion implantation: Ion implantation refers to the insertion of ions into the sample material. This implantation occurs during the sputtering process, where (gallium) ions become incorporated into the specimen. As a result, the local composition of the material within the interaction volume changes. Gallium implantation has profound effects, leading to modifications in the material's structure. These alterations extend to the mechanical, thermal, electrical, and optical properties of the milled sample. Ion implantation leads to the loss of FL in organic crystals and introduces undesirable refractive index inhomogeneity, the latter is undesirable for photonic-grade milled samples.

Amorphization: Exposure to ions can lead to material amorphization, disrupting the crystalline structure and causing swelling due to substantial atomic displacement in the affected region. The challenges associated with amorphization necessitate the use of low ion beam energies. This precautionary measure is crucial to prevent adverse impacts on crystalline structures, ensuring the integrity of the crystalline sample.

Redeposition: In the course of sputtering processes, particles expelled from a sample surface are prone to re-condensation on nearby surfaces, returning to a solid phase. This redeposition phenomenon leads to the undesired formation of debris, as some of the ejected atoms adhere back to their original surfaces. The use of low ion currents proves effective in minimizing redeposition. Moreover, applying protective layers serves as a significant measure to prevent excessive redeposition, ensuring cleaner and more precise material processing. Additionally, the strategic manipulation of ion milling configurations, particularly in terms of geometries, offers a valuable approach to minimizing the reattachment of particles.

5.1.2 Utilization of perylene single crystals for FIB milling

To establish organic crystals as reliable materials for PICs and to enable the creation of commercially viable all-organic microelectronic and nanophotonic components, it is imperative to demonstrate the following.

- Defect-free processing of crystals into diverse shapes (refer to section 5.1.1.2).
- Geometrical and dimensional precision.
- Reproducibility.
- Bulk-scale manufacturability.
- The universal applicability of the method to a wide range of organic crystals with various functional groups.

In 2019, my group reported findings on perylene (PER) single-crystal NLO resonators (refer to chapter 2, figure 2.19). Subsequently, in 2020, Bardeen *et al* published a study on the milling of PER single crystals into arbitrary shapes using FIB lithography (figures 5.2(a)–(d)) [18]. They employed an ultra-thin (4 nm) Au coating to prepare a conductive sample. Post-milling, the removal of the Au coating was achieved by immersing the sample in a KI/I_2 etchant briefly at room temperature. Notably, the authors highlighted the purpose of the Au layer in PER crystal was 'protecting it from electron beam damage'. Their study revealed an etching rate

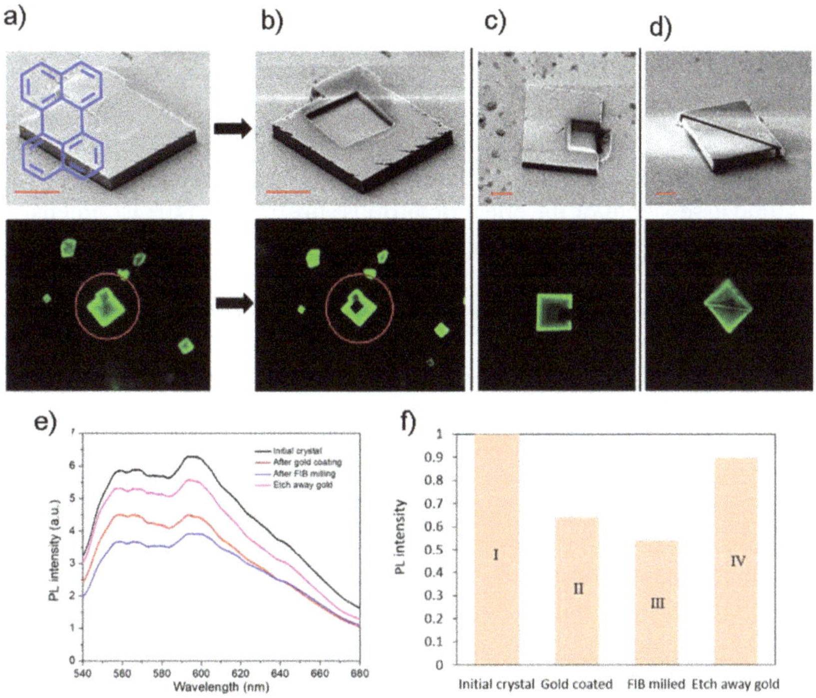

Figure 5.2. SEM (top) and FL (bottom) images of FIB-milled perylene (PER) crystals. A single PER crystal (a) before and (b) after FIB milling of a square in the center of the crystal. The inset in (a) shows the molecular structure of PER. (c) A PER crystal with a rectangular cut on the side that extends into the substrate, (d) a square crystal cut into two triangles. The scale bar for SEM images is 4 μm. The circled crystals in the FL image are the same as in the SEM images above. (e) FL spectrum of a PER crystal at each step. All the spectra are collected from a spot located within 2–4 μm of the FIB milling area. (f) Comparison of the integrated PL intensity at each step normalized to that measured at step I. Reprinted with permission from [18]. Copyright (2020) American Chemical Society.

of approximately 800 s μm^{-3}, utilizing a 0.05 nA ion beam at 30 kV. This methodology allowed for the milling of features with a spatial resolution on the order of 130 nm. Following the Au coating and FIB milling, a decrease in PL intensity by about 45% was observed. However, they observed a significant PL recovery, reaching 90% of the original intensity in the 'unmilled regions' of the crystal once the Au layer was etched away.

Nevertheless, our experimental plan encompassed several technical considerations.

- The inclusion of Au layer serves a dual purpose: safeguarding the organic crystals from *ion beam damage* and providing a *conductive layer for charge dissipation*.
- A crucial aspect involves probing the PL intensity in regions proximate to the milled area. This step is instrumental in identifying potential *ion beam damage triggered FL quenching*.
- The detection of optical resonant modes from the milled crystal cavities is a direct and vital confirmation of *refractive index homogeneity*. This, in turn, serves as evidence supporting the intrinsic purity of the crystal, ruling out any implications of ion implantation or crystal amorphization.
- Additionally, it is important to reduce the processing time of the crystals using an optimized beam current.

Therefore, the optimization of milling parameters, exploration of post-processing methods, and assessment of the effect of various substrates on the FL intensity of the organic microresonators become imperative.

5.1.2.1 Optimization of beam current

As the beam spot sizes are determined by the magnitude of beam currents, the aspect ratio of organic crystal under fabrication can also be tuned by varying beam currents. By modulating the ion beam current or deploying smaller spot sizes, the sputtering frequency can be rapidly and precisely regulated. With the use of low-current ion beams, high-quality and complex photonic modules can be attained at the expense of long milling duration. Shorter milling time severely affects the optical properties of crystals due to carbonization and ion implantation.

For instance, in 2011 Kim *et al* examined the effect of beam current on polymer thin films [19]. They systematically investigated FIB damage on thin polymer films (PS-*b*-PMMA) by varying the beam current and exposure time while maintaining a constant voltage (figure 5.3(a)). SEM images were captured for the fabricated holes in a PS-*b*-PMMA polymer sheet. At a beam current of 10 and 100 pA (0.01–0.1 nA), the morphology of the polymer remained intact (figures 5.3(b) and (c)). However, at beam currents of 300 pA, the SEM images revealed a distorted morphology and wrinkles in the polymer sheet (figure 5.3(d)). The authors attributed this damage to local heat generation during ion beam interaction with the polymer.

My group conducted a comprehensive study of FIB milling on organic microcrystals (figure 5.3 (e)) to achieve defect-free photonic modules [20]. Specifically, perylene microcrystals were exposed to various ion beam currents while maintaining

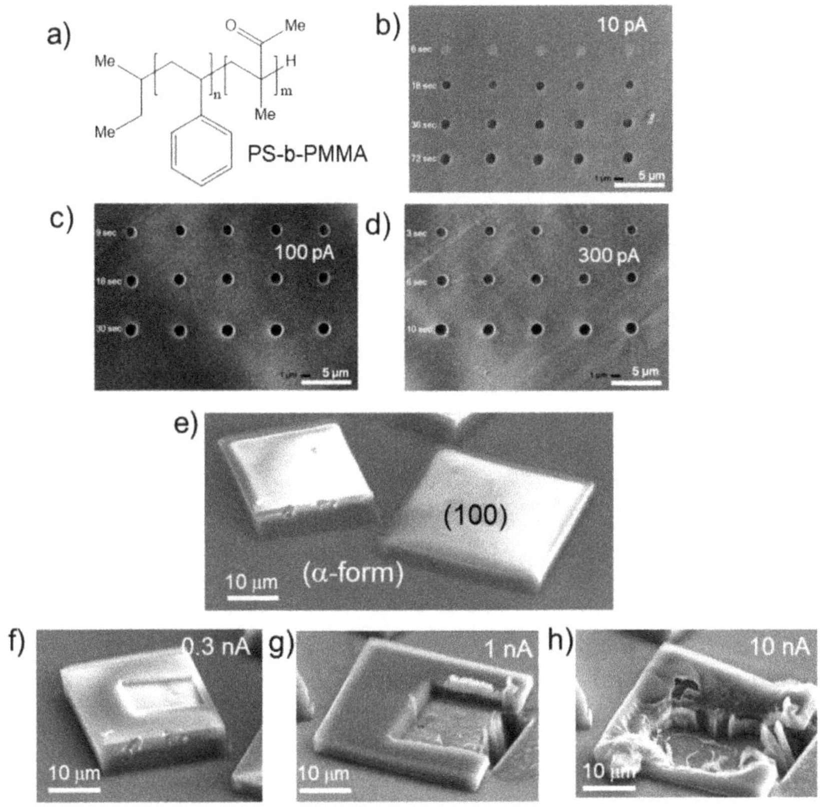

i)

Beam current	Depth	Time
300 pA	1 μm	10 min
1 nA	2 μm	17 min
10 nA	2 μm	5 min

Figure 5.3. (a) Chemical structure of PS-b-PMMA. SEM images of the polymer sheet when subjected to (b) 10 pA, (c) 100 pA, and (d) 300 pA beam currents to attain circular holes (acceleration voltage 30 kV). SEM images of perylene microcrystals (e) before and (f–h) after FIB milling. (i) The table shows the respective parameters incorporated for FIB milling. (b–d) Reprinted from [19], Copyright (2011), with permission from Elsevier.

a constant acceleration voltage. Initially employing FIB at 30 kV and 0.3 nA for approximately 15 min, we created a rectangular trench with smooth edges, reaching a depth of 1 μm (figure 5.3(f)). Subsequently, another microcrystal underwent 2 μm depth rectangular trench milling using a 1 nA beam current, requiring approximately 17 min (figure 5.3(g)). The crystal edges appeared smooth, with minimal redeposition near the trench edges. Subsequently, the same crystal underwent milling with a 10 nA beam for approximately 5 min, resulting in an irregular shape with significant redeposition (figure 5.3(h)).

Our experiments on PER microcrystals indicated that lower beam currents yield precise shapes but require longer milling times. Conversely, higher beam currents lead to more damage compared to lower beam currents. Consequently, a 1 nA beam current at 30 kV emerged as an optimal choice for FIB milling (figure 5.3(i)). This choice was based on a careful compromise between milling time and the desired smoothness of the final shape.

5.1.2.2 Effect of substrates [20]

Given that the perylene microcrystals are non-conductive, imaging in electron or ion microscopes led to the accumulation of static charge on and around the sample. Therefore, to study the effect of various substrates on FIB milling (accelerating voltage: 5 kV; beam current: 1 nA) of organic non-conducting crystals, indium tin oxide-coated polyethylene terephthalate (ITO-PET; a conductive material) and borosilicate glass substrate (a non-conductive material) substrates were compared. The ITO-PET used in this study (Sigma Aldrich, Product no. 639281) featured a 72 nm thick ITO coating on a PET substrate (127 μm thick). The refractive index of ITO and borosilicate glass coverslips at 632.8 nm is 1.85844 and 1.51446, respectively. The PET film has a refractive index of about 1.57–1.64, and for organic crystals, it is about 1.6.

Using the self-assembly technique, perylene microcrystals were grown on the two selected substrates: (i) on a conductive substrate (ITO-PET), and (ii) on a glass coverslip and then coated with a conductive gold layer.

Microcrystals on ITO-PET substrate: The ITO-PET substrate containing perylene microcrystals was kept in the FIB-SEM set-up and scanned using an electron beam. Five representative crystals (crystal-I to crystal-V) on the ITO-PET substrate were selected to fabricate various types of organic single-crystal resonators and waveguides. Initially, to create a circular disk resonator, a 25.5 × 27.2 μm^2 perylene single crystal (Crystal-I) was identified using SEM (figure 5.4(c)). Employing the predefined circular shape function and optimized beam parameters (1 nA and 30 kV), Crystal-I was milled into a circular disk with a diameter of 18 μm (figure 5.4(c)). Similarly, Crystal-II (25.6 × 31.5 μm^2) was milled into a circular ring with an outer diameter and width of 20 and 4.5 μm, respectively (figure 5.4(f)). Optical excitation of these circular modules with a continuous wave laser (405 nm, 4 mW power, 0.5 s integration time, and ten accumulations) resulted in the generation of yellowish FL at the excited position, which was subsequently outcoupled at the circular periphery. The FL spectrum collected at the periphery exhibited a bandwidth of ≈525–775 nm (figure 5.4(d) and (g)).

To create bent waveguides, perylene single crystals III and IV were selected (figure 5.4(h) and (i)). Crystal III underwent precise milling, resulting in the fabrication of two 90° bent waveguides, namely WG1 and WG2, with lengths of approximately 35 and 21 μm, respectively. The optical waveguiding nature of the milled waveguides was established from the controlled excitation of a portion of milled waveguides with a 405 nm continuous wave diode laser. Initially, excitation of the left terminal (T1) of WG1 resulted in the generation of yellowish FL (≈500–

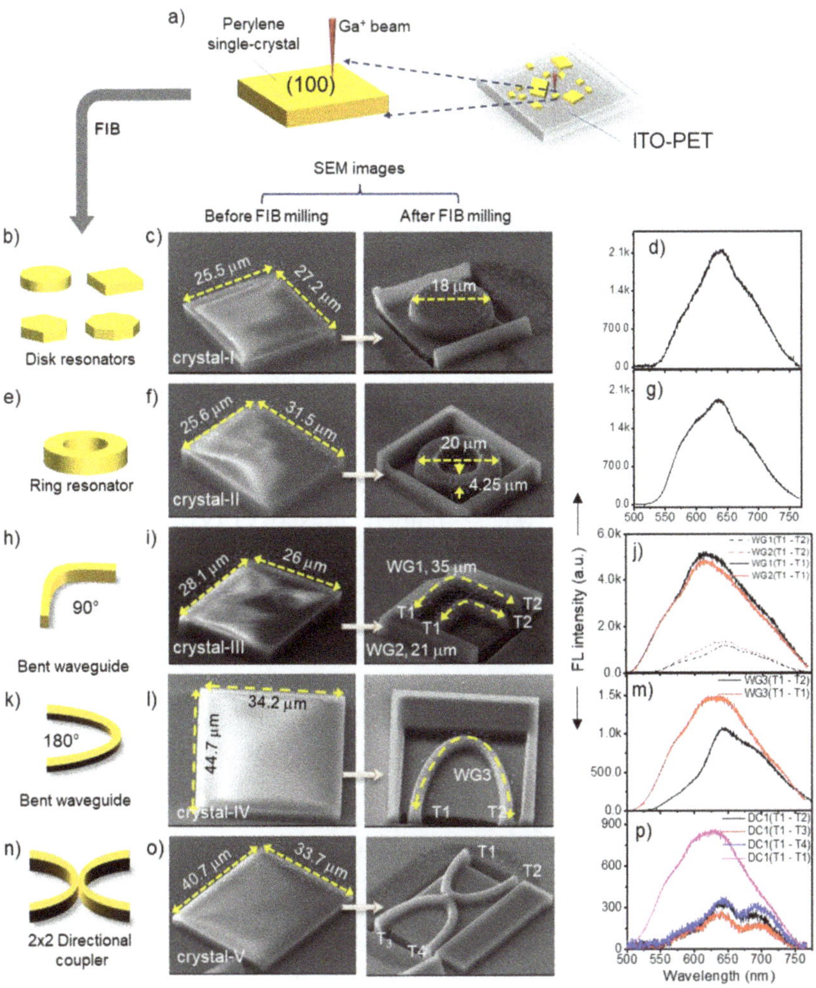

Figure 5.4. (a) Graphical representation of FIB milling of perylene microcrystals, on ITO-PET substrate, into various geometries namely, (b) disk resonators, (e) ring resonator, (h, k) bent waveguides, and (n) directional coupler. (c, f, i, l, o) FESEM images of selected perylene single crystals before and after FIB milling. (d, g, j, m, p) The corresponding FL spectra of fabricated optical components. Here, T1–T2 represent the excitation-detection positions of the respective component. Reprinted with permission from [20]. Copyright (2023) American Chemical Society.

775 nm collected using a 150× objective), and the same light propagated towards the other end of the waveguide (T2) as a reabsorbed signal (≈540–775 nm) due to overlap of perylene's FL with its absorption spectrum (figure 5.4(j)). WG2 displayed analogous waveguiding characteristics. Notably, the outcoupled FL intensity was increased in WG2 compared to WG1, attributed to the shorter optical propagation path (lower loss) in the former.

Likewise, crystal IV was milled to produce a 180° bent waveguide, WG3, with a length of 65.8 μm (figure 5.4(k) and (l)). Likewise, upon exciting the left terminal

(T1) of WG3, subsequently detection at the right terminal (T2) showed a FL band spanning from 550 to 775 nm (figure 5.4(m)).

Finally, a rectangular crystal denoted as crystal-V, measuring 40.7 by 33.7 μm, was meticulously shaped into a 2×2 directional coupler (DC1), as illustrated in figure 5.4(n) and (o). The optical excitation at one terminus (T1) of DC1 yielded FL spanning 525–775 nm (figure 5.4(p)). As anticipated from the DC's geometry, the propagating FL signal was detected at T2 and T4. Contrary to conventional DCs where no signal typically reaches T3 due to circuit geometry, the FL signal at T3 was observed. This anomaly might be attributed to a broader cross-section at the junction or the back reflection of light from T4 (figure 5.4(m)). The diverse line shapes in the spectra at T1–T4 can be ascribed to the molecular anisotropy exhibited by perylene crystals.

Microcrystals on borosilicate glass substrate: Given the non-conductive nature of both the glass substrate and perylene microcrystals, a 15 μA current was utilized to sputter-coat a thin layer of gold for 80 s. Following this procedure, the substrate was examined in SEM, and four crystals—namely, crystal-VI, VII–VIII, and -IX—were selected for subsequent analysis.

To produce four microdisks with distinct geometries (circular, square, hexagonal, and octagonal), all on a gold-coated single perylene crystal, crystal-VI with a large surface area of 22.89 × 22.93 μm^2 was selected (see figure 5.5(a) and (b)). Importantly, the gold coating of the microcrystals minimized the direct exposure of the entire microcrystal to the Ga ion beam during imaging. The fabrication process (30 kV acceleration voltage and a beam current of 1 nA) required a milling process lasting approximately 45 min to achieve distinct resonator geometries on a gold-coated perylene single crystal. Notably, these resonator geometries comprised a square with an edge length of 6.9 μm, a circular disk with a diameter of 6.9 μm, a hexagon with an edge length of 4.1 μm, and an octagon with an edge length of 3.3 μm (figure 5.5(c) and (d)). A parallel methodology was implemented for crystal-VII, possessing a surface area of 31.20 × 26.09 μm^2, resulting in the milling of three ring-shaped crystals with outer diameters of 10.8, 8.6, 6.3 μm, respectively and one square-shaped crystal with a side length of 8.6 μm (figure 5.5(g–j)).

After successful milling, the substrate containing the sample underwent treatment with Lugol's iodine solution to remove the gold layer. Subsequent confocal micro-spectroscopic studies were conducted on the fabricated structures using a 405 nm continuous wave diode laser. Exciting the central regions of the milled polygonal and ring-shaped crystals resulted in the emission of yellow FL (525–775 nm) at the excitation point and a bright FL at the crystal's periphery, signifying effective light confinement (figure 5.5(e), (f), (k), and (l)). The FL images distinctly revealed the inner and outer circular peripheries of the rings.

For the fabrication of a bent waveguide (WG4) and a 2×2 DC2 from perylene single crystals grown on a glass substrate, crystal-VIII and crystal-IX were chosen (left, figure 5.5(m), (n), (p), and (q)). Crystal-VIII, having dimensions of 21.8 × 18.7 μm^2, was subjected to FIB milling (30 kV and 1 nA) to produce a U-shaped waveguide (WG4) featuring a 180° bend (right, figure 5.5(m) and (n)). Similarly, crystal-IX with dimensions of 26.6 × 31.1 μm^2 was carved to create a 2×2 DC2

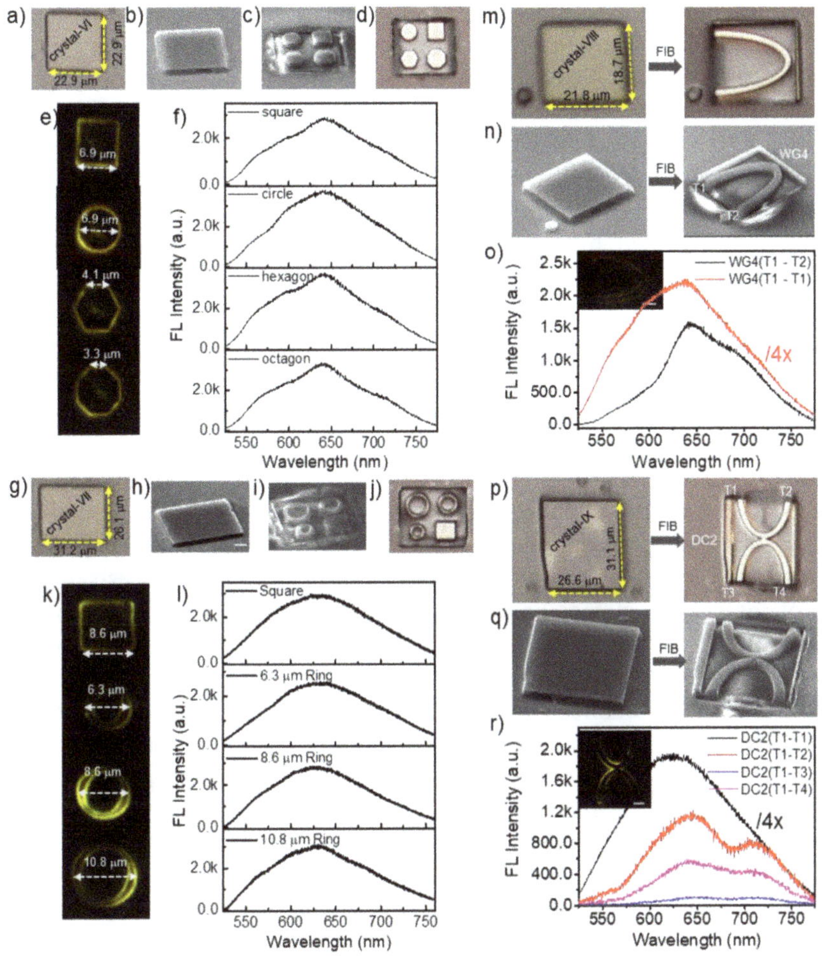

Figure 5.5. Perylene crystals on a glass substrate. (a, g) Optical microscope and (b, h) SEM images (tilt view) of selected perylene single crystals (crystal-VI and -VII). (c, i) SEM (tilt view) and (d, j) optical microscope images of FIB-milled selected perylene single crystals (crystal-VI, and -VII) into various shaped resonators. (e, k) The FL images and the (f, l) corresponding FL spectra of fabricated resonators. (m, p) Optical microscope and (n, q) SEM images (tilt view) of selected perylene single crystals (crystal-VIII and -IX) before and after FIB milling. (o, r) FL spectra of the fabricated WG4 and DC2. Here T1–T2 represent the excitation and detection positions of the respective fabricated geometries. Insets show the corresponding FL images. The scale bar is 5 μm. Reprinted with permission from [20]. Copyright (2023) American Chemical Society.

(right, figure 5.5(p) and (q)). Excitation of T1 of the WG4 with a 405 nm laser generated the yellowish FL in the range of 475–775 nm (red line in figure 5.5(o)). The transmitted FL signal recorded at T2 (150× objective with a 0.5-s integration time and data accumulation over ten cycles) showcased a notable suppression of the FL, attributed to the reabsorption effect (black line in figure 5.5(o)). Likewise, the optical excitation of terminal T1 on DC2 demonstrated the transmission of FL to the remaining three terminals, namely T2, T3, and T4 (figure 5.5(r)). Although the DC design was intended to have no output in terminal T3, the dimensions of the DC joint permitted some degree of light propagation toward the T3 terminal.

In contrast to the DC fabricated on ITO-PET, the FL intensity observed for DC on the coverslip exhibited a notable increase, exceeding double the intensity. This disparity can be attributed to the differing conditions under which the crystals were milled. Specifically, the crystals on the ITO-PET substrate were exposed directly to the Ga ion beam during the milling process. In contrast, the crystals on the glass substrate were shielded by a protective gold layer. Consequently, the extent of damage incurred directly by the crystals was significantly mitigated in the case of those coated with gold. This protective gold layer acted as a barrier, minimizing the detrimental effects of the ion beam, thereby resulting in comparatively higher FL intensity when compared to the crystals on ITO-PET.

The FL intensity of milled resonator geometries is relatively low, and notably, there is no apparent signature of optical modes due to resonances in the FL spectrum (figure 5.4(d), (g), (j), (m), and (p)) and figure 5.5(f), (l), (o), and (r)). It is crucial to note that the selection of the crystal (thickness) for milling and the beam current needs further optimization to preserve FL from the milled crystal. While a higher ion dose expedites the milling process, it also induces undesirable amorphization of the crystal surface and deep ion implantation, causing the area to turn black, leading to FL quenching and uneven refractive index of the crystal. The milling resolution is contingent on the ion beam diameter, with the beam current governing the latter. Smaller currents allow for higher milling resolution, enabling precision in targeting specific areas. The outcomes of these experiments suggest the gold coating helps mitigate the adverse effects of the high ion beam; however, there is a need for further optimization of milling parameters.

5.1.2.3 Fabrication of perylene single-crystal disk resonators

In our subsequent experiments, while keeping the acceleration voltage at 30 kV, we decreased the beam current from 1 to 0.4 nA for FIB milling. Two perylene single crystals, labeled I and II, grown on a borosilicate glass coverslip, were accurately milled into circular- and rectangular-shaped configurations, respectively (figure 5.6(a)) [2].

Prior to FIB milling, confocal micro-spectroscopy experiments were conducted in a transmission mode geometry to demonstrate the resonator characteristics of the selected single crystals. Perylene single crystal I, nearly square-shaped with dimensions of ≈5.2 × 5.3 μm^2, was excited on the (100) facet with a 405 nm laser (power: 0.05 mW; objective: 60×). The crystal exhibited a bright yellow FL at its four edge facets (left; figure 5.6(b)). The FL recorded at one of the crystal edges displayed a broad spectrum covering the bandwidth of ≈525–775 nm, featuring a series of pairs indicative of TM and TE modes. These sharp peaks or modes provided evidence of WGM resonances, highlighting the light-trapping characteristics of the single-crystal resonators. The broad FL background subtracted spectra (figure 5.6(b)) confirmed the presence of optical modes, with a calculated FSR value of approximately 8.10 nm. The resonator characteristics of the initially square-shaped perylene crystal,

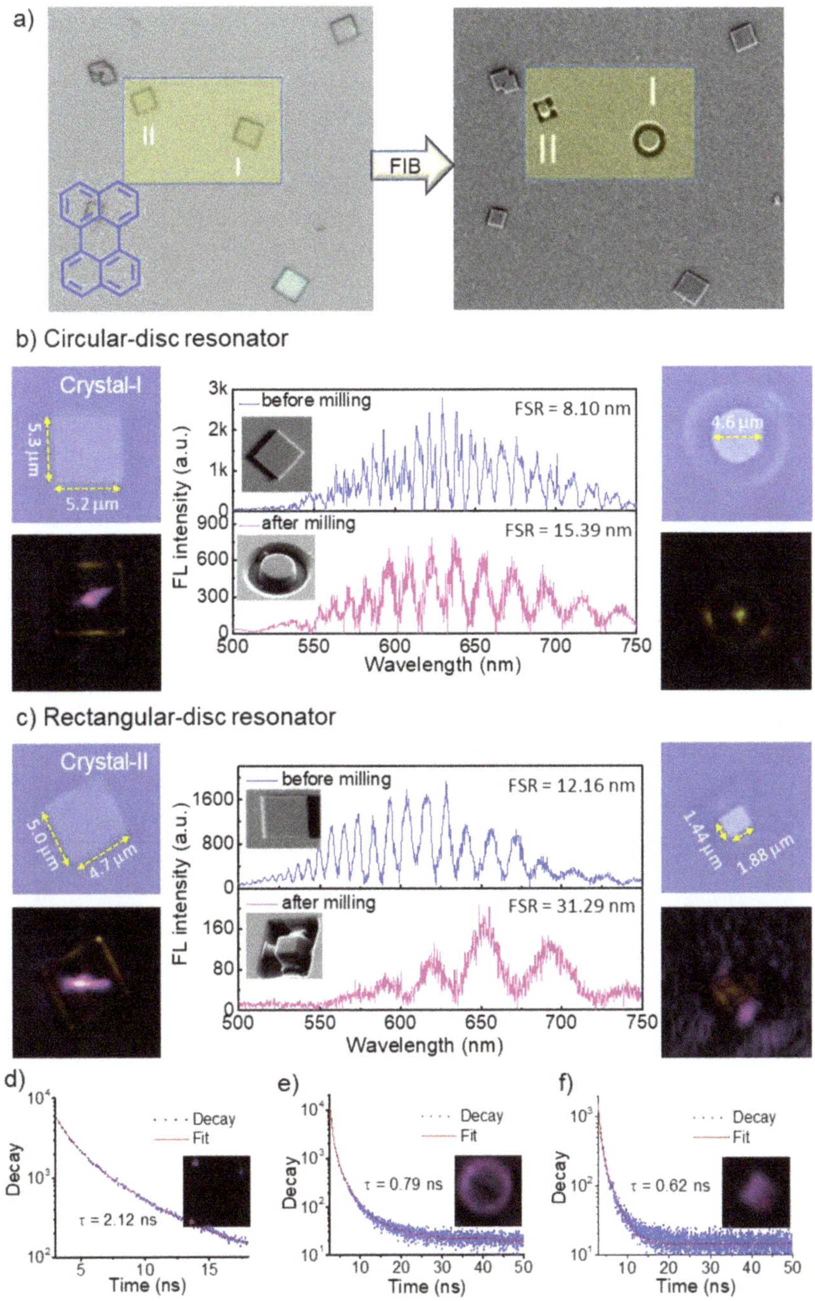

Figure 5.6. (a) SEM images of selected perylene single crystals before and after FIB milling. (b) Photonic studies on crystal-I before and after milling into a circular disk. (c) Photonic studies on crystal-II before and after milling into a rectangular disk. FL lifetime decay plots with estimated average lifetime values along with FL lifetime images of perylene crystals (d) before and (e, f) after milling, respectively. Scale bar 5 µm. Reprinted with permission from [2]. Copyright (2022) American Chemical Society.

as depicted in the microscopic image, were attributed to multiple round trips of light, facilitated by the crystal's four light-reflective facets.

Subsequently, after confirming the resonator traits, the crystal-containing substrate was gold-coated, and milling was performed orthogonal to the (100) facet utilizing a gallium ion beam. This milling process yielded a circular disk with a diameter of 4.6 μm (right; figure 5.6(b)). The gold coating was subsequently removed through washing with Lugol's iodine solution. Photonic experiments conducted on the circular disk unveiled an intense FL spectrum supporting relatively broad optical modes, characterized by an increased FSR of 15.39 nm (spectra: after milling; figure 5.6(b)). The observed optical modes were directly influenced by the geometry and size of the resonator, with the circular shape indicative of WGM resonances.

Likewise, crystal II, with dimensions approximately 5 × 4.7 μm^2, exhibited optical resonances in FL spectra upon focused excitation with a 405 nm laser (power: 0.05 mW). The calculated FSR value was determined to be 12.16 nm (figure 5.6(c)). Subsequent milling of crystal I into a smaller rectangular crystal, with dimensions of 1.88 × 1.44 μm^2, was followed by optical experiments revealing relatively broader modes with an increased FSR of 31.29 nm (figure 5.6(c)). The observed full-width at half-maximum (FWHM) of these modes and FSR values align with the inverse relationship between FSR and resonator dimensions given in equation (2.24).

FL lifetime imaging (FLIM) investigations were conducted on perylene crystals both before and after gallium ion beam milling (figures 5.6(d)–(f)). The milled circular disk resonator exhibited a distinct FL lifetime image with a high FL signal at the rim, indicative of circumnavigating light interactions at the crystal–air interface. However, the fabricated rectangular disk displayed a relatively uniform distribution of FL. The average lifetime decay values of milled crystal resonators are slightly lowered compared to crystal resonators before milling.

To investigate the size-dependent optical cavity effect of microcavities with the same height but varying diameters, a large perylene crystal (with a thickness of approximately 2.5 μm, as measured by AFM) was selected (figure 5.7(a)) [3]. The chosen single perylene crystal was deliberately broken into three pieces using an AFM cantilever tip force, and these broken pieces remained movable with the AFM cantilever tip (figure 5.7(b)). These characteristics allowed the manipulation of milled resonators and their integration with other optical modules. A shoe-shaped crystal (figure 5.7(c)) was selected, gold-coated, and subsequently milled into three different circular-shaped disk resonators (figure 5.7(d)). The color-coded FESEM images (figure 5.7(e–g)) exhibited the precise circular shape of the disk resonators (referred to as DR-1, DR-2, and DR-3) with diameters of 3.88, 4.83, and 5.74 μm. respectively. The smooth surface morphology of the fabricated microdisks was also evident.

Optical investigations were carried out by irradiating the microdisk's center with a 405 nm laser (power ≈0.1 mW). At the excitation point, yellow FL from perylene was emitted, lacking any apparent optical resonances in the spectrum. However, the FL spectrum recorded at the microdisk periphery, displaying a ring-shaped bright emission, revealed periodic sharp peaks indicative of WGMs. The optical modes in

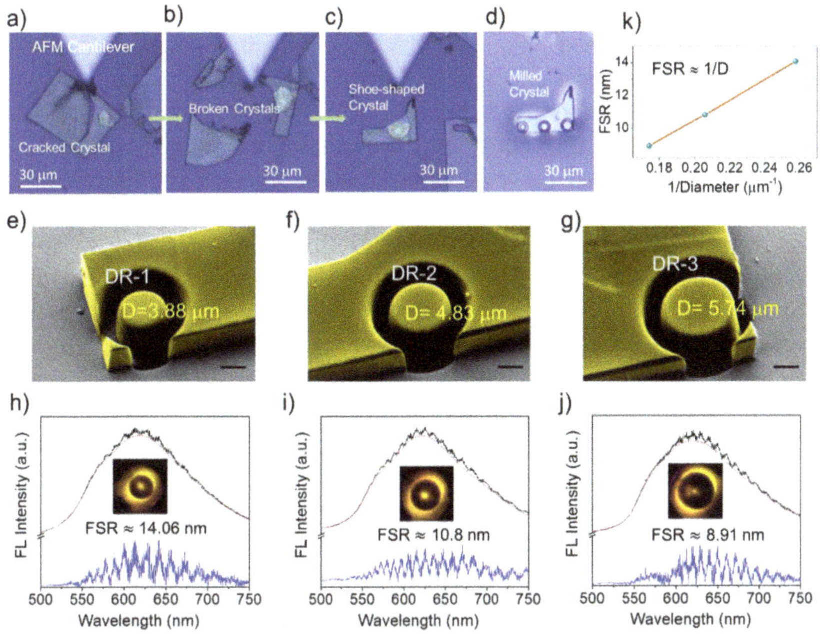

Figure 5.7. (a–c) Optical images depicting the separation of a broken shoe-shaped perylene single crystal via micromanipulation from a cracked crystal. (d) Optical image of a shoe-shaped single crystal milled into a series of disk resonators of varying diameters. (e–g) Color-coded FESEM images of three milled disk resonators revealing their diameters with the same crystal height. (h–j) FL spectra of disk resonators revealing optical modes and diameter-dependent FSR values. The spectra with blue lines show the FL background removed spectra. The insets show the FL images of optically excited disk resonators. (k) A plot of $1/D$ versus FSR for three disk resonators. Reproduced from [3] with permission from the Royal Society of Chemistry.

DR-1 to DR-3 were not well pronounced initially due to the presence of a gold coating on the disk surfaces [3].

Following the removal of the gold coating, spectra obtained at the periphery of the resulting disks (DR-1, DR-2, and DR-3) exhibited enhanced FL spectral intensity with clearly observable optical modes (figures 5.7(h)–(j)). The FSR values of DR-1 to DR-3 were measured as 14.06, 10.8, and 8.91 nm, respectively. This outcome underscores the preservation of the cavity effect in microdisks of varying diameters, as evidenced by decreased FSR values with an increase in cavity diameter, consistent with equation (2.24).

5.1.2.4 Fabrication of a perylene ring resonator and bent optical waveguide

The ring resonator is a very important component in OPICs. In the preceding chapters, the crucial role of ring resonators in ADF has been highlighted. Achieving a ring-shaped microcrystal with precise dimensions through conventional growth techniques such as crystallization, self-assembly, and sublimation poses practical challenges.

Building upon this foundation, my group demonstrated the utilization of the mechanophotonics approach to fabricating a single-crystal ring resonator (see section 4.3). This approach involved the transformation of a straight single crystal into a circular ring-shaped geometry by mechanical bending, addressing the practical challenges associated with conventional growth techniques such as crystallization, self-assembly, and sublimation.

However, the application of considerable mechanical force to achieve the desired curvature in a single crystal presents inherent challenges. A prominent issue is the substantial accumulation of strain in the ring resonator, with the potential risk of causing physical damage. In our continuous efforts to refine this technique and address strain-related concerns while elevating the reliability and performance of single-crystal ring resonators, we find that FIB milling proves to be a desirable solution. This precision milling method allows for controlled material removal at the nanoscale, offering a means to fine-tune the geometry of the single-crystal ring resonators.

The selection of a naturally grown single perylene crystal, initially measuring 8.4 × 8.2 μm^2, served as the foundation for the fabrication of a precisely sized circular ring-shaped microstructure through FIB milling (figure 5.8(a) and (d)). Initially, the chosen perylene crystal was first coated with a thin layer of gold. Subsequent FIB milling resulted in the successful creation of a ring-shaped microstructure, characterized by outer and inner diameters of 6.3 and 2.35 μm, respectively (figure 5.8(d) and (f)). Here, to refine the microstructure geometry and enhance its FL (figure 5.8(b) and (e)), we further reduced the beam current to 0.1 nA, while maintaining a constant acceleration voltage of 30 kV. Although this extension of milling time was implemented, it led to the creation of a ring-shaped structure with significantly improved FL. The enhancement observed can be attributed to the deliberate reduction in amorphization, achieved by minimizing gallium ion exposure on the unmilled crystal area.

Upon closer examination using FESEM and optical microscope images, the resulting ring-shaped perylene crystal displays a smooth and well-defined surface morphology (figure 5.8(f)). Initial photonic investigations conducted on the crystal, prior to the milling process, unveiled a WGM-type resonator effect with a FSR of 3.18 nm (see spectra before milling in figure 5.8(g), top).

After removing the gold coating, the FL spectrum recorded at the outer convex periphery of the ring resonator revealed relatively broad optical modes with an increased FSR of 20.75 nm, indicative of the resonator characteristics (see spectra after milling in figure 5.8(g), bottom). This observed change in FSR, post-milling, underscores the impact of resonator dimension and shape on the photonic properties, offering insights into the tuning and enhancement of the resonator's photonic behavior.

In the preceding chapter, we explored the fabrication of bent waveguides using pseudo-plastic 1D organic crystals, specifically employing the AFM cantilever tip-driven crystal micromanipulation technique. However, the presented manipulation method is not universal as all crystals cannot be mechanically bent. Further, the inherent limitations of flexible crystals must be recognized, as they exhibit mechanical deformability predominantly in one direction and rarely in two directions due to their molecular anisotropic nature. Additionally, the bent crystal, being derived from a naturally grown crystal, lacks uniform thickness. These

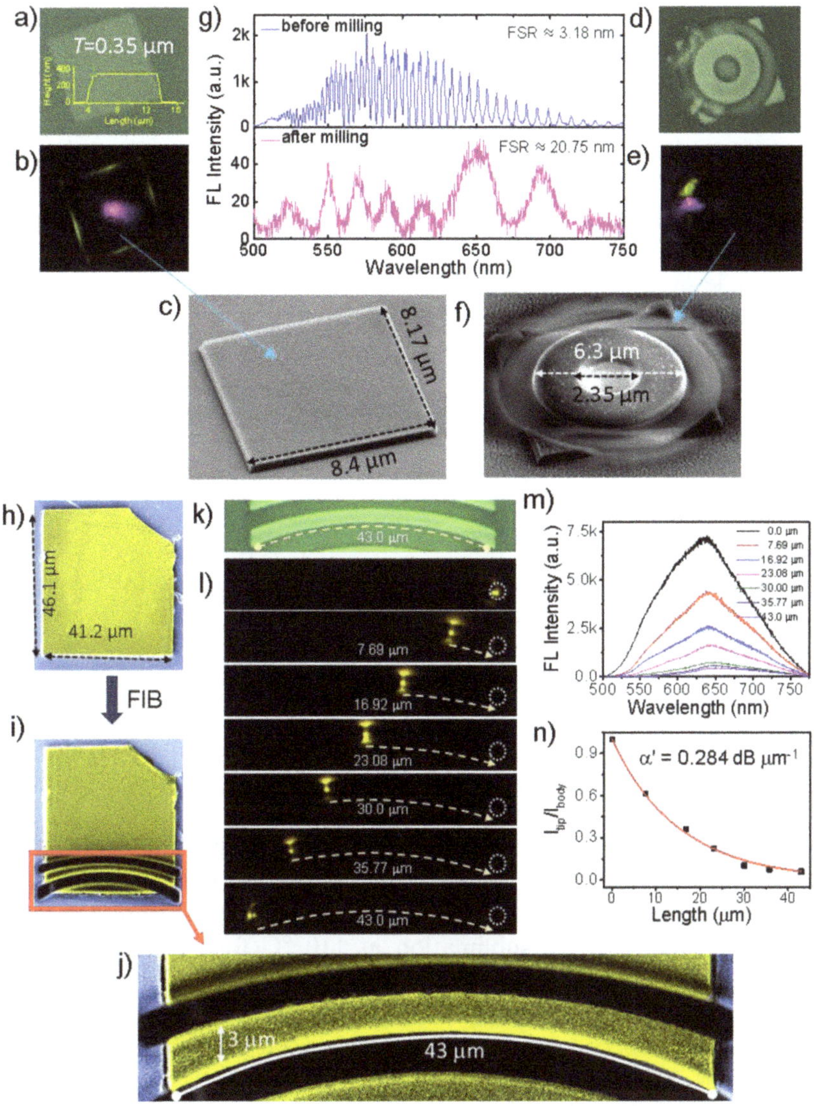

Figure 5.8. (a, d) Confocal optical, (b, e) FL microscopy, and (c, f) FESEM images of perylene microcrystal before and after milling into a ring resonator. The inset in (a) shows the AFM thickness profile. (g) The FL background subtracted spectra of perylene microcrystal before and after milling. (h, i) FESEM images of perylene microcrystal before and after milling into a bent waveguide, respectively. A zoomed-in view is shown in (j). (k) Confocal optical, (l) FL microscopy images of the bent waveguide. (m) The FL background subtracted spectra of a bent waveguide and (n) corresponding optical loss estimation plot. (d, e, g) Reproduced from [3] with permission from the Royal Society of Chemistry.

challenges emphasize the need for a nuanced approach in considering the potential applications of these crystals. Alternatively, FIB milling of 2D crystals provides curved geometries of uniform dimension without being constrained by molecular orientational preferences for flexibility.

To demonstrate our novel approach to creating a bent waveguide, a perylene microcrystal with dimensions of 46.1 × 41.2 μm^2 was milled on the widest (100) plane (see figure 5.8(h), (i), and (k)). This resulted in the successful creation of a bent waveguide with a curvature length of 43 μm (see figure 5.8(i) and (j)). Micro-spectroscopic studies on the fabricated bent waveguide, post-gold coating removal, showed distinct yellow FL upon excitation at the right terminal with a continuous wave 405 nm laser (black line in figure 5.8(l)). Shifting the excitation position slightly to the left resulted in a subtle reabsorbed signal (red line in figure 5.8(l)). Exciting the bent waveguide at four positions, with increasing distance between excitation and collection points, revealed a decrease in guided FL intensity as the path length increased with the optical loss of 0.284 dB μm^{-1} (figure 5.8(m) and (n)).

This alternative crystal processing technique using the ion beam again provides a pathway to surpass the limitations inherent in traditional methods for fabricating flexible crystal waveguides.

5.1.3 FIB milling of coumarin-153 single crystals

To meet the industrial PICs standard, there is a need to generalize the FIB milling technique to diverse molecular crystals. However, the specifics of universalizing this approach for chemically diverse organic microcrystals have yet to be investigated. Therefore, FIB milling was tested on a bright green FL organic molecular crystal with functional groups, namely, 2,3,6,7-Tetrahydro-9-(trifluoromethyl)-1H,5H,11H-[1]benzopyrano(6,7,8)quinolizin-11-one (coumarin-153 or C-153) (inset of figure 5.9(a)). C-153 is a laser dye characterized by a rigid molecular structure and a relatively large dipole moment of 6.55 D in the ground state.

Microcrystals of C-153 were obtained via sublimation at ambient pressure. C-153 is found in two polymorphs, namely, form I and form II. Form II of C-153 microcrystals, characterized by a broad FL spectrum ranging from 475 to 725 nm, was selectively created using sublimation at 110 °C for one minute. After sublimation, the sample was gradually cooled to room temperature, resulting in numerous microcrystals on the top coverslip. Optical microscope and FESEM images of C-153 microcrystals revealed smooth crystal facets (figure 5.9(a) and (b)).

The thickness of two representative coumarin-153 microcrystals (crystal-I: 9.8 × 9.4 μm^2; crystal-II: 9.9 × 8.3 μm^2) was measured using contact mode AFM, resulting in values of 1.18 μm for crystal-I and 0.91 μm for crystal-II. The color-coded FESEM images illustrate the rough edges formed during the contact mode AFM measurements (figure 5.9(b) and (f)). Subsequently, the crystals were gold-coated, and FIB milling (beam current: 0.1 nA) was conducted orthogonal to the (1–10) facet of crystal-I to fabricate a circular disk with a diameter of 7.5 μm. The color-coded FESEM images highlight the geometric perfection and clean finishing of the constructed microdisk (figure 5.9(c)). Similarly, a rectangular disk with dimensions of 6.3 × 5.5 μm^2 (length × width) was created from crystal-II, as depicted in figure 5.9(g). Notably, the gold-coated perylene crystals exhibit inertness to etchants, while coumarin-153 crystals display reactivity and damage the crystals.

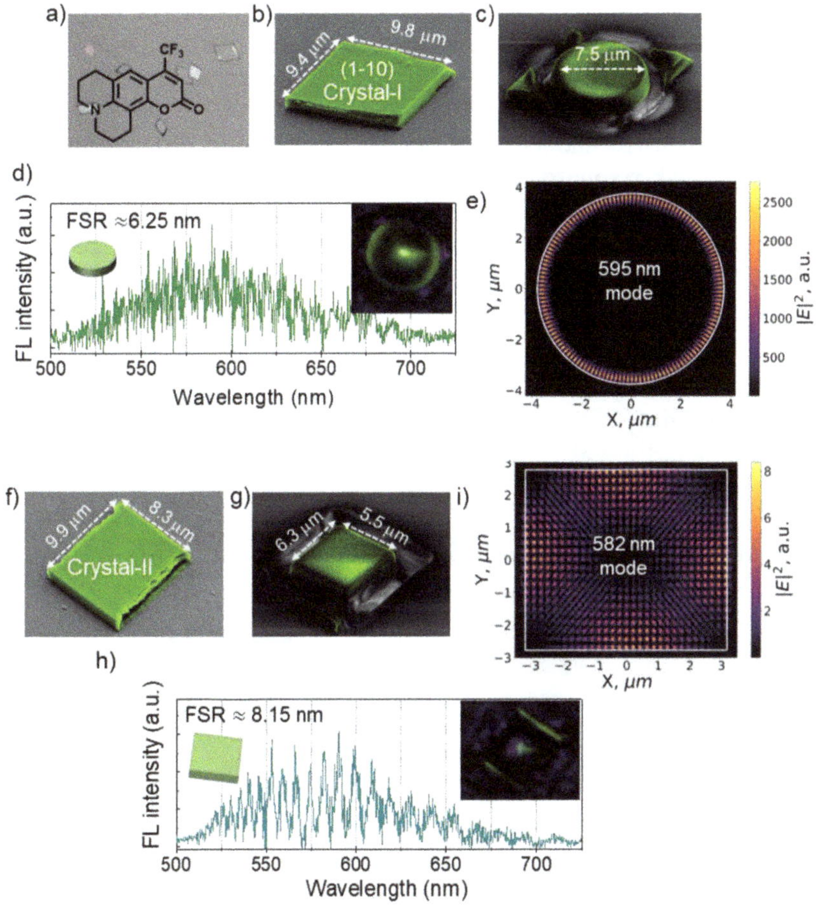

Figure 5.9. (a) Optical microscopic image of coumarin-153 microcrystals. The inset shows its molecular structure. (b, f) The color-coded FESEM images of selected coumarin-153 microcrystals, namely crystal-I and -II. (c, g) The color-coded FESEM and (d, h) FL background subtracted spectra of fabricated circular and rectangular disk resonators from crystal-I and -II, respectively. The insets show the corresponding FL images. Note: The rough edges in (b) and f) are due to the damage caused by the AFM-tip during contact mode topography imaging. (e, i) FDTD calculations displaying $|E_z|^2$ distribution for the 595 nm mode in the circular disk resonator and 582 nm mode in the rectangular disk resonator, respectively. (c–e, g–i) Reproduced from [3] with permission from the Royal Society of Chemistry.

Therefore, photonic investigations on the coumarin-153 resonators, without removing the gold coating, revealed sharp peaks in the FL spectra when excited with a 405 nm laser, attributed to the high quantum yield of coumarin-153. The FSR value for the crystal resonator was determined to be approximately 6.25 nm (figure 5.9(d)). Furthermore, for the representative 595 nm mode, FDTD calculations substantiate the distribution of the light's electric field around the periphery of the fabricated circular disk (figure 5.9(e)).

Similarly, without any post-processing to remove the gold layer, the fabricated rectangular disk demonstrated intense FL as well as optical resonances (figure 5.9(h)). The corresponding FSR value is determined to be 8.15 nm. FDTD calculations on the rectangular disk resonator revealed that the electric field at the wavelength of 582 nm is concentrated at the edges and not at the four corners (figure 5.9(i)).

For a comprehensive understanding of the electric field distribution of light, a smaller circular disk resonator was created from a coumarin-153 microcrystal by FIB milling (figure 5.10(b)). The original microcrystal featured a cross-section of 11 $\times$ 9 μm^2 and a thickness of 1.33 μm, as depicted in figure 5.10(a). Comparative color-coded FESEM images showcase the geometric perfection and smooth finishing of the disk-shaped crystal with a diameter of 2.5 μm (figure 5.10(b)). Photonic

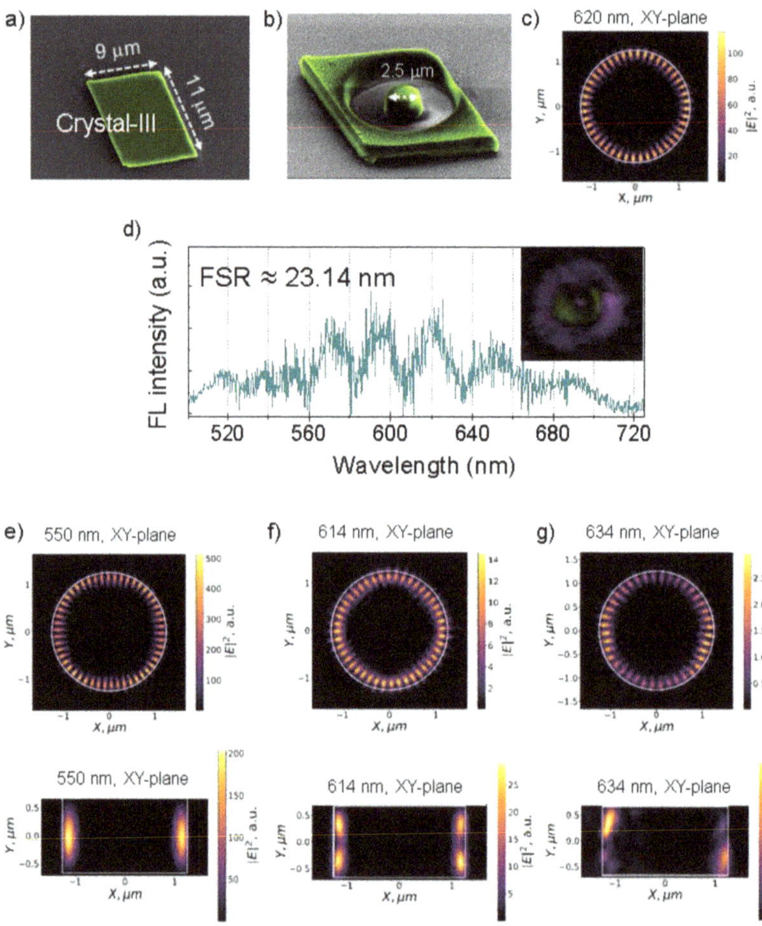

Figure 5.10. (a) The color-coded FESEM images of a selected coumarin-153 microcrystal. (b) FESEM and (c) FL background subtracted spectra of fabricated circular disk resonator from crystal-III. The insets show the FL images of microcrystals upon focused optical excitation. FDTD calculations display $|E_Z|^2$ distribution for (d) 620, (e) 550, (f) 614, and (g) 634 nm in the circular disk resonator of diameter ≈2.5 μm. (b–e) Reproduced from [3] with permission from the Royal Society of Chemistry.

investigations revealed the functionality of the fabricated circular disk (figure 5.10(b)). The calculated FSR values for the crystal, both before and after milling, were determined to be 4.8 nm and 23.14 nm, respectively (figure 5.10(d)). FL image of the crystal resonator exhibiting ring-shaped bright FL under optical excitation is presented in the inset of figure 5.10(d), inset.

For the produced single-crystal circular disk resonator (figure 5.10(d)), 2D and 3D FDTD numerical calculations were conducted for the 620, 550, 614, and 634 nm modes (figures 5.10(c), (e), (f), and (g)). The geometric analysis consistently reveals an intense electric field localized around the circular periphery in the 2D FDTD. The resonator demonstrates characteristics indicative of WGMs based on predicted azimuthal mode numbers of 22 and 25 for the peaks at 620 and 550 nm, respectively (figures 5.10(c) and (e)). The 3D FDTD calculations revealed the WGM with higher-order field distribution in a plane perpendicular to the mode circulation direction. The mode at 614 nm was found to have a lower Q-factor and a higher-order field distribution in a plane perpendicular to the mode circulation direction (figure 5.10(f)). Furthermore, the mode doesn't circulate in a plane parallel to the substrate, as seen in figure 5.10(g). These modes also have low Q-factors, but maybe they are simpler to identify than higher Q-factor modes because of their distinct radiation patterns. The field distribution cross-sections perpendicular to the substrate in the right panels and parallel to the substrate plane at half the resonator's height above the substrate in the left panels are shown in figure 5.10. The resonator borders are indicated with white lines.

Remarkably, C-153 disk resonators exhibit strong FL intensity and optical modes even without the removal of gold, owing to their relatively high quantum yield. Consequently, no additional post-processing treatment was performed for the coumarin-153 crystals. Another viable option is to grow coumarin-153 microcrystals on a conductive substrate (ITO-PET), which would alleviate the charging effects observed in electron/ion microscopy.

5.1.4 A tandem approach to fabricate a hybrid organic-add-drop filter (HOADF)

The construction of a HOADF, a pivotal optical component, is elucidated in the following section through the synergistic application of two distinct methodologies: mechanophotonics (micromechanical manipulation) and crystal photonics foundry (focused ion beam, FIB-assisted crystal milling) [21].

For the experiment, we opted for perylene (PER) disk resonators (DRs) produced through FIB milling, as detailed in figure 5.7 [21]. However, the transfer of milled optical components from one substrate to another poses a challenge for constructing monolithic and hybrid OPICs.

In this context, the mechanophotonics approach becomes crucial for integrating two or more photonic components directly at the OPIC construction site. Therefore, we envisioned leveraging flexible crystal waveguides and FIB-milled optical resonators for the development of an efficient small-scale HOADF.

For the filter fabrication, waveguides need to be integrated with these DRs. Therefore, the milled resonators must be physically separated from the parent crystal through post-mechanical micromanipulation using an AFM cantilever tip. However, due to the absence of stress-absorbing channels in the crystal structure, employing the AFM cantilever tip for DR manipulation may lead to mechanical damage. Hence, the AFM cantilever tip was carefully employed for the delicate tasks of moving, lifting, and dropping the parental crystal without causing physical impairment to the DRs, as depicted in figure 5.11(a). Despite careful handling, during the removal of the parent crystal, one of the fabricated disk resonators (DR3) also dislodged together with the parent crystal. The bright yellow FL (≈485–775 nm) of perylene DRs was coded as a λ_1(A) signal.

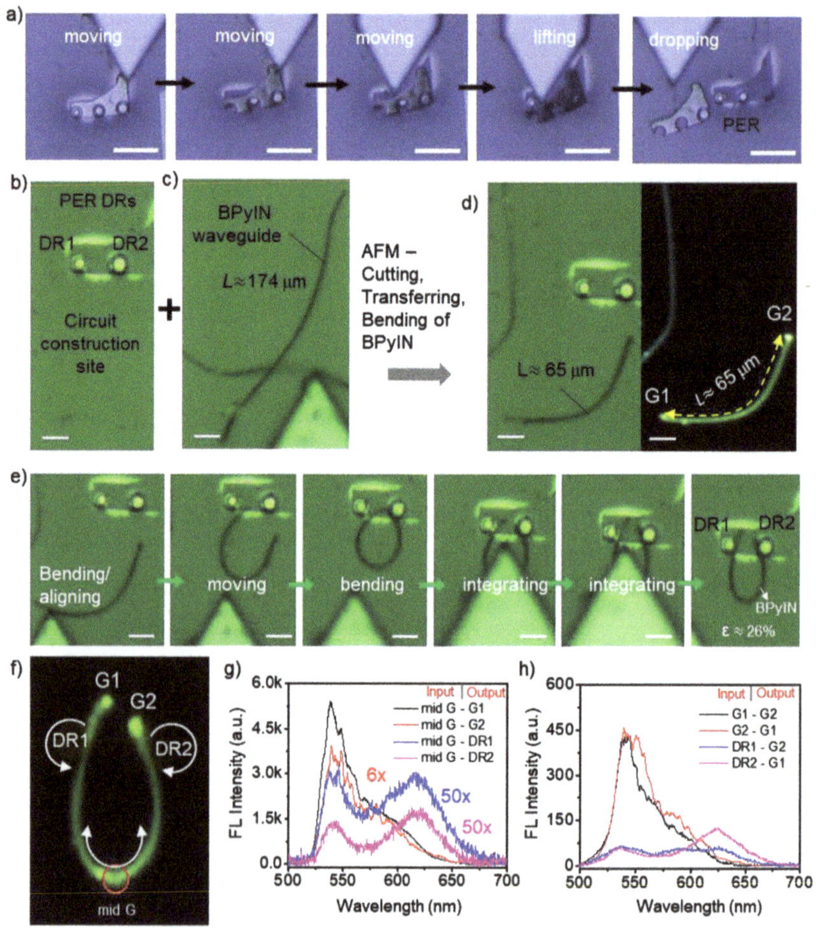

Figure 5.11. (a) Confocal optical microscopy images showing the micromechanical separation of the parent crystal from perylene DRs. Scale bars 30 μm. Confocal optical microscope images of (b) the circuit construction site. (c, d) AFM cantilever tip-based cutting and transferring of a BPyIN crystal waveguide closer to the circuit construction site. (e) Micromechanical integration of BPyIN waveguide with DRs to construct a DRs-coupled waveguide. (f) FL image of the DRs-coupled waveguide. FL spectra were recorded at various ports when excited at (g) the center of the BPyIN waveguide and (h) tips of BPyIN and and DRs (1 and 2). [21] John Wiley & Sons. (2023).

The next challenging task is to photonically couple DR-1 and DR-2 and to fabricate a novel HOADF. We selected BPyIN (see chapter 4, section 4.1.3) and BTD2CF_3 (see chapter 4, section 4.5) pseudo-plastic crystals. As one can see from the solid state optical data, there is a possibility for ET between two different crystals in physical contact viz. from BTD2CF_3 to PER, BPyIN to PER, PER to BTD2CF_3, and PER to BPyIN due to the spectral overlap of optical absorption of one crystal with the emission of another.

Firstly, for the physical integration of BPyIN with DRs, the BPyIN microcrystal was lifted with an AFM cantilever tip from the substrate containing BPyIN microcrystals. With caution, the lifted microcrystal was placed close to the perylene DRs containing substrate (figure 5.11(b)). Subsequently, to fit in the space between DR1 and DR2, this 173.8 μm long bent BPyIN waveguide was cut into a shorter waveguide of length, $L \approx 65$ μm (figure 5.11(c) and (d)). A 405 nm laser was used to excite the BPyIN waveguide to study its photonic performance in its bent state. Exciting one of the termini (G1; where G represents green FL) of the BPyIN microcrystal with a 405 nm laser resulted in a green FL band with a wavelength range of approximately 520–755 nm (signal: λ_2(A)). This FL guided to the opposite terminal (G2) as the $-\lambda_2$(A) signal due to reabsorption. Moreover, the spectra revealed optical modes arising from light confinement in the microcrystal waveguide.

The critical step in creating the HOADF was the fabrication of a U-shaped BPyIN waveguide which integrates with the disk resonators (DRs). The production of the HOADF brought forth several challenges:

- *Manual micromanipulation process*: The micromanipulation process is entirely manual, demanding attention and precision throughout the fabrication process to ensure seamless integration of crystals.
- *Extreme strain in the bent microcrystal*: The 180° bent microcrystal is susceptible to experiencing extreme strain during the fabrication process, necessitating careful handling to avoid crystal damage.
- *Operational area constraints*: The operational area is much smaller than the dimensions of the AFM cantilever, imposing spatial limitations. Any turbulence near the disk resonators (DRs) during cantilever movement may lead to potential damage during the micromanipulation process.

The micromanipulation steps, involving bending, moving, aligning, and integrating, constitute essential components of the integration process (figure 5.11(e)). The physical contact between the DRs and the U-shaped BPyIN waveguide is seen in the FL image of the BPyIN microcrystal waveguide coupled DRs circuit (figure 5.11(f)). The strain observed at the U-shaped BPyIN microcrystals was approximately 26%, surpassing the highest recorded values for flexible organic crystals, typically falling within the range of 3–11%.

To explore the optical communication between DRs and the bent waveguide, we excited the center of the BPyIN waveguide (denoted as 'mid G'; see figure 5.11(f)) with a 405 nm laser. As anticipated, both termini (G1 and G2) of the BPyIN waveguide exhibited outcoupling of green FL (figure 5.11(f)). Notably, the FL spectra at G1 and G2 revealed an additional band centered at 620 nm, alongside BPyIN's green FL ($\lambda_{max} \approx 538$ nm), corresponding to the PER emission maximum (figure 5.11(g)). This additional band can be attributed to the partial radiative excitation of PER DRs at the coupling region due to ET from BPyIN to PER, resulting in the simultaneous presence of two signals ($\lambda_{max} \approx 538$ nm and $\lambda_{max} \approx 620$ nm) in the DR-coupled waveguide.

This ET phenomenon was further validated by examining the spectra collected at DRs (1 and 2), where both BPyIN and PER FL bands were evident (figure 5.11(g), blue and pink lines). Consequently, there is a substantial contribution arising from the optical communication between PER DRs and the BPyIN waveguide in the operation of DRs-coupled waveguide. Similarly, the optical characteristics of the DRs-coupled waveguide were studied when input light was provided at G1, G2, DR1, and DR2 (figure 5.11(h)). Upon optically exciting DRs, the resulting spectrum at G1 or G2 includes a residual PER signal (see blue and pink spectra in figure 5.11(h)). This suggests that only a portion of the evanescently coupled PER FL is absorbed by BPyIN (via ET) to generate FL, while the unabsorbed light passively outcouples at G1 or G2.

Finally, the BTD2CF_3 microcrystal waveguide was air-lifted from another substrate to the circuit location and gently dropped close to the DRs-coupled waveguide circuit to create the envisioned hybrid OPIC, i.e., HOADF (figure 5.12(a)). In order to achieve HOADF, the BTD2CF_3 microcrystal was meticulously brought into evanescent contact with the DR1 using mechanical micromanipulation. The fabricated hybrid OPIC's physical contact between DR1 and BTD2CF_3 waveguide is evident in the color-coded FESEM image (figure 5.12(b)). The FESEM image measurements of the BPyIN and BTD2CF_3 microcrystals at the indicated locations showed that their respective thicknesses were 0.92 and 0.71 μm. Additionally, the thickness variations of the waveguide obtained via the self-assembly process are clearly visible in the FESEM image, whereas the FIB-milled DRs showed smooth morphology and well-defined dimensions.

The FL image of the assembled HOADF (figure 5.12(c)) vividly illustrated the diverse emissions from the three components. Assessing the photonic performance of the fabricated HOADF involved exciting one terminus (with a 405 nm laser) and analyzing the outputs at the other terminal. The BTD2CF_3 waveguide in HOADF was excited to generate a λ_3(A) (bandwidth ≈480–755 nm) signal at terminal Y1 (here, Y stands for yellow emission). The λ_3(A) signal traveled to the opposite end of BTD2CF_3, where the spectrum showed a slightly reabsorbed FL $-\lambda_3$(A) (bandwidth ≈520–755 nm). As it approached Y2, the $-\lambda_3$ (A) signal evanescently couples to DR1 and excites it via ET to produce λ_1 (A) (≈485–775 nm) signal. The generated signal is routed in an anticlockwise direction by DR1 and guided towards G1 in BPyIN waveguide. Because of the ET from perylene to BPyIN, a $-\lambda_2$(A)+λ_1(P) signal was detected at the G1 terminal, as observed in the DRs-coupled waveguide. Notably, because of DR1's controlled light guiding ability, light is not allowed to travel in the clockwise direction, and no optical signal is observed at G2 (figure 5.12(d) and (e)). As a result, hybrid OPIC serves as an add-drop filter and exhibits selective light guiding in the circuit.

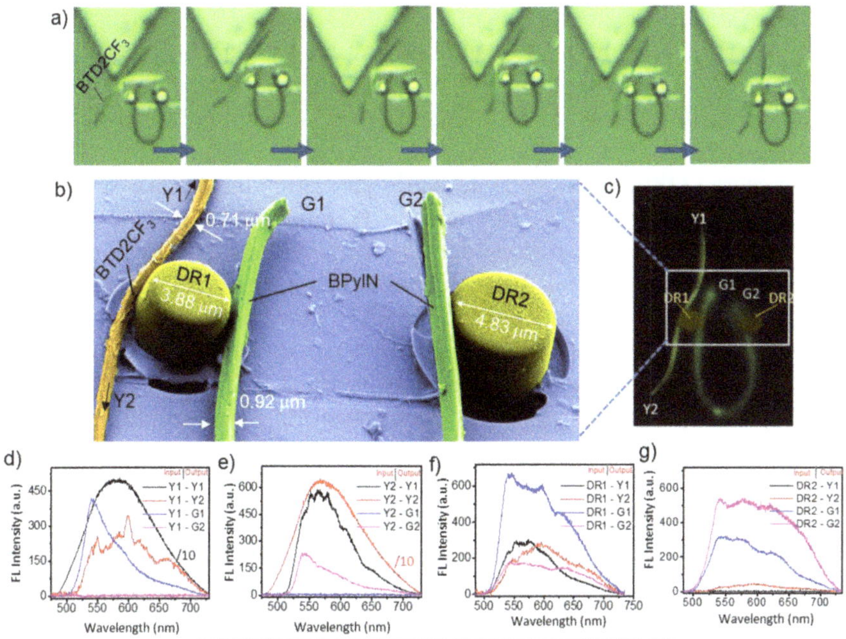

h)

Port		Output			
		Y1	Y2	G1	G2
Input (405 nm)	Y1	λ_3(A)	-λ_3(A)	-λ_2(A) + λ_1(P)	0
	Y2	-λ_3(A)	λ_3(A)	0	-λ_2(A) + λ_1(P)
	DR1	-λ_3(A)	-λ_3(A)	-λ_2(A) + λ_1(P)	-λ_2(A) + λ_1(P)
	DR2	0	-λ_3(A)	-λ_2(A) + λ_1(P)	-λ_2(A) + λ_1(P)

i)

Signal	λ_1(A)	λ_1(P)	λ_2(A)	-λ_2(A)	λ_3(A)	-λ_3(A)	0
Type	Active	Passive	Active	Reabsorbed	Active	Reabsorbed	NIL
Bandwidth (nm)	≈475-775	≈515-775	≈485-700	≈525-700	≈480-755	≈520-755	-

Figure 5.12. (a) Confocal optical microscope images depicting the micromechanical integration of $BTD2CF_3$ with the DR1. (b) The color-coded FESEM image of a portion of hybrid OPIC portraying the contact between DRs and waveguides. (c) FL microscope image of the fabricated hybrid OPIC using the FIB-milled perylene DRs and BPyIN and $BTD2CF_3$ microcrystals. (d) Input-dependent direction-specific optical outputs obtained from the FL spectra in the fabricated hybrid OPIC. (e–g) The FL spectra recorded at various outputs for input provided at Y1, Y2, DR1, and DR2, respectively. (h,i) The optical signals, bandwidths, and propagation mechanisms governing the functioning of hybrid ADF. [21] John Wiley & Sons. (2023).

Afterward, bright yellow FL of perylene (λ_1(A); ≈485–775 nm) was produced by excitation at DR1, and this FL was subsequently coupled into the $BTD2CF_3$ and BPyIN waveguides. As a result of ET from perylene to $BTD2CF_3$, a signal of $-\lambda_3$(A) (bandwidth ≈520–755 nm) was seen at Y1 and Y2. Maximum absorption of perylene FL is made possible by $BTD2CF_3$'s higher PLQY, and the λ_1(P) signal at Y1 and Y2 does not significantly contribute. Here, the light output at both Y1 and Y2 was detected because the point excitation in DR1 causes simultaneous scattering in all directions (figure 5.12(f)). However, as previously observed, a broad band at

G1 and G2 in the collected spectra corresponded to mixed signals from BPyIN and perylene, $-\lambda_3(A) + \lambda_1(P)$ because of simultaneous transduction of active and passive signals (figure 5.12(f) and (g)). The light input at DR2 also produced FL outputs at G1, G2, and Y2, but no signal at Y1 because of DR1's directional light guidance (figure 5.12(h) and (i)).

The engineered OPIC operates as an ADF, governing the directed propagation of light within the circuits. The DRs effectively guide optical signals in either a clockwise or anticlockwise direction. Comprehensive photonics investigations have reaffirmed the directional specificity inherent in the circuit geometry, underscoring the hybrid OPIC's optical functionality. This pioneering work unveils the latent potential of organic crystalline materials as an alternative platform to silicon-based materials, showcasing the precision fabrication of optical components. Crucially, it highlights the seamless integration of diverse photonic modules composed of different materials. The amalgamation of micromechanical manipulation techniques with crystal milling technology is expected to pave the way for automated, large-scale manufacturing of OPICs, a prospect eagerly anticipated by the materials science community.

References

[1] Pradeep V V and Chandrasekar R 2022 Micromanufacturing of geometrically- and dimensionally-precise molecular single-crystal photonic micro-resonators *via* focused ion beam milling arXiv:2203.14218v1 [physics.optics]

[2] Pradeep V V and Chandrasekar R 2022 Micromanufacturing of geometrically- and dimensionally-precise molecular single-crystal photonic micro-resonators via focused ion beam milling *Adv. Opt. Mater.* **10** 2201150

[3] Pradeep V V, Chosenyah M, Mamonov E and Chandrasekar R 2023 Crystal photonics foundry: Geometrical shaping of molecular single crystals into next generation optical cavities *Nanoscale* **15** 12220–6

[4] Mueller E W and Tsong T T 1969 *Field Ion Microscopy Principles and Applications* (New York: American Elsevier)

[5] Giannuzzi L A and Stevie F A 2005 *Introduction to Focused Ion Beams: Instrumentation, Theory, Techniques, and Practice* (Berlin: Springer)

[6] Thompson G B, Zalm P C and Miller R M 2007 *Focused Ion Beam Systems: Basics and Applications* (Cambridge: Cambridge University Press)

[7] Notte J A, Russell P E Jr and Notte J A (ed) 2007 *Focused Ion Beam Systems* (New York: Wiley)

[8] Stoykovich M P and Knickerbocker J U 2010 Dual-beam FIB/SEM for advanced micro-characterization of advanced materials *Microscopy Today* **18** 22–7

[9] Zhang X F, Ning Z and Fu Y Q 2016 Advances in FIB-SEM tomography for 3D imaging and analysis of biological specimens *Micron* **85** 8–20

[10] Larson D J, Prosa T J and Ulfig R M 2013 Advances in focused ion beam technology and applications *MRS Bulletin* **38** 324–32

[11] Patterson M 2012 Focused ion beam–scanning electron microscopy in cell biology *Methods Cell Biol.* **111** 325–42

[12] Smith N S, Notte J A, Weber M H, Larrabee R D and Swanson L W 1989 Focused ion beam technology for microfabrication and microanalysis *Microelectron. Eng* **9** 355–69
[13] Krohn V E and Ringo G R 1976 Ion source of high brightness using liquid metal *Appl. Phys. Lett.* **27** 479
[14] Platzman P M 1982 Liquid metal ion sources *J. Vac. Sci. Technol.* A **1** 1558–64
[15] Bischoff L, Mazarov P, Bruchhaus L and Gierak J 2016 Liquid metal alloy ion sources—an alternative for focused ion beam technology *Appl. Phys. Rev.* **3** 021101
[16] Hofmann T, Seidman D N and Robertson I M 2010 Sample damage and removal in focused ion beam milling of submicrometer GaN cross sections *J. Appl. Phys.* **107** 104906
[17] Nord M, Hlawacek G and Utke I 2013 Focused electron and ion beam induced processing at the nanoscale *Nano Today* **8** 640–65
[18] Li W, Baren J V, Berges A, Bekyarova E, H C and Bardeen C J 2020 Shaping organic microcrystals using focused ion beam milling *Cryst. Growth Des.* **20** 1583
[19] Kim S, Park M J, Balsara N P, Liu G and Minor A M 2011 Minimization of focused ion beam damage in nanostructured polymer thin films *Ultramicroscopy* **111** 191
[20] Pradeep V V, Ummethala G, Malladi S R K and Chandrasekar R 2023 Focused ion beam milling of perylene microcrystals into photonic modules: effect of substrate and the ion beam current *Cryst. Growth Des* **8** 5414
[21] Pradeep V V, Kumar A V and Chandrasekar R 2023 A tandem approach to fabricating a hybrid, organic-add-drop filter using single-crystal disk-resonators and pseudo-plastic crystal waveguides *Laser Photonics Rev.* **17** 2300552

www.ingramcontent.com/pod-product-compliance
Ingram Content Group UK Ltd.
Pitfield, Milton Keynes, MK11 3LW, UK
UKHW051855100926
8556IPUK00005B/632